Analysis of Polymers

AN INTRODUCTION

Related Pergamon titles of interest

P.G. BARKER
Computers in Analytical Chemistry

T.R. CROMPTON
The Analysis of Plastics

P.G. JEFFREY & D. HUTCHINSON
Chemical Methods of Rock Analysis, 3rd ed

L. PATAKI & E. ZAPP
Basic Analytical Chemistry

R. PRIBIL
Applied Complexometry

J. YINON & S. ZITRIN
The Analysis of Explosives

Related Pergamon Journals

Computers & Chemistry

European Polymer Journal

Journal of Pharmaceutical & Biomedical Analysis

Progress in Analytical Spectroscopy

Progress in Polymer Science

Tetrahedron

Tetrahedron Letters

Free specimen copy sent on request

Analysis of Polymers

AN INTRODUCTION

by

T. R. CROMPTON

PERGAMON PRESS

OXFORD · NEW YORK · BEIJING · FRANKFURT
SÃO PAULO · SYDNEY · TOKYO · TORONTO

U.K.	Pergamon Press plc, Headington Hill Hall, Oxford OX3 0BW, England
U.S.A.	Pergamon Press, Inc., Maxwell House, Fairview Park, Elmsford, New York 10523, U.S.A.
PEOPLE'S REPUBLIC OF CHINA	Pergamon Press, Room 4037, Qianmen Hotel, Beijing, People's Republic of China
FEDERAL REPUBLIC OF GERMANY	Pergamon Press GmbH, Hammerweg 6, D-6242 Kronberg, Federal Republic of Germany
BRAZIL	Pergamon Editora Ltda, Rua Eça de Queiros, 346, CEP 04011, Paraiso, São Paulo, Brazil
AUSTRALIA	Pergamon Press Australia Pty Ltd., P.O. Box 544, Potts Point, N.S.W. 2011, Australia
JAPAN	Pergamon Press, 5th Floor, Matsuoka Central Building, 1-7-1 Nishishinjuku, Shinjuku-ku, Tokyo 160, Japan
CANADA	Pergamon Press Canada Ltd., Suite No. 271, 253 College Street, Toronto, Ontario, Canada M5T 1R5

First edition 1989

Library of Congress Cataloging-in-Publication Data

Crompton, T.R. (Thomas Roy)
Analysis of polymers: an introduction/by T.R. Crompton.
p. cm.
Includes index.
1. Polymers and polymerization—Title.
QD139.P6C76   1988   547.7'046—dc19   88-19696

British Library Cataloguing in Publication Data

Crompton, T.R. (Thomas Roy)
Analysis of polymers.
1. Polymers. Chemical analysis
I. Title
547.7'046

ISBN 0-08-033942-5 Hardcover
ISBN 0-08-033936-0 Flexicover

Printed in Great Britain by A. Wheaton & Co. Ltd, Exeter

Contents

Preface

The aim of this book is to familiarize the reader with all aspects of plastic analysis. The book covers not only the analysis of the main types of plastics now in use commercially, but also the analysis of minor non-polymeric components of the plastics formulation, whether they be deliberately added, such as processing additives, or whether they occur adventitiously, such as moisture and residual monomers and solvents.

Practically all of the major newer analytical techniques, and many of the older classical techniques, have been used to examine plastics and their additive systems. As so many different polymers are now used commercially it is also advisable when attempting to identify a polymer to classify it by first carrying out at least a qualitative elemental analysis and possibly a quantitative analysis (Chapter 2) and then in some cases, depending on the elements found, to carry out functional group analysis (Chapter 3). If a simple qualitative identification of the plastic is all that is required then it is examined by fingerprinting techniques, as discussed in Chapter 4, in order to ascertain whether a quick identification can be made by comparing its infrared spectrum and pyrolysis–gas chromatography pattern with those of authentic specimens of known polymers.

Frequently, however, the identification of a polymer, especially co-polymers or terpolymers, is not as simple as this, and it is necessary to obtain a detailed picture of the microstructure of the polymer before identification can be made (Chapter 5). Techniques that might be used, in addition to elemental and functional group analysis, include spectroscopic techniques such as infrared, NMR, PMR, and systematic investigations by pyrolysis–gas chromatography. Examination for the type of unsaturation present, the nature of side-chain groups and end-groups, the presence of oxygenated groups such as carbonyl and whether they are macro or micro constituents will all assist in building up a picture of the polymer structure. In many cases, considerable experience and innovative skills are required by the analyst in order to successfully identify polymers by these techniques, and it is hoped that this book will assist the analyst in developing such skills.

The book gives an up-to-date and thorough exposition of the present state of the art of polymer analysis and, as such, should be of great interest to all those engaged in this subject in industry, university, research

establishments and general education. It is also intended for under-
graduate and graduate chemistry students and those taking courses
in plastics technology, engineering chemistry, materials science and
industrial chemistry. It will be a useful reference work for manufacturers
and users of plastics, the food and beverage packing industry, the
engineering plastics industry, plastic components manufacturers, and
those concerned with pharmaceuticals and cosmetics.

Before proceeding to the first two chapters, which deal, respectively,
with the determination of elements and functional groups, it would be of
interest in Chapter 1 to discuss briefly the various types of polymers used
commercially, and their properties and applications.

1

Types and properties of polymers

SYNTHETIC resins, in which plastics are also included, vary widely in their chemical composition and in their physical properties. The number of synthetic resins which can be made is vast; relatively few, however, have achieved commercial importance.

Well over 90% of all synthetic resins made today comprise no more than 20 different types, although there are certain variations to be found within each type. Synthetic resins are familiar to most people as plastics, but they have other uses, such as in the manufacture of surface coatings, glues, synthetic textile fibres, etc. The rapid growth of the synthetic resin industry has to a large extent been made possible by the fact that ample supplies of the necessary raw materials have become available from petroleum.

The synthetic resins may be divided into two classes, known respectively as 'thermosetting' and 'thermoplastic' resins, each class differing in its behaviour on being heated. The former do not soften; the latter soften, but regain their rigidity on cooling. Both types are composed of large molecules, known as macromolecules, but the difference in thermal behaviour is due to differences in internal structure.

The large molecules of the thermoplastics have a long-chain structure, with little branching. They do not link with each other chemically, although they may intertwine and form a cohesive mass with properties ranging from those of hard solids to those of soft pliable materials, in certain cases resembling rubber. On being heated, the chain molecules can move more or less freely relative to each other, so that, without melting, the material softens and can flow under pressure and be moulded to any shape. On cooling, the moulded articles regain rigidity. Some resins require the addition of liquid plasticizers to improve the flow of the plastic material in the mould. In such cases the moulded articles are usually softer and more flexible than the products made from the unplasticized resins.

The macromolecules of the thermosetting resins are often strongly-branched chains and are chemically joined by crosslinks, thus forming a complex network. On heating, there is less possibility of free movement, so that the material remains rigid.

Production of synthetic resins

Production of these resins also falls into two groups since there are, generally, two main types of chemical reaction by which they are made. These are polycondensation reactions and polymerization reactions.

Polycondensation reactions

In this type of reaction two or more chemicals are brought together and a reaction between them is initiated by using heat or a catalyst or both. The reaction proceeds with the elimination of water and the molecules are joined by chemical bonds to form macromolecules, either long-chain or crosslinked structures of the thermoplastic or thermosetting types, respectively. Many resins obtained by polycondensation are the thermosetting type.

In the manufacture of these resins the chemical reactions are arrested at an intermediate stage in which the resins are temporarily thermoplastic; they are set in their final shape by the application of heat and pressure. At this stage the interlinking of the molecules takes place.

Important thermosetting synthetic resins made by polycondensation, using petroleum chemicals as raw materials, include the phenol-formaldehyde ('Bakelite'), urea-formaldehyde, alkyd- and epoxy- types.

Resins with long-chain macromolecules obtained by polycondensation have thermoplastic properties. Polyesters ('Terylene') and polyamides (nylon) are examples of polycondensations. The synthetic fibre 'Terylene' (known as 'Dacron' in the U.S.A.) is a polyester formed by the reaction of ethylene glycol with terephthalic acid; the terephthalic acid is obtained from paraxylene by oxidation.

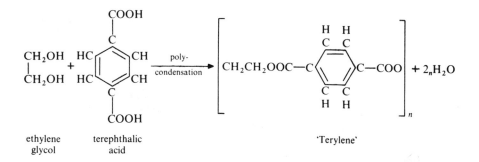

ethylene glycol   terephthalic acid   'Terylene'

Nylon type fibres (polyamides) are manufactured from adipic acid, which can be made from either cyclohexane or phenol. The adipic acid is condensed with hexa-methylene diamine, which is a derivative of adipic acid.

phenol — cyclohexanol — adipic acid

$$\begin{matrix} COOH \\ | \\ (CH_2)_4 \\ | \\ COOH \end{matrix} + \begin{matrix} NH_2 \\ | \\ (CH_2)_6 \\ | \\ NH_2 \end{matrix} \xrightarrow[\text{condensation}]{\text{poly-}} \left[NH(CH_2)_6NH{-}CO(CH_2)_4CO \right]_n + 2nH_2O$$

adipic acid + hexa-methylene diamine → nylon

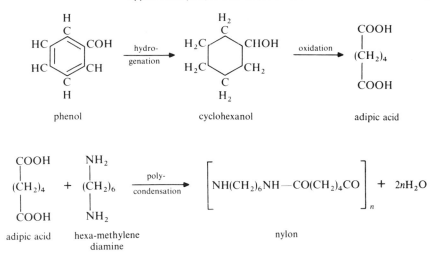

Polymerization reactions

Resins produced by polymerization reactions, known technically as high polymers, are rapidly increasing in number and in importance as compared with the polycondensation resins. High polymers are usually made by joining together into long chains a number of molecules which have the same kind of reactive points or groupings in their structure. These individual molecules are usually olefins or other compounds with double bonds, and are called 'monomers'. The molecule of the polymer often contains hundreds of monomer units.

The manufacture of high polymers therefore takes place in two stages: first, the production of the monomer, or repeating chemical unit; and second, the polymerization to a resin.

Thus, if we take the preparation of polyvinyl chloride as an example we have:

1st Stage

$$\underset{\text{ethylene}}{CH_2{=}CH_2} + \underset{\text{chlorine}}{Cl_2} \rightarrow \underset{\text{dichloro ethane}}{CH_2Cl{-}CH_2Cl} \rightarrow \underset{\text{vinyl chloride (monomer)}}{CH_2{=}CHCl} + \underset{\text{hydrochloric acid}}{HCl}$$

2nd Stage

$$\underbrace{CH_2{=}CHCl + CH_2{=}CHCl + CH_2{=}CHCl + \cdots}_{n \text{ molecules vinyl chloride}} \xrightarrow{\text{polymerization}}$$

$$\underbrace{{-}CH_2CHCl{-}CH_2{-}CHCl{-}CH_2{-}CHCl{-}CH_2CHCl{-}}_{\text{polyvinyl chloride}}$$

In some cases it is possible to form polymers from two or even three monomers which may differ from one another in chemical form and yet be capable of linking end-to-end to form mixed monomer chains. These are known as 'copolymers', and such polymers form the basis of the most important types of synthetic rubber.

Further examples of polymerizations:

Styrene butadiene copolymer

$$CH=CH_2 + CH_2=CH-CH=CH_2$$
$$\underset{\text{styrene}}{|\ Ph} \qquad \underset{\text{butadiene}}{}$$

$$-CH-CH_2-CH_2-CH=CH-CH_2-$$
$$|\ Ph$$

styrene–butadiene copolymers

Polymethylmethacrylate

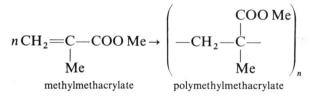

$$n\,CH_2=C-COO\,Me \rightarrow \left(-CH_2-\underset{Me}{\overset{COO\,Me}{C}}- \right)_n$$

methylmethacrylate   polymethylmethacrylate

Buna N rubber

$$\underset{\text{acrylonitrile}}{CH_2=CHCN} + \underset{\text{butadiene}}{CH_2=CH-CH=CH_2} \rightarrow$$

$$-CH_2-\underset{CN}{CH}-CH_2-CH=CH-CH_2-$$

acrylonitrile–butadiene copolymer

Vinyl chloride–vinylidene chloride copolymers

$$\underset{\text{vinyl chloride}}{CH_2=CH-Cl} + \underset{\text{vinylidene chloride}}{CH_2=CCl_2} \rightarrow$$

$$-CH_2-CCl_2-CH_2-CHCl-$$

Polyvinylacetate

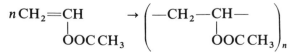

$$n\,CH_2=\underset{OOC\,CH_3}{CH} \rightarrow \left(-CH_2-\underset{OOC\,CH_3}{CH}- \right)_n$$

vinyl acetate   polyvinylacetate

Polyacrylonitrile

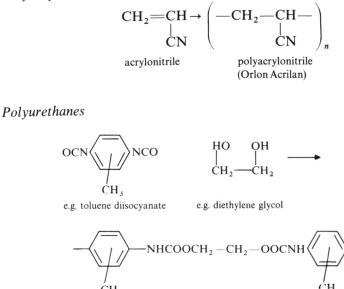

$$CH_2{=}CH{\rightarrow}\left(-CH_2{-}CH-\right)_n$$
$$\quad\quad CN \quad\quad\quad\quad\quad CN$$

acrylonitrile   polyacrylonitrile
(Orlon Acrilan)

Polyurethanes

e.g. toluene diisocyanate   e.g. diethylene glycol

$-\text{NHCOOCH}_2-\text{CH}_2-\text{OOCNH}-$

Numerous high molecular weight polymers are produced commercially. The properties and some of the uses of these are discussed below under separate headings. Many more copolymers exist than polymers and many, but not all, of these are produced in relatively small quantities for more specialized applications. Some of the more important copolymers are included; mechanical, physical, electrical and chemical properties of the important commercially produced polymers are compared in Table 1. Data on the properties of metals are included for comparison. Table 2 tabulates some of the uses to which these plastics are put.

TABLE 1   *Physical and chemical properties of polymers*

	Specific gravity ASTM D 792	Impact strength ASTM D 256 (J/12.7 mm)	Tensile strength ASTM D 638,651 (MN m⁻²)	Elongation in tension ASTM D 638 (%)	Modulus of elasticity in tension ASTM D 747 (MN m⁻² × 10⁻²)	Flexural strength ASTM D 790 (MN m⁻²)	Compressive strength ASTM D 695 (MN m⁻²)
Polycondensation types							
Phenol formaldehyde							
(a) Unfilled	1.25–1.3	0.13–0.24	48–55.3	1.0–1.5	51.8–69	83.9–103.6	69–203
(b) Woodflour/cotton flock filled	1.32–1.45	0.16–0.4	44.9–58.8	0.4–0.8	55.3–83.9	58.8–83.9	152–276
Urea formaldehyde							
Cellulose filled	1.47–1.52	0.17–0.24	41.4–89.7	0.5–1.0	104	69–110.5	172–241
Melamine formaldehyde							
(a) Unfilled	1.48	—	—	—	—	76–96.6	276–311
(b) α-cellulose filled	1.47–1.52	0.16–0.24	48–89.7	0.6–0.9	89.7	69–110.5	172–297
Polyesters							
(a) Resin	1.1–1.4	0.5	42–91	1.5–2.5	—	59–152	91–280
(b) Dough moulding compound	1.7–2.3	1.4–2.3	28–70	—	—	84–140	140–210
(c) Sheet moulding compound	1.7–2.6	7.0–10.0	56–140	—	—	67–172	105–210
Epoxides							
(a) Rigid	1.0–3.2	0.2–0.62	34.5–83.9	5–10	13.8–41.4	69–138	104–203
(b) Flexible	1.2	0.43–1.02	6.9–27.6	10–100	—	20.3–69	—
Nylons							
(a) Type 6	1.13	0.68–2.45	70.5–83.9	90–320	10.4–24.8	55.3–110.5	48.4–96.6
(b) Type 6/6	1.14	0.68–1.36	48–83.9	60–300	17.9–27.6	55.3–95.3	48.4–110.5
(c) Type 11	1.04–1.05	1.2	55–60	300–350	—	69.0	55
(d) Type 12	1.01–1.02	1.4	55	150–300	—	73.5	—
Polymerization types							
Polyethylenes							
(a) Low density	0.91–0.93	No Break	6.9–15.9	90–650	1.18–2.42	No Break	Excessive cold flow
(b) High density	0.941–0.965	1.02–8.15	21.4–38	50–800	5.53–10.4	13.8–20.3	16.5
Polypropylene	0.9–0.91	0.27–4.25	29–38	50–600	8.92–13.8	34.5–55	58.8–69

Material							
Polystyrenes							
(a) Conventional	1.04-1.11	0.13-0.34	34.5-83.9	1.0-2.5	27.6-41.4	83.9-117.5	79.4-110
(b) Toughened	0.98-1.1	0.27-2.05	17.2-48	7-60	17.2-31	27.6-69	27.6-62.2
Styrene acrylonitrile	1.075-1.1	0.24-0.34	65.7-83.9	1.5-3.5	27.6-38.7	96.6-131	96.6-117
Acrylonitrile/butadiene/styrene	0.99-1.1	2.04-8.15	17.2-62	10-140	6.9-28.3	24.8-93.2	17.2-76
vinyl polymer							
(a) Rigid polyvinyl chloride	1.38-1.4	0.68-2.04	58.8	2-40	24.2	93.2	55.3
(b) Rigid vinyl chloride/vinyl acetate	1.37-1.45	0.34-0.68	51.9-58.8	200-450	↕	83.9	—
(c) Rubber modified PVC	1.35	10.2 (Unaged)	41.4	—	41.4	83.9	89.7
Polyacetals	1.425	1.56 extrusion / 0.95 injection	69	15 injection / 75 extrusion	28.3	97.4	ca 124
Polycarbonates	1.2	8.15-10.7	58.8-65.7	60-100	22	76-89.7	76
Ionomers	0.93-0.96	No Break	28-35	—	1.8-2.1	2100	—
Polyphenylene oxide	1.06-1.1	1.5	50-65	20-60	—	90-95	102-104
Polyphenylene sulphide	1.6-1.9	—	74-131	0.5-1.25	—	102-185	70-145
Acrylics	1.17-1.2	0.2-0.43	48-76	3-10	31	89.7-117.5	83.9-138
Ethylene vinylacetate	0.925-0.95	No Break	10-18	750-900	—	20-26	—
Fluorinated polymers							
(a) Fluorinated Ethylene Propylene	2.14-2.17	No Break	17.2-24.2	250-600	3.45-4.8	11.05	19.7
(b) Polytetrafluoroethylene	2.1-2.2	1.70-2.72	17.2-41.1	125-175	3.45-6.22	No Break	4.84-12.4
(c) Polytrifluorochloroethylene	2.1-2.15	1.70-2.45	31.8-39.4	—	13.1-20.3	24.1	13.8
Cellulosic plastics							
(a) Cellulose acetate	1.23-1.34	0.27-3.53	13.1-58.8	6-70	4.49-27.6	13.8-110.5	15.2-248
(b) Cellulose acetate butyrate	1.15-1.22	0.54-4.27	17.9-47.7	40-90	3.45-13.8	12.4-64.3	14.5-152
(c) Cellulose propionate	1.18-1.24	0.13-6.78	13.8-50.5	30-100	4.14-14.8	26.9-76	21.4-152
Metals and alloys							
(a) Brass (Cu-Zn, 70/30)	8.5	45.8	320	65-70	965*	—	—
(b) Mild steel (0.6% Carbon)	7.87	24	459	45	2030*	—	760-897
(c) Carbon steel (0.4% Carbon)	7.85	3.4	505	28	2000*	—	ca 1360
(d) Aluminium (99% Pure)	2.82	13.4	83.9	15-30	710*	—	—
(e) Duralumin (Al-Si-Cu-Mg)	2.8	Variable	247	25-29	690*	—	—
(f) Phosphor bronze (Cu-Sn-P)	8.98	Variable	402	15	1100*	—	—

*Calculated from tensile strength and elongation data

Table 1 Contd.

	Hardness (Rockwell) ASTM D 785	Dissipation (power) factor ASTM D 150 (10^6 Hz)	Dielectric constant ASTM D 150 (10^6 Hz)	Dielectric strength ASTM D 149 (Short time 0.125" thick V mm^{-1} × 10^{-2})	Volume resistivity ASTM D 257 (Ohm cm^{-1} at 23°C and 50% RH)	Thermal. expansion ASTM D 696 (mm mm^{-1} °C^{-1} × 10^{-5})	Specific heat (kJ kg^{-1})
Phenol formaldehyde							
(a) Unfilled	M124–M128	0.015–0.03	4.5–5.0	118–158	10^{11}–10^{12}	2.5–6	1.6–1.76
(b) Woodflour/cotton flock filled	M100–M120	0.03–0.07	4.0–7.0	79–168	10^9–10^{13}	3–4.5	1.45–1.68
Urea formaldehyde							
Cellulose filled	M115–M120	0.25–0.35	6.4–6.9	118–158	10^{12}–10^{13}	2.7	1.68
Melamine formaldehyde							
(a) Unfilled	—	—	—	—	—	—	—
(b) α-Cellulose filled	M110–M125	0.027–0.045	7.2–8.2	118–158	10^{12}–10^{14}	4	1.68
Polyesters							
(a) Resin	—	0.006–0.026	2.8–4.1	132–200	10^{13}–10^{16}	5.5–10	—
(b) Dough moulding compound	—	0.007–0.02	5.2–6.4	136–168	10^{14}–10^{15}	1.1–5	1.04
(c) Sheet moulding compound	—	0.015–0.024	4.2–5.8	152–180	10^{14}–10^{16}	2	1.04
Epoxides							
(a) Rigid	—	0.01–0.02	3.0–4.0	138–177	$> 10^{15}$	5–9	—
(b) Flexible	—	0.01–0.02	3.0–4.0	118–158	$> 10^{15}$	5–9	—
Nylons							
(a) Type 6	R103–R118	0.02–0.13	3.0–7.0	173–201	10^{12}–10^{15}	8–13	1.68
(b) Type 6/6	R108–R118	0.02–0.06	3.6–6.0	152–185	0.45–4×10^{14}	10–15	1.68
(c) Type 11	R108	0.011–0.022	3.2	168	4.3×10^{13}	15	2.4
(d) Type 12	R106	0.03	3.1	180	2.5×10^{15}	10.4	2.1
Polymerization types							
Polyethylenes							
(a) Low density	D41–D46 (Shore)	< 0.0005	2.25–2.35	181–276	$> 10^{16}$	16–18	2.3
(b) High density	D60–D70 (Shore)	< 0.0003	2.25–2.35	> 316	$> 10^{16}$	11–13	2.22–2.3
Polypropylene	R85–R110	0.0002–0.0003	2.25–2.3	> 316	$> 10^{16}$	11	1.93

	Hardness						
Polystyrenes							
(a) Conventional	M65–M90	0.0001–0.005	2.4–3.1	197–276	$>10^{13}$	6–8	1.34–1.45
(b) Toughened	R50–R100, M35–M70	0.0004–0.002	2.4–3.8	118–237	$>10^{13}$	3.4–21	1.34–1.45
Styrene acrylonitrile	M80–M90	0.007–0.01	2.75–3.1	158–197	$>10^{13}$	6–8	1.34–1.43
Acrylonitrile/buta-diene/styrene	R30–R118	0.007–0.026	2.7–4.75	122–162	0.5×10^{13} 2.7×10^{16}	6–13	1.38–1.68
Vinyl polymers							
(a) Rigid polyvinyl chloride	R110		3.0	167	10^{16}	5	0.84–2.1
(b) Rigid vinyl chloride/vinyl acetate	—	0.006	3.0–3.5	555	↕	7	0.84–2.1
(c) Rubber modified PVC	R105	0.14	—	435	10^{13}	5	0.84–2.1
Polyacetals	M94–R118	0.004	3.7	197	6×10^{14}	8.1	1.45
Polycarbonates	M80–R125	0.01	2.6	158	2.1×10^{16}	7	1.26
Ionomers	D50–D65 (Shore)	0.0019	2.4–2.5	360–440	10^{16}	12	2.3
Polyphenylene oxide	R115–R119	0.0009	2.64	88–220	10^{17}	5.2–6	1.33
Polyphenylene sulphide	R120–R123	—	3.8–6.6	—	2×10^{15}	—	—
Acrylics	M85–M105	0.02–0.04	2.2–3.2	177–213	$>10^{14}$	5–9	1.47
Ethylene vinyl acetate	D27–D36 (Shore)	0.03–0.05	2.6–3.2	248–312	1.5×10^{8}	16–20	2.8
Fluorinated polymers							
(a) Fluorinated ethylene propylene	R25	0.0003	2.1	197–237	10^{19}	8.3–10.5	1.18
(b) Polytetrafluoroethylene	D50–D65 (Shore)	0.0002	2.1	158–237	10^{19}	9–22, varies with temperature	1.05
(c) Polytrifluorochloroethylene	R110–R115	ca 0.01	2.5	209–237	10^{18}	0.7	0.93
Cellulosic plastics							
(a) Cellulose acetate	R35–R125	0.01–0.10	3.2–7.0	102–144	10^{10}–10^{13}	8–16	1.26–1.76
(b) Cellulose acetate butyrate	R31–R116	0.01–0.04	3.2–6.2	98–158	10^{10}–10^{12}	11–17	1.26–1.68
(c) Cellulose propionate	R20–R120	0.01–0.04	3.4–3.6	118–177	10^{12}–10^{15}	11–17	1.26–1.68
Metals and alloys	Brinell: Hardness						
(a) Brass (Cu–Zn, 70/30)	ca 100	—	—	—	—	1.99	0.379
(b) Mild steel (0.6% Carbon)	92	—	—	—	—	1.26	0.463
(c) Carbon steel (0.4% Carbon)	135	—	—	—	—	1.12	0.463
(d) Aluminium (99% Pure)	22	—	—	—	—	2.4	0.965
(e) Duralumin (Al–Si–Cu–Mg)	82	—	—	—	—	2.25	0.965
(f) Phosphor bronze (Cu–Sn–P)	110	—	—	—	—	1.78	0.379

TABLE 1 *Contd.*

	Thermal conductivity ASTM C177 (W m⁻¹°C⁻¹ × 10⁻³)	Heat distortion temperature ASTM D648 (°C* 4.5 MN m⁻² 18 MN m⁻²)	Softening Point (vicat) (°C)	Burning Rate ASTM D635 (in/min)	Low temperature performance	Odour
Phenol formaldehyde						
(a) Unfilled	1.26–2.52	115–127	—	—	—	None
(b) Woodflour/cotton flock filled	1.68–2.94	127–171	—	—	Good	None
Urea formaldehyde						
Cellulose filled	2.94–4.2	132–138	—	—	Good	None
Melamine formaldehyde						
(a) Unfilled	—	147	—	—	—	None
(b) α-Cellulose filled	2.94–4.2	205	—	—	Good	None
Polyesters						
(a) Resin	3.4–3.6	70–100	—	—	Good	None when cured
(b) Dough moulding compound	9.4–10.6	205	—	—	Good	None when cured
(c) Sheet moulding compound	—	205	—	—	Good	None when cured
Epoxides						
(a) Rigid	1.68–2.1	Up to 299	—	—	Good	None
(b) Flexible	1.68–2.1	Up to 60	—	—	Very Good	None
Nylons						
(a) Type 6	—	127–171*	M.P. 215	—	Good	None
(b) Type 6/6	2.18–2.43	149–182*	M.P. 264	—	Good	None
(c) Type 11	4.2	144–155	185	—	Good	None
(d) Type 12	2.9	155	175	—	Good	None
Polymerization types						
Polyethylenes						
(a) Low density	3.37	40–50*	85–87	1.02–1.06	Good	None
(b) High density	4.63–5.22	60–82*	120–130	1.02–1.06	Good	None
Polypropylene	1.38	90–110*	150	0.75–0.83	Fair	None
Polystyrenes						
(a) Conventional	1.1–1.38	65–113	82–103	—	Poor	None
(b) Toughened	0.42–1.26	64–93	78–100	—	Poor/Fair	None

Styrene acrylonitrile	1.21	65–113	85–103	—	Poor	Slight Smell
Acrylonitrile/buta-diene/styrene	0.62–3.62	74–107	85	0.6–1.0	Poor	None
vinyl polymers						
(a) Rigid polyvinyl chloride	1.47	74*	82	0.04–0.4	Good	None
(b) Rigid vinyl chloride/vinyl acetate	1.55	63–77†		0.04–0.4	Good	None
(c) Rubber modified PVC	1.89	71*	78		Good	None
Polyacetals	2.3	169*	175	1.0–1.1	Good	None
Polycarbonates	1.93	138–143	165		Excellent	None
Ionomers	2.4	38–43	68–76		Good	None
Polyphenylene oxide	5.2–6	110–130	125–148	0.9–1.1	Good	None
Polyphenylene sulphide		217–260				None
Acrylics	1.68–2.52	71–95	80–98	0.6–1.3	Good	None
Ethylene vinyl acetate	0.3	60–64	64	1.02–1.06	Very Good	None
Fluorinated polymers						
(a) Fluorinated ethylene propylene	2.1	121*	M.P. 285–295		Excellent	None
(b) Polytetrafluoroethylene	2.52		Trans Pnt 372		Excellent	None
(c) Polytrifluorochloroethylene	—	91–144*	Dec Pnt >400		Good	None
Cellulosic plastics						
(a) Cellulose acetate	1.68–3.37	43.5–99*	70		Good	Slight smell
(b) Cellulose acetate butyrate	1.68–3.37	46–108*	70		Good	Smell of rancid butter
(c) Cellulose propionate	1.68–3.37	43.5–121*		1.0–1.3	Good	Slight smell
Metals and alloys		Melting Point	Numerical flame spread ratings cannot directly reflect behaviour under actual fire conditions and can be no more than a very general indicator. Performance of test pieces may be radically different from that of plastic products in fire situations and all polymers will burn with varying degrees of intensity under suitable conditions.		Impact strength rises as temperature falls	
(a) Brass (Cu-Zn 70/30)	1210	935			Good	
(b) Mild steel (0.6% Carbon)	631	1515			Good	
(c) Carbon steel (0.4% Carbon)	505	1495			Good	
(d) Aluminium (99% Pure)	2180	648			Good	
(e) Duralumin (Al-Si-Cu-Mg)	1470	583			Good	
(f) Phosphor bronze (Cu-Sn-P)	840	930			Good	

TABLE 1 Contd.

	Specific volume ASTM D792 (ml kg⁻¹)	Refractive index ASTM D542 (ηD)	Clarity transparency	Effect of sunlight	Machining qualities	Mouldability	Mould Shrinkage (a) Injection (b) Compression (mm. mm⁻¹)
Phenol formaldehyde							
(a) Unfilled	800–771	1.5–1.7	Transparent to translucent	Surface darkens; loss of gloss	Fair	Fair-Good	b 0.01–0.02
(b) Woodflour/cotton flock filled	756–645	—	Opaque	General darkening	Excellent	Fair-Good	b 0.004–0.009
Urea formaldehyde							
Cellulose filled	682–659	1.54–1.56	Transparent to opaque	Colour fade to grey	Fair	Excellent	b 0.006–0.014
Melamine formaldehyde							
(a) Unfilled	678	—	Opalescent	(a) Colours fade: surface deterioration	—	Good	b 0.011–0.012
(b) α-Cellulose filled	682–659	—	Translucent	(b) Yellowing	Fair	Excellent	b 0.005–0.01
Polyesters							
(a) Resin	910–715	1.52–1.58	Transparent to opaque	Yellowing	Good	—	cast 0.08
(b) Dough moulding compound	690–435	—	Opaque	Subject to deterioration	Good	Good	b 0.001–0.012
(c) Sheet moulding compound	590–400	—	Opaque	Subject to deterioration	Good	Good	b 0.001–0.004
Epoxides							
(a) Rigid	832	1.6	Transparent to opaque	Slight yellowing	Very Good	Very Good	—
(b) Flexible	832	—	Transparent to opaque	Slight yellowing	Good-Poor	Very Good	—
Nylons							
(a) Type 6	886–875	—	Translucent to opaque	Embrittles; improved by stabilizers	Excellent	Excellent	a 0.009
(b) Type 6/6	921–875	1.53	Translucent to opaque		Excellent	Good	a 0.015
(c) Type 11	960–950	1.52	Translucent		Very Good	Good	a 0.012
(d) Type 12	990–980	—	Translucent to opaque		Very Good	Good	a 0.003–0.015
Polymerization types							
Polyethylenes							
(a) Low density	1091–1061	1.51	Near transparent to opaque	Surface crazing; embrittles; improved by stabilizers/pigments	Fair	Excellent	a 0.02–0.035
(b) High density	1061–1037	1.54	Translucent to opaque	As for Polyethylenes	Excellent	Good	a 0.02–0.035
Polypropylene	1119–1048	1.49	Transparent to Translucent		Excellent	Excellent	a 0.015–0.025
Polystyrenes							
(a) Conventional	953–896	1.59–1.6	Transparent	Yellowing and loss of strength; unsuitable for outdoor use	Fair-Good	Excellent	a 0.002–0.006
			Translucent		Good	Excellent	b 0.001–0.008
							a 0.002–0.008

Material		(refractive index)	Appearance	Effect of sunlight	Weathering	Machining	Water absorption (%)
Styrene acrylonitrile	932–911	1.57	Transparent	Slight yellowing	Good	Good	a 0.002–0.005
Acrylonitrile/butadiene/styrene	1012–911	—	Translucent	Yellowing; some embrittlement	Good	Good	a 0.003–0.008; b —
Vinyl polymers							
(a) Rigid polyvinyl chloride	728	1.52	Transparent to opaque	Slight in suitably stabilized grades	Excellent	Fair–Good	a 0.001
(b) Rigid vinyl chloride/vinyl acetate	939–691	1.55	Transparent to opaque		—	Fair–Good	↕
(c) Rubber modified PVC	742	—	Transparent to opaque	Slight; varies with formulation		Fair–Good	a 0.05
Polyacetals	706	1.48	Translucent to opaque	Powdering; embrittles; improved by stabilizers	Excellent	Good	a 0.025 av
Polyethylenes	832	1.584	Transparent	Crazing; loss of impact	Excellent	Good	a 0.006–0.008
Ionomers	1060	1.51	Transparent	Embrittles; improved by stabilizers	Fair–Good	Good	a 0.003–0.02
Polyphenylene oxide	945–910	—	Opaque	Colours may fade	Very Good	Very Good	a 0.005–0.007
Polyphenylene sulphide	—	—		—	—	—	—
Acrylics	857–835	1.49	Transparent	Virtually unaffected	Excellent	Excellent	a 0.002–0.008; b 0.001–0.004
Ethylene vinyl acetate	1070	—	Transparent to opaque	Embrittles; improved by stabilizers	Fair	Good	a 0.007–0.012; b —
Fluorinated polymers							
(a) Fluorinated ethylene propylene	466–459	1.34	Translucent	Unaffected	Good	Good	a 0.001–0.005; b 0.003–0.005
(b) Polytetrafluoroethylene	475–433	1.35	Translucent to opaque	Unaffected	Good	Excellent (special technique)	a —; b 0.03–0.06
(c) Polytrifluorochloroethylene	475–463	1.43	Transparent to opaque	Unaffected		Fair	ab 0.003–0.007
Cellulosic plastics							
(a) Cellulose acetate	815–746	1.46–1.5	Transparent	Slight	Excellent	Excellent	a 0.002–0.005
(b) Cellulose acetate butyrate	867–821	1.46–1.49	Transparent	Slight	Excellent	Excellent	a 0.002–0.005
(c) Cellulose propionate	846–815	1.46–1.49	Transparent	Slight	Excellent	Excellent	a 0.001–0.005
(c) Cellulose propionate	846–815	1.46–1.49	Transparent	Slight	Excellent	Excellent	b 0.003–0.005
Metals and alloys							
(a) Brass (Cu–Zn, 70/30)	126	—			Ratings have been assessed from exposure of standard specimens in a temperate climate. As the effect of sunlight is influenced by many factors these ratings can give only a general guide, and should not be considered as an absolute indicator of outdoor performance.	Excellent though the cutting speed and/or lubricant is of course important	—
(b) Mild steel (0.6% Carbon)	127	—					—
(c) Carbon steel (0.4% Carbon)	134	—					—
(d) Aluminium (99% Pure)	395	—					—
(e) Duralumin (Al–Si–Cu–Mg)	359	—					—
(f) Phosphor bronze (Cu–Sn–P)	113	—					—

TABLE 1 Contd.

	Water absorption ASTM D 570 (1/8" thickness % 24 hours)	Permeability (ml cm^{-2} s^{-1} mil^{-1} cm Hg^{-1} at 25°C × 10^{-3})			Resistance (chemical) ASTM D 543 (See key at foot of table)	Resistance (solvent) ASTM D 543 (See key at foot of table)
		H$_2$O vap	O$_2$	CO$_2$		
Phenol formaldehyde						
(a) Unfilled	0.1–0.2	—	—	—	A/B, d/c, a/b, a/c	Unaffected
(b) Woodflour/cotton flock filled	0.3–1	—	—	—	A/B, A/C, a/b, A/c	Unaffected
Urea formaldehyde						
Cellulose filled	0.4–0.8	—	—	—	d/B, d/C, a/b, a/c	Unaffected
Melamine formaldehyde						
(a) Unfilled	0.3–0.5	—	—	—	A/B, A/C, U/b, U/c	Unaffected
(b) α-cellulose filled	0.1–0.6	—	—	—	A/B, A/C, U/b, U/c	Unaffected
Polyesters						
(a) Resin	0.15–0.6	—	—	—	a/B, A/C, U/b, A/c	Attacked by most solvents A/t, A/K, a/T
(b) Dough moulding compound	0.06–0.28	—	—	—	U/B, A/C, U/b, a/c	a/t, a/K, a/T
(c) Sheet moulding compound	0.1–0.15	—	—	—	U/B, A/C, U/b, a/c	
Epoxides						
(a) **Rigid**	—	—	—	—	U/B (mineral), U/b U/C, U/c	U/h, U/H, a/t, a/k
(b) Flexible	—	—	—	—	Increased attack compared with rigid	U/p
Nylons						
(a) Type 6	0.9–3.3	2800	0.12	0.5	A/B, U/C, U/b, U/c	Unaffected
(b) Type 6/6	0.4–1.5	—	—	—	A/B, U/C, U/b, U/c	Unaffected
(c) Type 11	0.4	166	1.6	7	A/B, U/C, U/b, U/c	a/t
(d) Type 12	0.25	36.3	2.4	10	A/B, U/C, U/b, U/c	a/t
Polymerization types						
Polyethylenes						
(a) Low density	<0.015	420	15	55	A/B, U/C, U/b, U/c	S/T above 60°C
(b) High density	<0.01	60	—	13	A/B, U/C, U/b, U/c	Unaffected below 80°C
Polypropylene	<0.01	160	4	12	A/B, U/C, U/b, U/c	Unaffected below 80°C
Polystyrenes						
(a) Conventional	0.03–0.4	4700	4	35	A/B, U/C, A/b, U/c	S/e, h, T, t, K
(b) Toughened	0.1–0.3	—	7.2	49.6	A/B, U/C, A/b, U/c	S/e, h, T, t, K

Styrene acrylonitrile	0.2–0.3	—	—		A/B, U/C, A/b, U/c	S/e, t, K
Acrylonitrile/butadiene/styrene	0.1–0.3	—	—		A/B, U/C, A/b, U/c	S/e, S/t, S/K
vinyl polymers						
(a) Rigid polyvinyl chloride	0.05	0.4	0.7	630	U/B, U/C, U/b, U/c	S/e, S/K, A/T
(b) Rigid vinyl chloride/vinyl acetate	0.08	—	—	—	U/B, U/C, U/b, U/c	S/e, S/K, A/T
(c) Rubber modified PVC	0.1	—	—	—	U/B, U/C, U/b, U/c	S/e, S/K, A/T
Polyacetals	0.12	—	—	—	A/B, A/C, variable bc	Unaffected
Polycarbonates	0.3	—	—	—	a/B, a/c, U/b, U/c	Unaffected
Ionomers	0.1–1.4	—	18	1260	A/B, a/b, a/C, a/c	Slightly attacked by most solvents
Polyphenylene oxide	0.07–0.08	—	—	—	U/B, U/C, U/b, U/c	a/c, a/T, u/p, U/K
Polyphenylene sulphide	0.05–0.06	—	—	—	—	—
Acrylics	0.3–0.4	—	—	—	A/B, A/C, a/b, a/c	S/K, S/e, S/t
Ethylene vinyl acetate	0.05–0.13	132	59.2	2800	A/B, U/b, U/C, U/c	A/t, A/T
Fluorinated polymers						
(a) Fluorinated ethylene propylene	0	—	—	—	U/B, U/C, U/b, U/c	Unaffected
(b) Polytetrafluoroethylene	0	—	—	—	U/B, U/C, U/b, U/c	Unaffected
(c) Polytrifluorochloroethylene	0	—	—	—	U/B, U/C, U/b, U/c	a/t in some cases
Cellulosic plastics						
(a) Cellulose acetate	1.9–6.5	27–28	3–4	29,000–47,000	d/B, d/C, a/b, a/c	S/K, S/e, S/H
(b) Cellulose acetate butyrate	0.9–2.2	—	—	—	d/B, d/C, a/b, a/c	S/K, S/e, S/H
(c) Cellulose propionate	1.2–2.8	—	—	—	d/B, d/C, a/b, a/c	S/K, S/e, S/H
Metals and alloys						
(a) Brass (Cu-Zn, 70/30)	—					
(b) Mild Steel (0.6% Carbon)	—					
(c) Carbon Steel (0.4% Carbon)	—					
(d) Aluminium (99% Pure)	—					
(e) Duralumin (Al-Si-Cu-Mg)	—					
(f) Phosphor Bronze (Cu-Sn-P)	—					

In interpretation of this section, which acts as a guide to chemical and solvent resistance, note that considerations of temperature are all-important.

s = slightly soluble	d = decomposed	p = paraffins
S = soluble	h = higher alcohols	b = dilute acids
i = insoluble	P = alcohols	B = strong acids
U = unaffected	T = aromatics	c = dilute alkalis
a = slightly affected	t = chlorinated hydro-carbons	C = strong alkalis
A = attacked		K = ketones
		e = esters

Example: d/C = decomposed by strong alkalis.

TABLE 2 *Uses of polymers*

Polycondensation types		
Phenol formaldehyde	(a) Unfilled	(a) Adhesives, laminates, pulp mouldings, particle board.
	(b) Woodflour/cotton flock filled	(b) Bottle tops, electrical parts, fuse boxes, meter cases, heat-resistant close-tolerance mouldings, toilet seats, restricted in colours obtainable, dark coloured plastic ashtrays.
Urea formaldehyde	Cellulose filled	As for, cellulose filled melamine formaldehyde but unsuitable for dinner ware. U/F resins are used for similar applications to those shown under unfilled phenol formaldehyde, white electrical plugs.
Melamine formaldehyde	(a) Unfilled	(a) Usually occurs in laminate form as surfacing for tables, etc.
	(b) α-cellulose filled	(b) Noted for durability, hardness and good electrical properties, suitable for appliance housings, dinnerware, closures, writing equipment, clock housings, knobs, handles, lighting fixtures, appliances, instruction panels.
Polyesters	(a) Resin	(a) Unreinforced resin for buttons, surface coatings, embedding and potting and nut locking. Filled resin for imitation marble, flooring, pipe joints, mortars and body stoppers.
	(b) Dough moulding compound	(b) Protective housings, connectors, cowls, guards and ducts. Components often replace metal, offering non-corrosion, durability, good electrical performance and high strength.
	(c) Sheet moulding compound	(c) Outlets in the electrical, building, motor engineering and furniture industries that compete on a cost basis with die castings and sheet metal fabrications owing to ease of moulding complicated shapes and short moulding cycles.
Epoxides		Chemically resistant paints, adhesives, tools, PVC stabilizers, electrical insulation, chemical- and wear-resistant jointless flooring, road coatings, cements, laminates, powder coatings, stopping compounds, repair kits, printed circuits, filament wound pipes, tanks and pressure vessels.
Nylons	(a) Type 6	(a) Moulded mechanical parts, gear wheels, bushings, sliding parts for storm windows, automobile and refrigerator door closures, mixer valves, switch housings, grommets, cable clamps, pipes, tubing, filaments, aerosol bottles, stockings, clothing, zips, curtain fittings.
	(b) Type 6/6	(b) As for Type 6.
	(c) Type 11	(c) Electro/mechanical components such as cams, housings, guides, terminal blocks, transformers and bobbins. Tubing and hose for automotive use, reinforced hose, technical components in copiers, calculators, machinery and tools.
	(d) Type 12	(d) Injection moulded parts for automotive, electrical and electronic and precision machine industries. Semi-rigid or flexible tubing for fuel lines, air brakes, pressurised air lines. Cable and wire sheathing, chill roll and blown film. Powder coatings.

Polymerization types	
Polyethylenes (a) Low density	(a) Housewares (bottles, bowls, buckets, containers), closures for squeeze tubes, spouts for detergent cans, carboys, shoe parts, toys, packaging film, garment bags, sheet, piping for domestic, industrial and agricultural use, cable and wire insulation and sheathing, paper, cellulose and foil coating, carpet backing, monofilament, cold water storage tanks.
(b) High density	(b) As for low-density polyethylene. Material have greater rigidity, specially suitable for large carrying cases, housings, closures, appliance parts, packaging film, sheet, piping, carboys, containers, bottles.
Polypropylene	Domestic, hospital and laboratory ware, textile, automotive, electrical and industrial usages. Containers and closures, crates, toys, blown containers. Film fibre for baler twine, ropes, sacks and carpet backing. Fibre for carpet face yarns. Extruded sheet, pipe, film and filament.
Polystyrenes (a) Conventional	(a) Packaging (disposable and others), dishes and utensils, refrigerator parts, emblems, signs, displays, toys, novelties, combs, brush-backs, radio and TV cabinets, lighting fixtures, rigid containers, housewares, reels/spools for film/tape, appliance panels, handles and switches.
(b) Toughened	(b) Wheels, helmets, valve parts, refrigerator parts, bobbins, radio cabinets, electric fan blades, toys, housewares, containers, battery cases, refrigerator linings and trays, yoghurt pots, footwear.
Styrene acrylonitrile / Acrylonitrile/ butadiene/ styrene	Cups, tumblers, trays and general table, kitchen and picnic ware, toothbrush handles, refrigerator components, radio knobs and scales, lenses, cosmetic items, hi-fi covers and cases, packaging. Shoe heels, telephone handsets, housings for consumer durables, food containers, luggage, refrigerator liners, safety helmets, radio cabinets, tote boxes, car fascia panels, instrument clusters, boat hulls, furniture.
Vinyl polymers (a) Rigid polyvinyl chloride	(a) Extrusion of piping, profiles and sheet in applications requiring chemical inertness and scuff-resistance combined with light weight. Plastisols for toys, leathercloth, etc. Plastic guttering, high clarity bottles, formed packaging trays, moulded containers.
(b) Rigid vinyl chloride/acetate	(b) Similar applications to those for PVC but widely used in the manufacture of calendered sheet used for toys, novelties, wall coverings, displays, templates, etc. Gramophone records.
(c) Rubber modified PVC	(c) Same as for PVC.
Polyacetals	Load bearing mechanical parts, small pressure vessels, aerosol containers, gears, business machine parts, appliances, automobile, engineering and industrial products.
Polycarbonates	Business machine parts, camera components, electrical apparatus, sterilisable ware, draughtsman's instruments, lamp covers, safety helmets, tail-lights, de luxe housewares, engineering and industrial components, sterilisable transparent feeding bottles for babies.
Ionomers	Shoe heel tips, tool handles, hammer and mallet heads, bottles, skin packaging, coating, toys, shoe soles, shoe stiffeners, meat packaging, flexible packaging, packaging of wine and fruit juices.
Polyphenylene oxide	TV set components, valve bases, switches, housings, parts for domestic appliances, meter cases, machine housing, computer and camera parts, automotive grilles, ducts, light housing, instrumentation parts.

TABLE 2 *Contd.*

Polymer	Applications
Polyphenylene sulphide	Engineering plastic, high heat and chemical resistance applications. Non-flammability applications.
Acrylics	Automatic parts, control knobs, dials and handles, meter cases, lenses, pens and pencils, brush-backs, hospital equipment, display material, signs, light fittings, inspection panel covers, windscreens, machine guards, skylights, some telephones, sanitary ware, TV tube implosion guards.
Polyvinylacetate	Foodstuff packaging film.
Ethylene vinyl acetate	Flexible extrusions, tubing and hose, sachets, sheathing, cable coverings. Closures, gaskets, handle grips, shoe soles, teats, disposable gloves, box liners, packaging film, greenhouse film, inflatable toys.
Polyoxyalkylene glycols	Waterproofing of paper wrappings, wood preservation.
Polyacrylonitriles	Synthetic film (e.g. Acrilan).
Polyurethanes	Thermal insulation panels, sealant material, e.g. battery containers.
Polysulphones	Engineering plastic, replacement for stainless steel.
Fluorinated polymers (a) Fluorinated ethylene propylene	Coil formers, terminal blocks, valve holders, wire insulation, electronic components, encapsulations and fluidised bed coatings, non-stick valves.
(b) Polytetrafluoroethylene	Gaskets, packings, valves, sintered metal bearings, rigid and flexible pipes, membranes, wire insulations, electronic engineering applications, non-stick coatings for kitchen utensils, heat sealing equipment and confectionery machinery, tanks.
(c) Polytrifluorochloroethylene	Extruded sheet, profile and film, electronic parts, gaskets, pump sealants, dispersion coatings, liquid level indicators of particular use where resistance to aggressive chemicals is needed.
Cellulosic plastics (a) Cellulose acetate	Toys, beads, cutlery handles, electrical parts, knobs, steering wheels, shoe heels, packaging sheeting, toothbrushes, cosmetics, windows in window cartons.
(b) Cellulose acetate butyrate	Moulded or extruded parts for metallization (reflectors etc.), outdoor signs, automobile tail-light covers, tool handles, toothbrushes, pipe inspection traps, piping.

2

Determination of elements

VARIOUS elements occur in polymers, metallic and non-metallic, such as chlorine in polyvinylchloride. These elements can be divided into three categories:

1. Elements which are a constituent part of the monomers used in polymer manufacture, such as nitrogen in acrylonitrile used in the manufacture of, for example, acrylonitrile–butadiene–styrene terpolymers.
2. Elements which occur in substances deliberately included in the polymer formulation such as, for example, zinc stearate which is incorporated in some polystyrene formulations.
3. Elements which occur as adventitious impurities in the polymer. For example, during the manufacture of polyethylene by the low-pressure process polymerization catalysts such as titanium halides and organo-aluminium compounds are used, and the final polymer would be expected to, and indeed does, contain traces of aluminium, titanium and chlorine residues.

Generally, speaking elements covered by type (1) occur in the percentage range, and those covered by types (2) and (3) occur in the parts per million range.

For a variety of reasons, ranging from control during the manufacturing process to the identification of unknown polymers, it is necessary to be able to determine accurately the elements in polymers, and many different types of techniques have been evolved for this purpose. Generally speaking the techniques used come under two broad headings: (1) adaptations of classical techniques, and (2) instrumental techniques such as X-ray fluorescence and neutron activation analysis.

2.1 Classical techniques

The classical techniques are generally based on one of four possible approaches to the analysis: (a) dry ashing of the polymer with or without an ashing aid, (b) fusion of the polymer with an inorganic compound to effect solution of the elements, (c) acid digestion techniques and, finally, (d) bomb digestion techniques.

2.1(a) Dry ashing methods

Dry ashing without ashing aid

This technique, which simply involves combustion of the polymer under controlled conditions in a platinum crucible – followed by solution of the residual ash in a suitable aqueous reagent prior to final analysis by spectrophotometry, is of limited value. It has been used for the determination of parts per million sodium and lithium in polyolefins. A quite complicated and lengthy ashing programme is necessary in this technique to avoid losses of alkali metal during the ignition.

Time from start   0–1 hour,   heat to 200°C
                  1–2 hours,   hold at 200°C
                  3–5 hours,   heat to 450°C
                  5–8 hours,   hold at 450°C

The disadvantages of such techniques are immediately apparent. Dry ashing in platinum has been found to give reasonably good results in the determination of low concentrations of vanadium in ethylene–propylene; 10 g of polymer is ashed in platinum by charring on a hot plate followed by heating over a Mekér burner. Dilute nitric acid is added to the residue and any residue in the crucible dissolved by fusion with potassium persulphate. The vanadium is determined spectrophotometrically by the 3, 3′-diaminobenzene method.[1] The table below compares results obtained by this method with those obtained by neutron activation analysis, which in this case can be considered to be an accurate reference method. Quite a good agreement was obtained between the two methods from samples containing above 1 ppm vanadium.

Vanadium (ppm)

Sample	Dry ashing	Neutron activation
A	10.2	9.9 $\pm$ 0.2
B	14.0	14.1 $\pm$ 0.1
C	14.6	15.6 $\pm$ 0.3
D	0.5	0.14 $\pm$ 0.01
E	13.0	14.8 $\pm$ 0.2
F	0.9	0.27 $\pm$ 0.01
G	15.2	18.8 $\pm$ 0.3
H	18.2	17.9 $\pm$ 0.3

Dry ashing with ashing aid

It has been shown[2,3] in studies using a radioactive copper isotope that, when organic materials containing copper are ashed, losses of up to some 10% of the copper will occur due to retention in the crucible, and this could not be removed by acid washing. Virtually no retention of copper in the silica crucible occurred, however, when copper was ashed under the same conditions in the absence of added organic matter. This was attributed to reduction of copper to the metal by organic matter present, followed by partial diffusion of the copper metal into the crucible wall. This effect also occurs with elements other than copper, and would be responsible for the return of low metal recoveries in polymers. One method of reducing or eliminating such losses is to ignite the polymer after it has been initially mixed with an inorganic salt solution such as magnesium nitrate which, upon ignition, provides a matrix of magnesium oxide in which the element being determined is retained and thereby prevented from being lost in the crucible material. The metallic oxide residue is then dissolved in a suitable acid and analysed by conventional techniques. To obtain intimate contact between polyethylene powder and the ashing acid during the preliminary mixing it was necessary to dissolve the magnesium nitrate in a 70:30 v/v mixture of distilled water and redistilled methyl alcohol. Following ignition at 600°C any residual carbon particles are dissolved by digestion with 1:3 concentrated sulphuric–nitric acid and the copper determined by the sodium diethyl dithiocarbamate spectrophotometric procedure. Distinctly higher copper determinations were obtained on polyolefins by the procedure involving the use of an ashing

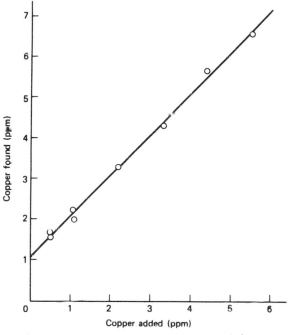

FIG. 1. Recovery of added copper to polyethylene.

aid than are obtained without an ashing aid, or by the use of a molten potassium bisulphate fusion technique (see section 2.1(b)) to take up the polymer ash. This is confirmed in Fig. 1, which shows copper recoveries obtained by the above technique when known additions of copper are made to a polyethylene whose original copper content was 1.1 ppm.

Low-temperature ashing

Low-temperature ashing furnaces are available commercially. The principle behind these is that the oxygen in the atmosphere over the sample to be ashed is activated by exposure to an electric discharge and/or ultraviolet light. This activated oxygen combusts the polymer at a lower temperature then it would combust under normal ashing conditions, with the result that, in addition to a more rapid ashing, risks of losses of the element are reduced. The method has been used for the determination of trace metals in polypropylene,[4] polystyrene,[4] polyvinyl-chloride[4] and polyethylene terephthalate.[4]

2.1(b) Fusion methods

Fusion with sodium carbonate

This method is very useful for the fusion of polymers which, upon ignition, release acidic vapours – for example polyethylene containing traces of chlorine, or polyvinylchloride, both of which, upon ignition, release anhydrous hydrogen chloride. To determine chlorine accurately in the polymer in amounts down to 5 ppm this acid must be trapped in a solid alkaline reagent such as sodium carbonate. In this method polyethylene is mixed with pure sodium carbonate and ashed in a muffle furnace at 500°C. The residual ash is dissolved in aqueous nitric acid, and then diluted with acetone. This solution is titrated potentiometrically with standard silver nitrate. Alternatively, chloride can be determined by another appropriate method.

Procedure – determination of chlorine in polyolefins

Accurately weigh 5 g of polymer into a metal crucible and cover the polymer with a layer of 2 g of sodium carbonate. Place the crucible in a cold muffle furnace and allow the temperature to increase gradually to 500°C, maintaining this temperature for 4 h. Allow the crucibles to cool, then dissolve the residue in 10 ml of distilled water. Transfer this solution with crucible washings to a beaker (30 ml) and add five drops of screened methyl orange indicator solution. Neutralize the mixture by dropwise addition of 30% nitric acid to the purple–red-coloured end-point. Add a further 10 drops of 30% nitric acid solution to the beaker and then add 30 ml of acetone. Titrate the resultant solution potentiometrically with silver nitrate solution (N/100). Preferably using an automatic titrator equipped with a silver measuring electrode and a glass reference electrode. Carry out a blank determination exactly as described above, omitting only the sample addition.

Calculation   ppm (w/w) chlorine in polymer

$$= \frac{(T_S - T_B) \times N \times 35.46 \times 10^3}{W}$$

where T_B = titration of silver nitrate (ml) in blank determination
  T_S = titration of silver nitrate (ml) in sample determination
  N = normality of silver nitrate solution
  W = weight of polymer sample in grams.

Table 3 compares results for chlorine determinations in polyethylene obtained by this method with those obtained by an accurate independent referee method – in this case X-ray fluorescence spectroscopy (see later). The averages of results obtained by the two methods agree satisfactorily within ± 15% of each other. The superior reproducibility of the chemical method of analysis is believed to be due to the fact that a considerably higher weight of sample is taken for analysis than is used in the X-ray method of analysis. Thus, whereas in the chemical method one obtains the average chlorine content of 5 g of polymer, in the X-ray method the amount analysed is only a few microns thickness of the disc of polymer taken for analysis. Any local variations in the chlorine content of the polymer will be picked up much more readily, therefore, by the X-ray method of analysis.

TABLE 3   *Comparison of chlorine contents by X-ray method and chemical method (averages in parentheses)*

X-ray on discs	Chemical method on same disc as used for X-ray analysis	Chemical method on powder
865, 815	700	786, 761
(840)		(773)
535, 570	606	636, 651
(552)		(643)
785, 675	598	650, 654
(730)		(652)
625, 675	600	637, 684
(650)		(660)
895, 870	733	828, 816
(882)		(822)

Fusion with sodium bisulphate

Sodium bisulphate is another useful reagent for the fusion of polymer samples prior to the determination of metals. It has the advantage of forming a very high-temperature melt in which the oxides of many metallic elements dissolve and are put in a chemical form suitable for subsequent analysis. The technique has been used, for example, to determine values down to one part per million of titanium, iron and aluminium in dried low-pressure polyolefins. The ignition takes place in clean silica crucibles over 4 h at 800°C. All that is then necessary to do is to dissolve the cooled residue in warm dilute sulphuric acid. The three metals can be determined by classical spectrophotometric techniques involving the use of Tiron, thioglycollic acid and aluminun, or by other suitable techniques.

Although, nowadays, these analyses would possibly be carried out by modern physical techniques such as X-ray fluorescence, there are no doubt circumstances (such as small one-off batches of samples) where the sodium bisulphate technique is still of value to the analyst. The technique would also be applicable to the solution of polymers preparatory to the application of techniques such as atomic absorption spectrometry.

Fusion with sodium peroxide

Sodium peroxide is another useful reagent for the fusion of polymer samples preparatory to analysis for metal such as zinc, and non-metals such as chlorine[5] and bromine. In this method the polymer is intimately mixed either with sodium peroxide in an open crucible[6] or with a mixture of sodium peroxide and sucrose in a micro-Parr bomb. After acidification with nitric acid chlorine can be determined complexometrically using EDTA.[6] In a method for the determination of traces of bromine in polystyrene in amounts down to 100 ppm bromine a known weight of polymer is mixed intimately with pure sodium peroxide and sucrose in a micro-Parr bomb which is then ignited. The sodium bromate produced is converted to sodium bromide by the addition of hydrazine sulphate

$$2NaBrO_3 + 3NH_2NH_2 = 2NaBr + 6H_2O + 3N_2$$

The combustion mixture is dissolved in water and acidified with nitric acid. The bromine content of this solution is determined by potentiometric titration with standard silver nitrate solution. The organic bromo content of the polymer can then be calculated from the determined bromine content.

2.1(c) Acid digestion methods

Apart from the classical Kjeldahl digestion procedure for the determination of organic nitrogen, acid digestion of polymers has found little application. One of the problems is connected with the form in which the polymer sample exists. If it is in the form of a fine powder, or a very thin film, then digestion with acid might be adequate to enable the relevant substance to be quantitatively extracted from the polymer. Thus, reliable nitrogen contents can be obtained for fine polymer powders using the Kjeldahl digestion procedure utilizing a mixture of concentrated sulphuric acid, potassium sulphate and an oxidation catalyst. However, low nitrogen results would be expected for polymers in larger granular form, and for the analysis of such samples classical microcombustion techniques are recommended.

Similar considerations apply in the case of metals. Antimony, present as the trioxide, has been accurately determined in a concentrated hydrochloric acid extract of polypropylene powder.[7] Arsenic in the same form has been determined in acrylic fibres by hot extraction with a mixture of concentrated nitric, perchloric and sulphuric acid.[8] In this latter procedure arsenic is separated from interfering elements in the acid digest, and determined by atomic absorption spectrophotometry as the volatile arsenic hydride. Arsenic recoveries of between 94 and 106% were obtained by this procedure, which was capable of determining arsenic in acrylic fibres in amounts down to 0.04 ppm with a relative standard deviation of 7%.

Another reason why the Kjeldahl sulphuric acid digestion procedure is not usually applicable to the digestion of polymer powders, and certainly not to polymer granules prior to the determination of metals, is that the Kjeldahl digestion catalyst, usually a mercury, copper on selenium salt, would in all likelihood interfere in methods for the determination of traces of metals in digests.

2.1(d)   Oxygen flask combustion method

This well-known technique has found several applications in the analysis of polymers, including the determination of sulphur in polyolefins, chlorine in polyvinyl chloride,[9] chlorine in polyolefins and in chlorobutyl rubber,[10] and phosphorus in polyolefins.[11]

Oxygen flask combustion methods have been used to determine traces of chlorine in polyolefins at levels between 0 and 500 ppm. The Schoniger oxygen flask combustion technique requires a 0.1 g sample and the use of a 1-litre conical flask. Chlorine-free polyethylene foil is employed to wrap the sample, which is then supported in a platinum wire attached to the flask stopper. Water is used as the absorbent. Combustion takes place at atmospheric pressure in oxygen. The chloride formed is potentiometrically titrated in nitric acid/acetone medium in a small beaker using 0.01 N silver nitrate solution added from a syringe microlitre burette.

To determine sulphur in amounts down to 500 ppm in polyolefins the sample is wrapped in a piece of filter paper and burnt in a closed conical flask filled with oxygen at atmospheric pressure. The sulphur dioxide produced in the reaction reacts with dilute hydrogen peroxide solution contained in the reaction flask to produce an equivalent amount of sulphuric acid. The sulphuric is estimated by visual titration with N/1000 or N/100 barium perchlorate using Thorin indicator.

$$H_2SO_3 + H_2O_2 = H_2SO_4 + H_2O$$

The repeatability of the method is $\pm 40\%$ of the determined sulphur content at the 500 ppm sulphur level, improving to $\pm 2\%$ at the 1% level. Chlorine and nitrogen concentrations in the sample may exceed the sulphur concentration several times over without causing interference. Fluorine does not interfere unless present in concentrations exceeding 30% of the sulphur content. Phosphorus interferes even when present in moderate amounts. Metallic constituents also interfere when present in moderate amounts.

2.2 Instrumental techniques

Instrumental techniques for determining elements in polymers can be broadly divided into two categories:

1. Non-destructive techniques: these include techniques such as X-ray fluorescence and neutron activation analysis, in which the sample is not destroyed during the analysis.
2. Destructive techniques: these are techniques in which the sample is decomposed by a reagent such as those discussed in the previous section, and then the concentration of the element in the extract is determined by a physical technique such as atomic absorption spectrometry, emission spectrography, flame photometry, polarography or any one of the increasing repertoire of physical methods of analysis.

These techniques are now briefly discussed, with some examples of their application to particular problems.

2.2(a) Non-destructive techniques

X-ray fluorescence spectrometry

X-ray fluorescence spectrometry has been used extensively for the determination of traces of metals and non-metals in polyolefins and other polymers. The technique has also been used in the determination of major metallic constituents in polymers, such as cadmium selenide pigment in polyolefins.

The X-ray fluorescence spectrometric method has several advantages over other methods. The analysis is non-destructive; specimen preparation is simple, involving compressing a disc of the polymer sample for insertion in the instrument; measurement time is usually less than for other methods; and X-rays interact with elements as such, i.e. the intensity measurement of a constituent element is independent of its state of chemical combination. However, the technique does have some drawbacks, and these are evident in the measurement of cadmium and selenium. For example, absorption effects of other elements present, e.g. the carbon and hydrogen of the polyethylene matrix, excitation of one element by X-rays from another, e.g. cadmium and selenium mutually affect one another. The technique has been applied to the determination of metals in polybutadiene, polyisoprene and polyester resins.[12] The metals determined were iron, cobalt, nickel, copper and zinc. The samples were ashed and the ash dissolved in nitric acid prior to X-ray analysis. No separation schemes are necessary and concentrations as low as 10 ppm can be determined without inter-element interference. A solution technique was used (rather than examination of the solid compressed polymer) to circumvent inter-element and matrix effects. Nitric acid was used because, unlike the other mineral acids, it does not have a strong absorbing effect on the X-ray fluorescence of these metals. In general the theoretical and found values agree within $\pm 10\%$. The determination of chromium tended to be the most erratic, relative to the other metals. Possibly the reason for the good recoveries after dry ashing found by these workers was due to the relatively low maximum temperature, 550°C, that was used.

X-ray fluorescence spectrography has been applied very successfully, industrially, to the routine determination in hot-pressed discs of polyethylene and polypropylene of down to a few parts per million of the following elements: chlorine, bromine, titanium, aluminium, sodium, potassium, calcium, magnesium and vanadium.

TABLE 4 *Determination of chlorine by X-ray fluorescence procedure*

A: Polymer not treated with alcoholic potassium hydroxide before analysis, ppm chlorine. X-ray fluorescence on polymer discs. Average of two discs (A)	B: Polymer treated with alcoholic potassium hydroxide before analysis, ppm chlorine. X-ray fluorescence on polymer discs. Average of two discs (B)	Difference between average chlorine contents obtained on potassium hydroxide-treated untreated samples (B)–(A)
510	840	330
422	552	130
440	730	290
497	650	153
460	882	422

An interesting phenomenon was observed in applying the method to the determination of parts per million of chlorine in hot-pressed discs of low-pressure polyolefins. In these polymers the chlorine is present in two forms, organically bound and inorganic, titanium–chlorine compounds resulting as residues from the polymerization catalyst used in this process. The organic part of the chlorine is determined by X-ray fluorescence without complications. However, during hot processing of the discs there is a danger that some inorganic chlorine will be lost. This can be completely avoided by intimately mixing the powder with alcoholic potassium hydroxide, then drying at 105°C before hot pressing into discs. The results (Table 4) illustrate this effect. Considerably higher total chlorine contents were obtained for the alkali-treated polymers.

Neutron activation analysis

This is another non-destructive technique capable of determining a wide range of elements – for example chlorine in polyolefins, metals in poly-methylmethacrylate,[13] total oxygen in polyethylene–ethylacrylate and polyethylene–vinyl acetate copolymers[14] and total oxygen in polyolefins. The advantages of the technique are extreme sensitivity and freedom from interference effects. A disadvantage is that, involving as it does the use of a nuclear reactor, the samples would usually have to be sent away for analysis. However, the technique is extremely useful, and in many cases the results obtained can be considered as reference values for those materials, and these data are of great value when these samples are analysed by alternative methods in the originating laboratory.

To illustrate this some work will be discussed on the determination of parts per million of sodium in polyolefins. During the manufacture of high-density polyethylene and polypropylene, sodium-containing alkalis are added to the polymer to neutralize residual acidity originating from the catalyst system. It was found that replicate sodium contents determined on the same sample by a flame photometric procedure were frequently widely divergent. Neutron activation analysis offers an independent non-destructive method of checking the sodium contents which does not involve ashing. Both methods are discussed below.

In the flame photometric procedure the sample is dry-ashed in a nickel crucible and the residue dissolved in hot water before determining sodium by evaluating the intensity of the line emission occurring at 589 nm.

Neutron activation analysis (flux 10^{12} neutrons $cm^{-1} s^{-1}$) for sodium was carried out on polyethylene and polypropylene moulded discs containing up to about 550 ppm sodium which had been previously analysed by the flame photometric method. The results obtained in these experiments (Table 5) show that significantly higher sodium contents are usually obtained by neutron activation analysis, and this suggested that

TABLE 5   *Interlaboratory variation of flame photometric sodium determinations in polyolefins (sodium content, ppm)*

Neutron activation analysis		Flame photometry (moulded discs)
Powder	Moulded discs	
Polyethylene		
207	211, 204	35, 165, 140
177	175, 172	100, 140, 148
266	267, 263	85, 210, 221
203	197, 191	70, 160, 150
Polypropylene		
165	151, 161	50, 130, 133
198	186, 191	95, 173
322	333, 350	95, 138

TABLE 6   *Comparison of sodium determination in polyolefins by neutron activation analysis, emission spectrography and flame photometry (sodium, ppm)*

Sample description	By neutron activation analysis	By emission spectrography	By flame photometry
Polyethylene	99, 96, 99	95	60, 76, 55
Polyethylene	256, 247, 256	258, 259	160, 178, 271
Polyethylene	343, 321, 339	339, 287	250, 312
Polyethylene	213, 210, 212	218, 212	140, 196
Polypropylene	194, 189, 192	209, 198	80, 158, 229
Polypropylene	186, 191, 198	191, 191	95, 173

sodium is being lost during the ashing stage of the flame photometric method. Sodium can also be determined by a further independent method, namely emission spectrographic analysis, which involves ashing the sample at 500°C in the presence of an ashing aid consisting of sulphur and the magnesium salt of a long-chain fatty acid (compared to dry-ashing at 650–800°C without an aid as used in the flame photometric procedure.[15-17] In Table 6 results by neutron activation and emission spectrography are shown to agree well with each other, thereby confirming that low results can be obtained by flame photometry when a simple ashing process is employed. Unlike the neutron activation technique, emission spectrography involved a preliminary ashing of the sample at 500°C, thereby involving the risk of sodium losses, despite this, the technique gives correct results. The losses of sodium in the flame photometric ashing procedure were probably caused by the maximum ashing temperature used exceeding that used in the emission spectro-

TABLE 7   *The effects of modification of ashing procedure on the flame
photometric determination of sodium (sodium, ppm)*

		By flame photometry		
By neutron activation	By emission spectrography	original (ashed between 650°C and 800°C)	Dope ash at 500°C	Direct ash at 500°C
99, 96, 99	95	60, 75, 55	100	75
256, 247, 259	258, 259	160, 178, 271	225	208
343, 321, 339	339, 287	250, 312	282	265
213, 210, 212	218, 212	140, 196	210	191
194, 189, 192	209, 198	80, 158, 229	196	169
186, 191, 198	191, 191	95, 95, 173	193	173

graphic method by some 150–300°C. This suggested that it might be
possible to obtain more reliable sodium contents by flame photometry if
the magnesium AC dope ashing procedure could be used in conjunction
with flame photometry.

The results in Table 7 show clearly that flame photometry following
dope ashing at 500°C gives a quantitative recovery of sodium. Direct
ashing without an ashing aid at 500°C causes losses of 10% or more of
the sodium, whilst direct ashing at 800°C causes even greater losses.

2.2(b)   Destructive techniques

Atomic absorption spectroscopy

This is a very useful technique for the determination of metals. As the
sample had to be applied to the instrument in the form of a solution,
the polymer has first to be digested to provide a solution suitable for
analysing by one of the dissolution techniques discussed earlier. In
addition to this, certain elements (such as arsenic, antimony, mercury,
selenium and tin) can, after producing the soluble digest of the polymer,
be converted to gaseous metallic hydrides by reaction of the digest with
reagents such as stannous chloride or sodium borohydride, and these
hydrides can be determined by atomic absorption spectrometry.

$$As_2O_3 + 3SnCl_2 + 6HCl = 2AsH_3 + 3H_2O + 3SnCl_4$$
$$NaBH_4 + As_2O_3 \qquad \rightarrow 2AsH_3$$

To illustrate, let us consider a method developed for the determination
of trace amounts of arsenic in acrylic fibres containing antimony oxide
fire-retardant additive.[18] The arsenic occurs as an impurity in the
antimony oxide additive and, as such, its concentration must be controlled
at a low level.

In this method a weighed amount of sample (6 mg–1 g) is digested with 3 ml each of concentrated nitric, perchloric and nitric acids and digested strongly until the sample is completely dissolved. Pentavalent arsenic in the sample is then reduced to trivalent arsenic by the addition of titanium trichloride dissolved in concentrated hydrochloric acid.

$$As^{5+} + 2Ti^{3+} \rightarrow As^{3+} + 2Ti^{4+}$$

The trivalent arsenic is then separated from antimony by extraction with benzene, leaving antimony in the acid layer. The trivalent arsenic is then extracted with water from the benzene phase. This solution is then treated with a mixture of hydrochloric acid, potassium iodide and stannous chloride in a glass reaction vessel to convert trivalent arsenic to arsine (AsH_3), which is swept into the atomic absorption spectrophotometer. Arsenic is then determined at the 193.7 μm absorption line. Recoveries between 96 and 104% are obtained by this procedure in the 0.5–1.0 μg arsenic range.

Electron probe microanalysis

This is a technique for identifying the nature of metallic inclusions in polymer film and sheet. The sample is bombarded with a very narrow beam of X-rays of known frequency and the back-scattered electron

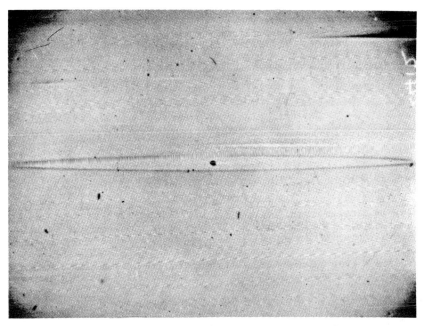

FIG. 2.   Lens in polypropylene film. Magnification × 18.

radiation is examined. An image is produced of the distribution of elements of any particular atomic number. The technique can be illustrated by its application to the elucidation of a phenomenon known as lensing in polypropylene film.[19] When the film in question was drawn down to between 1 and 2 thou thickness an imperfection became apparent (Fig. 2) Electron probe microanalysis was used a direct check on whether or not a higher concentration of chlorine is associated with a lens than is present in the surrounding lens-free polymer. The polypropylene film was first coated with a thin layer of copper, to keep the sample cool by conduction during electron bombardment. Of the lenses examined, a number were seen to have a speck at the centre. Figure 3 shows a back-scattered electron image of such a speck; part of the lens is also seen. The specks examined were shown to contain the elements sodium and chlorine. Figure 4a, b and c show respectively the image in back-scattered electrons

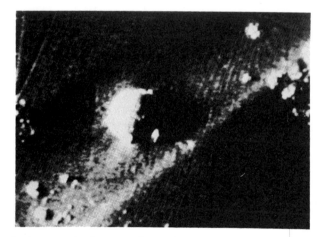

FIG. 3.   Back-scattered electron microanalysis of a speck within a lens in polypropylene

a) ELECTRON   **b) CHLORINE**   **c) SODIUM**
    IMAGE IN BACK   **DISTRIBUTION**   **DISTRIBUTION**
    SCATTERED
    ELECTRONS

FIG. 4.   Electron probe microanalysis of a speck within a lens in polypropylene.

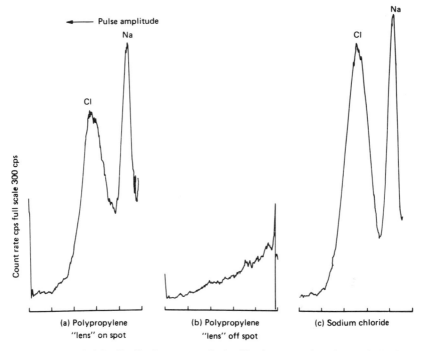

FIG. 5. Pulse height distribution curves obtained in electron probe microanalysis of polypropylene lenses.

of a speck within a lens, the emitted chlorine radiation, and the emitted sodium radiation. The chlorine and sodium distribution in Fig. 4b and c show as a ring round the circumference of the speck.

The X-rays emitted by the sample in electron probe microanalysis can be detected and counted by a proportional counter. In Fig. 5 are shown three pulse height distribution curves. Curve 5a was obtained with the electron probe positioned on the sodium and chlorine containing ring shown in Fig. 4b and c. Curve 5b was obtained with the electron probe displaced from the ring shown in Fig. 4b and c, i.e. on a nonlensed part of the polypropylene film. The curve in 5 was obtained with the electron probe positioned on a crystal of pure sodium chloride. Comparison of Fig. 5a and b demonstrates that chlorine and sodium only occur in high concentrations in the speck occurring in a lens, and that these elements are not generally distributed through the polymer. Comparison of Fig. 5a and c shows that the material which surrounds the central speck within the lens is not very different from pure sodium chloride.

3

Determination of functional groups

As in the case of elemental analysis functional groups can occur in polymers over a wide range of concentration ranging from a few parts per million, as occurs for example in the case of end-groups, or micro-unsaturation to the percentage range. The occurrence of two or three double bonds per thousand carbon atoms in, for example, polyethylene can affect intrinsic polymer properties and can certainly help to distinguish between polyethylene manufactured by different manufacturers using different processes. As such, this type of determination falls within the province of microstructure, which is discussed in Chapter 5. At the other end of the concentration scale a copolymer of, for example, ethylene and vinyl acetate will contain between 1 and 90% of either monomer. Analysis of polymers and copolymers in this concentration range is considered below.

A wide range of physical and chemical techniques have been employed in such analyses. The application of these techniques to the determination of particular functional groups is discussed under two main headings:

1. Chemical methods comprising techniques based on halogenation, titration, saponification values, procedures based on phthalation, acetylation, etc., hydrogenation and colorimetric procedures.
2. Physical methods comprising procedures based on infrared, Raman and nuclear magnetic resonance spectroscopy, saponification and, finally, pyrolysis or alkali fission of the polymer followed by gas chromatography of the volatile products produced.

3.1 Chemical methods

Many of the techniques which are applicable to the determination of functional groups in organic compounds are also applicable to polymers. Some examples of these techniques are discussed below.

3.1(a) Titration methods

Iodine monochloride procedure for the determination of unsaturation in styrene–butadiene copolymers

Styrene–butadiene copolymers contain residual double bonds which enable the butadiene content of the copolymer to be determined

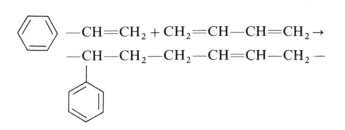

In this procedure the polymer is reacted with an excess of standard iodine monochloride dissolved in glacial acetic acid (Wijs reagent)

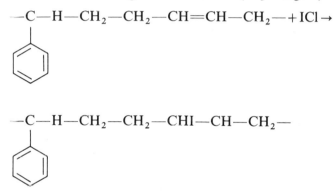

After completion of the reaction, excess iodine monochloride is reacted with potassium iodide and the liberated iodine estimated by titration with standard sodium thiosulphate.

$$ICl + KI = KCl + I_2$$

The double bond content of the original polymer can be then calculated from the measured consumption of iodine monochloride.

Crompton and Reid[20] have described procedures for the separation of high-impact polystyrene into the free rubber plus rubber grafted polystyrene plus copolymerized rubber and a gel fraction, and for using the iodine monochloride procedure to estimate total unsaturation in the two separated fractions. To separate a sample into gel and soluble fractions it was first dissolved in toluene. Only gel remains undissolved. Methanol is then added, which precipitates the polystyrene–rubber graft, ungrafted rubber, and polystyrene. Any styrene monomer, soap or lubricant remain in the liquid phase, which is separated from the solids and rejected. The toluene solubles are separated from the solid gel by centrifuging and made up to a standard volume with toluene. The gel is then dried *in vacuo* and weighed. Both the gel and toluene-soluble fractions are reserved

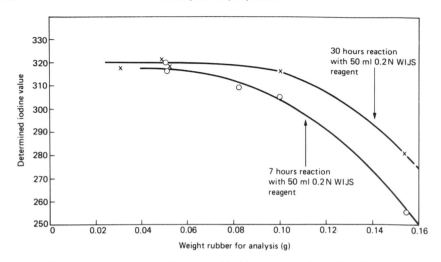

FIG. 6.   Influence of excess iodine monochloride reagent and reaction time on the determination of the iodine value of rubber.

for determination of unsaturation. To determine unsaturation in styrene–butadiene rubbers with good accuracy using the iodine monochloride procedure it was found necessary to contact the sample with chloroform for 15 h before reaction with iodine monochloride. Figure 6 shows a plot of sample size against the determined iodine value; it may be seen that, even with a 30 h reaction period, a constant iodine value (ca. 320) is obtained only when the sample size is 0.05 g or less, i.e. a five-fold excess of iodine monochloride reagent.

The solid gel, separated from a high-impact polystyrene by solvent extraction procedures, is completely insoluble in chloroform and in the iodine monochloride reagent solution. A contact time with chloroform of 90 h with a 75 reaction period with reagent is required.

Crompton and Reid[20] used these procedures to study the distribution of rubber added in several laboratory preparations of high-impact polystyrene containing 6 wt% of a styrene–butadiene rubber and 94% styrene, i.e. theoretical 4.1% butadiene. The results in Table 8 show the way in which the added unsaturation of 4.1% butadiene distributes between the gel and soluble fractions. The butadiene content of the separated gel remains fairly constant, in the 20–25% region, regardless of the quantity of gel present in the sample. As the gel content increases, therefore, so more of the rubber becomes incorporated into the gel and less remains as free rubber or soluble graft. The recovered unsaturation lies mainly in the 90–95% region, indicating that loss of unsaturation due to grafting or crosslinking reactions occurs only to the extent of some 5–10%.

TABLE 8 *Distribution of butadiene between soluble and gel fractions obtained from polystyrenes containing different amounts of gel*

Gel content of sample (wt%)	Butadiene content isolated gel (wt%)	Soluble graft butadiene content A (calculated on original sample) (wt%)	Gel butadiene content B (calculated on original sample) (wt%)	Total butadiene content (A + B) (calculated on original sample) (wt%)	Amount of original rubber unsaturation in the sample $C = (A + B) \times 100\%$
—		3.5	—	3.5	85
4.7	19.5	2.8	0.9	3.7	90
5.6	16.2	2.9	0.9	3.8	93
8.9	23.3	1.5	2.1	3.6	88
11.8	20.0	1.5	2.4	3.9	95

Albert[21] has compared determinations of butadiene in high-impact polystyrene by an infrared method and by the iodine monochloride method described by Crompton and Reid.[20] The infrared method is based on a characteristic absorbance in the infrared spectrum associated with the transconfiguration in polybutadiene:

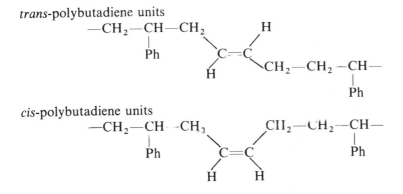

trans-polybutadiene units

cis-polybutadiene units

Since different grades of high-impact polystyrene may contain elastomers with different *trans*-butadiene contents, calibration curves based on the standard rubber are not always suitable for analysing these products. The results obtained by the two methods for several high-impact grades are compared in Table 9. The rubber content of high-impact polystyrene sample 1 determined by titration is lower than the value obtained by the infrared method. This is expected on interpolymerized polymers because of crosslinking, which reduces the unsaturation of the rubber. The other polymers (except sample 3), appear to contain diene 55-type rubber of high (*trans*-butadiene content, since reasonable agreement was obtained

TABLE 9   *Rubber content of high-impact polystyrenes (based on PBD)*

Sample	Polybutadiene, %w (iodine monochloride method)	Polybutadiene, %w (IR method)
Standard: 6.0%w diene 55	6.2	—
Standard: 12.%w diene 55	12.2	—
Standard: 15.%w diene 55	14.8	—
High-impact polystyrene 1	8.6	9.7
High-impact polystyrene 2	5.6	5.8
High-impact polystyrene 3	9.0	1.2
High-impact polystyrene 4	11.2	11.4
High-impact polystyrene 5	5.8	5.9

between the iodine monochloride and infrared methods. High-impact polystyrene 3, however, must contain a polybutadiene of high *cis*-content to explain the low (1.2%w) amount of rubber found by the infrared method compared to the 9.0% found by the titration method.

The iodine monochloride method has been used for a variety of polymers. These polymers include those which are highly unsaturated, such as polybutadiene and polyisoprene,[552–555] and polymers having low unsaturation such as butyl rubber,[556] and ethylene–propylene–diene terpolymer. Considerable work has been done investigating the side-reactions of iodine monochloride with different polymers.[556] These side-reactions are substitution and splitting out rather than the desired addition reaction.

Bromination procedure for unsaturation

In acidic medium potassium bromide and potassium bromate produce bromine stoichiometrically, and this is the basis of a titration method,[24] which has been used to determine double bonds in polymethylacrylate. After bromination, excess bromine is estimated by the addition of potassium iodide and estimated the iodine produced by titration with standard sodium thiosulphate to the starch end-point.

$$KBrO_3 + 5KBr + 6HCl = 3Br_2 + 3H_2O + 6KCl$$

$$\sim\left(\begin{array}{c} CH_3 \\ | \\ CH_2{=}C{-} \\ | \\ COOCH_3 \end{array}\right)_n + Br_2 \rightarrow \sim\left(\begin{array}{c} CH_3 \\ | \\ CH_2Br{-}CBr{-} \\ | \\ COOCH_3 \end{array}\right)_n$$

$$2KI + Br_2 = 2KBr + I_2$$

Titration procedure for the determination of carboxyl groups in acrylic copolymers

Most methods for the determination of carboxyl groups in polymers are based on titration techniques including, for example, the following copolymers, acrylic acid–itaconic acid[38], acrylic acid–ethyl acrylate[39] and maleic acid-styrene[40]. High-frequency titration has been applied[41] to the analysis of itaconic acid–styrene and maleic acid–styrene copolymers and ethyl esters of itaconic anhydride–styrene copolymers. The method can also be used to detect traces of acidic impurities in polymers and in the identification of mixtures of similar acidic copolymers. Titration indicates that the acid segments in the copolymers of itaconic acid–styrene, and maleic acid–styrene, and the homopolymer polyitaconic acid, act as dibasic acids. The method has a sensitivity that permits identification and approximate resolution of two carboxylate species in the same polymer, e.g.:

$$\text{polyitaconic} \quad \left(-CH_2=\underset{\underset{\displaystyle COOH}{|}}{\overset{\overset{\displaystyle CH_2COOH}{|}}{C}} \right)_n + nK^+$$

$$\rightarrow \left(-CH_2=\underset{\underset{\displaystyle COOK}{|}}{\overset{\overset{\displaystyle CH_2COOK}{|}}{C}} \right)_n + nH^+$$

High-frequency titration gives a precise location of the inflection points related to the polymer carboxyl groups, and is a sensitive method for the determination of the freedom of the copolymer samples from monobasic acid impurities (comonomer acids), since mixtures of copolymer acids with monobasic and dibasic acids show definite inflection points that can be related to the individual carboxylate species present.

A titration curve (Fig. 7) is shown for a monomethyl ester of an itaconic acid–styrene copolymer.

Potentiometric titration provides a method of investigating changes of conformation undergone by polyelectrolytes in solution, since the environment of the dissociating groups is dependent on the conformation of the polymer chain helix-coil transitions of polyacids. Thus, precise potentiometric titration of solutions of high molecular weight polyacrylic acid at constant ionic strength indicate the presence of such conforma tional transition.[13,14]

Figure 8 shows the titration results for polyacrylic acid plotted as $pH + \log((1 - \alpha)/\alpha)$ versus α (degree of dissociation), as points connected

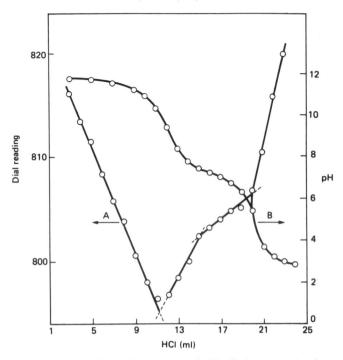

Fig. 7. High-frequency (A) and potentiometric (B) displacement titration of the monosodium salts of the monomethyl esters of poly(itaconic acid-costyrene). Titration of 0.2345 g 57:43 anhydridestyrene copolymer + MeOH (heat) + excess NaOH with 0.1286 N HCl.

by full curves. The four curves at the different ionic strengths all show the same features. The first short region, labelled A in the figure, is probably due to some instability in the solution, such as aggregation preceding precipitation. This region extends to higher values of α at the higher ionic strengths. The second region, B, represents the ionization of the first conformation of the polymer; the third, C, the conformational transition; and the fourth, D, the ionization of the second conformation. The first conformation, which exists at the lower degree of dissociation, has presumably the more tightly coiled structure, and is denoted PAA(a) (PAA-polyacrylic acid). The second conformation, stable at high degree of dissociation, and less tightly coiled, is denoted PAA(b). The four curves of PAA(a) in Fig. 8 have been extrapolated (dashed curves) semi-empirically to zero α, and they meet there at a value of $pH + \log((1 - \alpha)/\alpha)$ of 4.58, which is the value of $pK'o$, the intrinsic dissociation constant of the polyacid for ionic strengths of 0.06 and 0.11 found previously.[44] This extrapolation is made with the help of plots of pH versus $\log((1 - \alpha)/\alpha)$ shown in Fig. 9. These plots, though almost linear over the whole range,

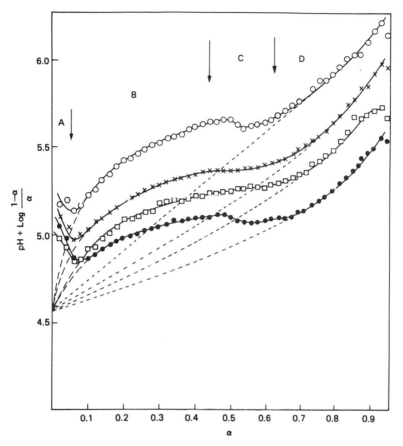

FIG. 8.   Dependence of the function $pH + \log((1-\alpha)/\alpha)$ on α for poly(acrylic acid) at various ionic strengths u: ($\bigcirc$) $u = 0.02$; ($\times$) $u = 0.065$; ($\square$) $u = 0.11$; ($\bullet$) $u = 0.20$.

as previously reported by Mandel and Leyte,[45,46] are not quite so, and regions A, B, C and D can be distinguished here also. Linear extrapolation of region B, which represents PAA(a), to higher values of $\log((1-\alpha)/\alpha)$ was used to obtain the extrapolations of the PAA(a) curves of Fig. 8 to zero α.

Acetylation and phthalation procedures for the determination of hydroxyl groups

Chemical methods for the determination of hydroxyl groups in polymers are based on acetylation[48,49] phthalation[49] and reaction with phenyl isocyanate[50,49] or, when two adjacent hydroxy groups and present in the polymers by reaction, with potassium periodate.[48]

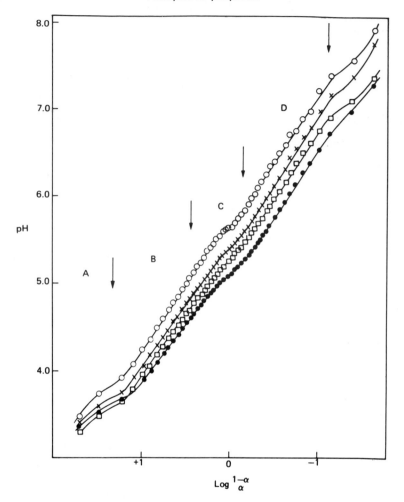

FIG. 9.   pH versus $\log((1-\alpha)/\alpha)$ for poly(acrylic acid) at various ionic strengths u; ($\bigcirc$) $u = 0.02$; ($\times$) $u = 0.065$; ($\square$) $u = 0.11$; ($\bullet$) $u = 0.20$.

The reactions on which these methods are based are as follows:

Acetylation:

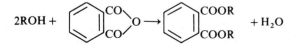

Phthalation:

Reaction with phenyl isocyanate:

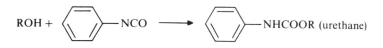

$$ROH + \text{}—NCO \longrightarrow \text{}—NHCOOR \text{ (urethane)}$$

Reaction with potassium perodate:

$$\underset{\underset{\displaystyle H}{\displaystyle |}}{\overset{\overset{\displaystyle OH}{\displaystyle |}}{R—C}}—\underset{\underset{\displaystyle H}{\displaystyle |}}{\overset{\overset{\displaystyle OH}{\displaystyle |}}{C}}—R' + HIO_4 \rightarrow RCHO + R'CHO + H_2O + HIO_3$$

In all these methods an excess of standardized reagent is added and, at the end of the reaction, unconsumed reagent is estimated. The concentration of hydroxyl groups can then be calculated from the amount of reagent consumed.

Acetylation and phthalation procedures

Determination of hydroxyl groups in polyoxyalkylene glycols

In these procedures the hydroxyl-containing polymer is reacted with an excess of a standard non-aqueous solution of acetic anhydride or phthalic anhydride, sometimes in the presence of a catalyst such as p-toluene sulphonic acid. After the formation of the ester is complete, an excess of water is added to convert excess anhydride to the free carboxylic acid. The acid is titrated with aqueous or alcoholic standard potassium hydroxide to the phenol phthalein end-point, to determine unconsumed acid. A blank run is carried out in which the sample is omitted. The hydroxyl content of the polymer can then be calculated from the difference between the sample and blank titrations.

Stetzler and Smullin[49] found that for polypropylene glycols the classical phthalation procedure gave consistently low results, as did the perchloric acid catalysed acetylation procedure developed by Fritz and Schenk.[51] In addition to its greater intrinsic accuracy, the advantages claimed for the Stetzler and Smullin[49] toluene p-sulphonic acid catalysed procedure over phthalation include shorter reaction time and lower reaction temperature, also good reproducibility, indicating no reaction with the ether groups.

Table 10 compares hydroxyl values obtained by three methods. It is seen that in every case the acid-catalysed acetylation method gives a higher result than that obtained by phthalation, the average difference between the two methods being 2.5% of the determined value. It will be seen from the table that the agreement between the p-toluene sulphonic acid method and the phenyl isocyanate method is good.

TABLE 10 *Hydroxyl values obtained on polypropylene glycol by catalysed acetylation, phthalation and phenyl isocyanate reaction*

Mol. wt.	p-toluene sulphonic acid catalysed acetylation method. Average hydroxyl value (mg KOH/g)	Phthalation method. Average hydroxyl value (mg KOH/g)	Reaction with phenyl isocyanate. Average hydroxyl value (mg KOH/g)
700	353	346	
1260	276	265	
5000	34.9	34.0	34.9
5000	34.3	33.8	35.1
2890	58.3	57.0	

The Stetzler and Smullin[49] method is applicable to polyoxyethylene, polyoxypropylene and ethylene oxide tipped glycerol/propylene oxide condensates. Compounds of this type in the molecular weight range 500 to 5000 can be analysed by this procedure.

Reaction with toluene diisocyanate

As mentioned above, organic isocyanates react with hydroxyl groups to produce a urethame as follows:

$$ROH + R'NCO \rightarrow R'NHCOOR$$

In many practical situations it is necessary to be able to distinguish between primary and secondary hydroxyl groups in polymers. Thus, the reaction product of a glycerol–propylene oxide condensate tipped with ethylene oxide would contain both types of hydroxyl group:

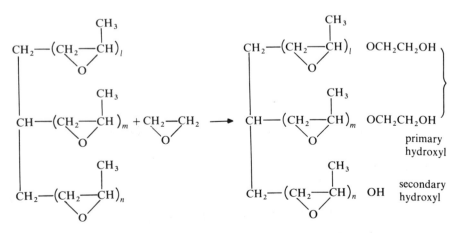

Reaction rate differences of primary and secondary hydroxyl groups with phenyl isocyanate are the basis of the methods described below, for carrying out this determination.[50,52]

Determination of primary and secondary hydroxyl groups in ethylene oxide tipped glycerol–propylene oxide condensates

Crompton[52] has developed a kinetic method for the evaluation of the reactivity of polyols in the range 3000–5000. This procedure may be used to determine the degree of ethylene oxide 'tipping' produced by the addition of ethylene oxide to glycerol/propylene oxide condensates. The method was developed to measure the relative reactivity with isocyanates of such condensates as a function of their ethylene oxide tipping content. The method is based on the observation that primary hydroxyl groups react with phenyl isocyanate to form a urethane faster than do secondary hydroxyl groups. Hence a 'tipped' polyol will react more completely in a given time with an equivalent amount of phenyl isocyanate than an 'untipped' polyol reacting under the same conditions. The greater the amount of ethylene oxide tipping the greater its rate of reaction with phenyl isocyanate.

The reaction is carried out under standard conditions in which a calculated weight of the polyol (depending on its hydroxyl number) is reacted under standard conditions with an excess of a standard toluene solution of phenyl isocyanate in the presence of a basic catalyst.

Unconsumed phenyl isocyanate is then reacted with excess standard potassium hydroxide

$$PhCNO + 2KOH = K_2CO_3 + PhNH_2$$

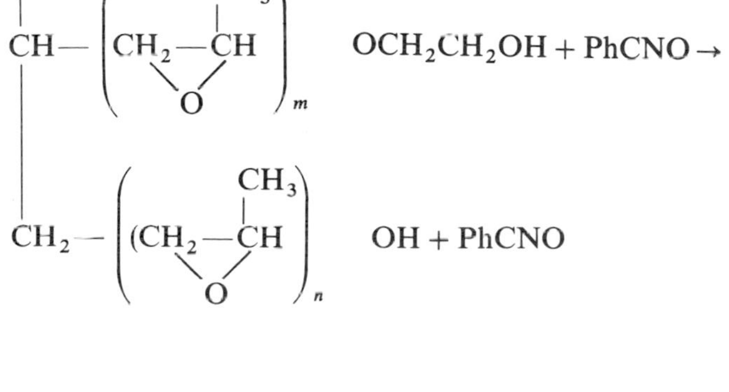

Excess sodium hydroxide is estimated by titration with standard acetic

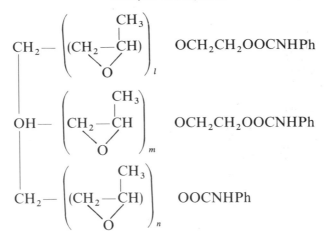

acid to the phenol phthalein end-point. A blank run is performed in which the sample is omitted. From the difference between the sample and the blank titrations it is possible to calculate the hydroxyl content of the original polymer:

1 mole hydroxyl groups ≡ 1 mole phenyl isocyanate ≡ 2000 ml N KOH

Reactivity is calculated by a kinetic procedure in which the percentage of the original phenyl isocyanate addition, which reacts with the sample in a given time, is taken as an index of its reactivity. A calibration graph can be prepared in which the determined phenyl isocyanate reactivity is plotted against the ethylene oxide content, and this enables a determination to be made of the ethylene oxide content of unknown samples from the calibration graph by interpolation.

Crompton applied this method to a range of glycerol/propylene oxide adducts containing various accurately known amounts of ethylene oxide tipping, up to 5.3 moles (Table 11). Figure 10 shows the reaction curves

TABLE 11   *Application of reactivity method to standard ethylene oxide tipped polyols*

Sample identi-fication	Approximate molecular weight	Hydroxyl number mg KOH/g polyol	Ethylene oxide tipping moles ethylene oxide/ mole glycerol (by weight addition)	Reactivity, i.e. percentage of original phenyl isocyanate addition consumed in:	
				60 min reaction	100 min reaction
A	3000	59.0	0.0	25.2	35.2
B	5000	34.9	0.0	27.1	36.8
C	5000	34.3	3.0	41.5	46.6
D	5000	35.9	3.5	44.4	51.0
E	5000	36.0	4.3	47.5	54.1
F	5000	33.3	5.3	51.9	57.4

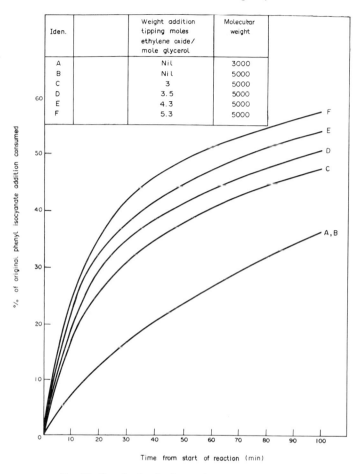

Iden.		Weight addition tipping moles ethylene oxide/ mole glycerol	Molecular weight
A		Nil	3000
B		Nil	5000
C		3	5000
D		3.5	5000
E		4.3	5000
F		5.3	5000

Fig. 10. Standard polyol samples – reaction curves.

for each of the standard samples analysed. It is seen that increasing the ethylene oxide tip content of a polyol leads to a distinct increase in the reactivity of the polyol with phenyl isocyanate. As expected, 'untipped' polyols which are relatively free from primary hydroxyl groups, i.e. curves A and B, react comparatively slowly with phenyl isocyanate. Decinormal solutions of 'untipped' glycerol/propylene oxide condensates of molecular weight 3000 and 5000 had an identical rate of reaction with phenyl isocyanate (Fig. 10 curves A and B). Thus the rate of reaction with phenyl isocyanate of the terminal isopropanol end-groups in polyols is independent of molecular weight in the molecular weight range 3000 to 5000 and depends only on proportions of primary and secondary hydroxyl end-groups present.

A calibration curve is prepared (Fig. 10) by plotting moles ethylene oxide per mole of glycerol for the range of standard tipped polyols of known ethylene oxide content against percentage of original phenyl isocyanate addition consumed after 60 and 100 min, i.e. P60% and P100%. This curve can be used to obtain from P60% and P100% data obtained for tipped glycerol for propylene oxide polyols of unknown composition their tipped ethylene oxide contents (in moles ethylene oxide).

Saponification procedure for determination of ester groups

Saponification procedures can be applied to the determination of ester groups in polymers. A copolymer of ethylene and vinyl acetate has the following structure which, upon hydrolysis with excess standard in the presence of potassium hydroxide/p-toluene sulphonic acid catalyst[63-64] reacts as follows:

$$-(CH_2-CH_2-CH_2-CH-)_n + nKOH \rightarrow$$
$$\underset{OOCCH_3}{|}$$

$$-(CH_2-CH_2-CH-CH-)_n + nCH_3COOK$$
$$\underset{OH}{|}$$

Excess potassium hydroxide is then determined by titration with standard acetic acid, and hence the vinyl acetate content of the polymer is calculated from the amount of potassium hydroxide consumed.

3.1(b)   Hydrogenation methods

Unsaturation in polymers is usually measured by physical techniques, as discussed later. This is especially so in the case of low levels of unsaturation, or in instances where a distinction has to be made between different types of unsaturation. Hydrogenation techniques have been used, however, to measure higher levels (0.5–5.0 mole%) of total unsaturation in polymers, a good example of which is the determination of terminal unsaturation in polystyrene a oligomers[25-27] (low molecular weight polymers), e.g. polystyrene dimer

$$2PhCH=CH_2 \rightarrow PhCH_2-CH=CH-CH_2Ph$$

3.1(c)   Derivatization methods

An example of this is the determination of low concentrations of carbonyl groups in PVC and vinyl chloride–vinyl acetate copolymers.[76] In this method the carbonyl groups in the polymer are reacted with 2,4-dinitrophenylhydrazine to produce the corresponding phenylhydrazone.

Excess reagent is washed away from the polymer, which is then digested with concentrated sulphuric acid to convert the bound hydrazone to ammonium sulphate, which is then estimated using Nessler's reagent.

Spectrophotometric methods

Direct spectrophotometry in the visible and ultraviolet region has been used to determine low concentrations of functional groups in the surface of polymer films. Thus Kato[77] has followed the regeneration of carbonyl groups from 2, 4-dinitrophenylhydrazones formed on the surface of irradiated polystyrene films by absorption measurements at 378 nm.

3.2 Physical methods

3.2(a) Nuclear magnetic resonance spectroscopy

This technique has been used extensively for the determination of unsaturation and carboxyl, ester and carbonyl groups in polymers.

Unsaturation

An advantage of NMR is that it is capable of distinguishing between the different types of unsaturation that can occur in a polymer.

Nuclear magnetic resonance spectroscopy has been used to determine unsaturation in acrylonitrile–butadiene–styrene terpolymers,[30] ethylene–propylene–diene terpolymers[31] and 1, 2-polybutadiene.[32]

Regarding acrylonitrile–butadiene–styrene terpolymers,[30] NMR is capable of determining ungrafted butadiene rubber in solvent extracts of these polymers.

About 60–80% of the butadiene in the entire sample was present as ungrafted rubber. Using the compositional analysis of the grafted and ungrafted rubber, and the amount of ungrafted rubber extracted, one can calculate the composition of the graft. Figure 11 shows a typical NMR spectrum of the grafted material. No aromatic protons of styrene or acrylonitrile protons are seen in the NMR spectra. The vinyl content of the polybutadiene is about 20%.

Polymerization of ethylene and propylene results in a saturated copolymer. In order to vulcanize this rubber, some unsaturation has to be introduced. This is commonly done by adding a few per cent of non-conjugated diene (termonomer) such as dicyclopentadiene, 1, 4-hexadiene, or ethylidene norbornene, during the polymerization. Since only one of the double bonds of the diene reacts during polymerization, the other is free for vulcanization. The amount of unsaturation left in the ethylene propylene diene terpolymer is of great interest, because the vulcanization properties will be affected.

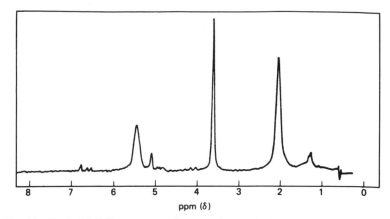

Fig. 11. Typical NMR spectrum of extracted ungrafted rubber from Samples 1 through 6. Signal at 5.4 ppm – olefinic protons of 1,4-polybutadiene. Signal at 5.1 ppm – olefinic protons of 1,2-polybutadiene. Signal at 3.5 ppm – dioxane (internal standard). Signal at 2.0 ppm – aliphatic protons of 1,4-polybutadiene. Signal at 1.2 ppm – methylene protons of the soap. Signals between 6 and 7 ppm-impurities in hexachlorobutadiene.

Sewell and Skidmore[31] used time-averaged NMR spectroscopy at 60 MC/s to identify low concentrations of non-conjugated dienes introduced into ethylene–propylene copolymers to permit vulcanization. Although infrared spectroscopy[33] and iodine monochloride unsaturation methods[34] have been used to determine or detect such dienes, these two methods can present difficulties. Identification of the incorporated third monomer is not always practical by infrared spectroscopy at the low concentrations involved, and in the high-resolution NMR spectra of these terpolymers the presence of unsaturation is not usually detected, since the signals from the olefinic protons are of such low intensity that they become lost in the background noise. The spectra obtained by time-averaged NMR are usually sufficiently characteristic to allow identification of the particular third monomer incorporated in the terpolymer. Moreover, as the third monomer initially contains two double bonds, differing in structure and reactivity, the one used up in copolymerization may be distinguished from the one remaining for subsequent use in vulcanization. Therefore information concerning the structure of the remaining unsaturated entity may be obtained. Table 12 shows the chemical shifts of olefinic protons of a number of different third monomers in the copolymers.

The cyclooctadiene and dicyclopentadiene terpolymers have olefinic protons with the same chemical shift, 4.55τ, and so these cannot be differentiated by this technique, but may be distinguished by the use of iodine monochloride. The hexadiene type of terpolymer may be identified by its olefinic resonance at 4.7τ.

TABLE 12

Third monomer	Chemical shift, τ
Cyclooctadiene-1, 5	4.55
Dicyclopentadiene	4.55
1, 4-Hexadiene	4.7
Methylene norbornene	5.25 and 5.5
Ethylidene norbornene	4.8 and 4.9

These three monomers have what appears as a single olefinic resonance in the terpolymer. On the other hand, the two norbornadiene types of monomer each show two characteristic resonances. In the methylene norbornene terpolymer the olefinic resonances arise from two protons, each giving a separate signal; whereas in the ethylidene norbonene terpolymer this is only one proton, the signal of which appears as a doublet. In view of these considerations it is more difficult to detect the olefinic resonance in the latter instance.

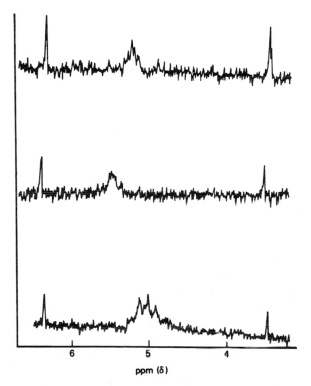

FIG. 12. Time-averaged NMR spectra of EPDM containing one of the following termonoments: upper spectrum: 1, 4-hexadiene; middle spectrum: dicyclopentadiene; lower spectrum: ethylidene norbornene.

TABLE 13   *Determination of termonomer in ethylene–*
propylene diene terpolymers: comparison of methods
(data shown as weight per cent termonomer)

NMR	Lee, Koltoff and Johnson[36]	Kemp and Peters[37]	Termonomer
7.3	5.0		Dicyclopentadiene
1.1	1.6		1,4-hexadiene
5.7	5.9	9.0	Ethylidene norbornene
2.8	3.6	4.5	Ethylidene norbornene
1.7	2.3	4.8	Ethylidene norbornene
4.6	5.4	6.0	Ethylidene norbornene

Altenau et al.[35] applied time-averaging NMR to the determination of low percentages of termonomers such as 1,4-hexadiene, dicyclopentadiene and ethylidene norbornene in ethylene–propylene termonomers. They compared results obtained by NMR and the iodine monochloride procedure of Lee et al.[36] The chemical shifts and splitting pattern of the olefinic response were used to identify the termonomer. Figure 12 show the time-averaged NMR spectra of ethylene–propylene terpolymers containing various dienes.

Table 13 compares the amount of termonomer found by the NMR method of Altenau et al.[35] and by iodine monochloride procedures.[36,37] The termonomers were identified by NMR and infrared.

Table 13 shows that the data obtained by the NMR method agree more closely with the Lee, Kolthoff and Johnson iodine monochloride method[36] than with the iodine monochloride method of Kemp and Peters.[37] The difference between the latter two methods is best explained on the basis of side-reactions occurring between the iodine monochloride and polymer because of branching near the double bond.[36] The reason for the difference between the NMR and Lee, Kolthoff and Johnson[36] methods is not clear. The reproducibility of the NMR method was ± 10–15%.

Carboxyl groups

Johnson et al.[47] found that, in titrating copolymers of methylacrylate and methacrylic acid with standard base to determine carboxyl groups, a number of deficiencies were encountered. The presence of up to 5% water and unreacted monomer both led to underestimations of the acid content. Additionally, titration was not applicable to polymers of molecular weight over one million, or high acid content, because of a tendency to reprecipitate during the titration. For this reason they investigated the application of proton NMR. By using the integral of the ester methoxy protons and combining this result with the total integral for CH_2 and CH_3 protons (the overlap between CH_2 and CH_3 resonances was enough at 100 MHz to prevent separate determination of these integrals), the copolymer composition could be ascertained. However, it was necessary to carry out the determinations at 100°C or higher to obtain resolution sufficient for reliable integrals. An additional problem was that the reaction solvents (toluene and hexane) and comonomers had resonances that overlapped those of the CH_2 and CH_3 of the copolymers, introducing considerable inaccuracy in the total CH_2–CH_3 integral. For this reason they investigated the applicability of ^{13}C NMR. Because of the greater spectral dispersion and narrower resonance lines obtained with ^{13}C NMR relative to proton NMR, problems associated with resonance overlap can be resolved. Excellent agreement was obtained between carboxyl values obtained by this procedure and conventional titration in 1:1 ethanol:water with standard potassium hydroxide to the phenol phthalein end-point over the acid content range 13–100%.

Ester groups

NMR spectroscopy has been used for the determination of isophthalate in polyethylene terephthalate isophthalate dissolved in 5% trichloroacetic acid. The NMR spectra of these polymers were measured on a high-resolution NMR spectrometer at 80°C. A singlet at 7.74 ppm is due to the four equivalent protons attached to the nucleus of the terephthalate unit. The complicated signals which appear at 8.21, 7.90, 7.80, 7.35, 7.22 and 7.10 τ are due to the four protons attached to the nucleus of the isophthalate unit. The content of the isophthalate unit can be calculated from the integrated intensities of these peaks.

NMR has also been used to determine ethyl acrylate in ethyl acrylate–ethylene and vinyl acetate–ethylene copolymers.[75] Measurements were made on 10% solutions in diphenyl ester at elevated temperature. Resolution improved with increasing temperature and lower polymer concentration in the solvent.

Figure 13 shows NMR spectra for an ethylene–ethyl acrylate copolymer. Spectra indicate clearly both copolymer identification and monomer ratio. A distinct ethyl group pattern (quartet, triplet), with the methylene quartet shifted downfield by the adjacent oxygen, is observed. The oxygen effect carries over to the methyl triplet which merges with the aliphatic methylene peak. No other ester group would give this characteristic pattern. The area of the quartet is a direct and quantitative measure of the ester content. All features in Fig. 13 are consistent with an identification of an ethylene-rich copolymer with ethyl acrylate. Ethyl acrylate contents obtained by NMR (6.0%) agreed well with those obtained by PMR (6.2%) and neutron activation analysis for oxygen (6.1%).

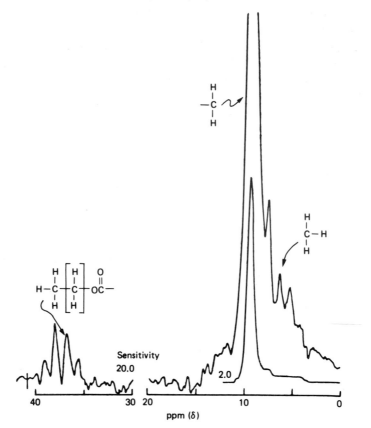

FIG. 13.   NMR spectra. Poly(ethylene–ethyl acrylate).

Carbonyl groups

Various physical techniques such as infrared and NMR spectroscopy have been described for the determination of very low concentrations of carbonyl groups in polymers. This is discussed in Chapter 5.

3.2(b)   Infrared spectroscopy

Infrared spectroscopy is useful both for microstructural studies on polymers (see Chapter 5) and for the determination of major structural groups in polymers, as discussed below.

Composition of olefin copolymers

In addition to polyethylene and polypropylene a wide range of olefin comonomers are produced which consist of copolymers of C_2 to C_8

olefins. A case in point is a copolymer of ethylene and butene-1 containing up to 10% butene-1. The preparation of calibration standards presents a difficulty in infrared methods for analysing such copolymers.

Physical blends of the two homopolymers, polyethylene and poly-butene-1 will not suffice, as these have a different spectrum to a true copolymer with the same ethylene–butene ratio. An excellent method for preparing such standards is to copolymerize blends of ethylene and ^{14}C-labelled butene-1 of known activity. From the activity of the copolymer determined by scintillation counting its butene-1 content can be calculated.

Standards prepared by this method are suitable for the calibration of the more rapid infrared method, which involves measurements of the characteristic absorption of the ethylbranches at 769 cm^{-1} (13 μ).[83] Absorbance at 769 cm^{-1} (13 μ) is directly proportional to the concentration of ethyl branches up to 10 per 1000°C.

Brown et al.[84] showed that pyrolysis of ethylene–propylene copolymers at 450°C produces derivatives that are rich in unsaturated vinyl and vinylidene groups, similar to the pyrolysis of natural rubber and styrene–butadiene rubber mixture,[85] which produces vinyl groups derived from the butadiene part of the molecule and the vinylidene groups from the methyl branches of the isoprene units. This unsaturation exhibits strong absorption in the infrared region. The ratio of the absorption of the vinyl groups to that of vinylidene groups varies with the mole fraction

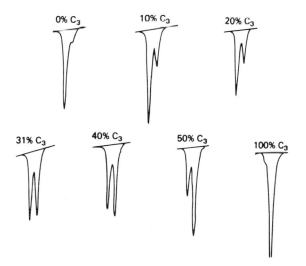

FIG. 14.   Typical infrared spectra of pyrolysates obtained from raw ethylene–propylene copolymers. The position of the absorption peaks from left to right are near 909 cm^{-1} and 889 cm^{-1} respectively, for each pair and the nominal compositions are indicated in mole% propylene.

of propylene in saturated ethylene–propylene copolymers.[86] Making use of this ratio they developed an analytical method for determining propylene in both raw and vulcanized ethylene–propylene copolymers. The vinyl group absorbs at about 909 cm^{-1} (11.00 μ) and the vinylidene at about 889 cm^{-1} (12.25 μ).[86,87]

Figure 14 shows some typical spectra obtained on the pyrolysates in the region of 950 cm^{-1} (10.5 μ) to 850 cm^{-1} (11.76 μ). The values of the ratio, R ($\times$ 100) range from 9.977 to 0.0290, respectively, for 0 to 100 mole% propylene for the raw samples and from 5.440 to 0.0431, respectively, for 10 to 100 mole% propylene for the vulcanized samples. The common logarithm of the ratio, R, can be represented by a linear function of the mole% of propylene in the copolymer.

Table 14 lists the results, expressed as common logarithms of $100R$, for unvulcanized samples.

Nitrile groups

An infrared method has been described[78] for the compositional analysis of styrene–acrylonitrile copolymers. In this method relative absorbance between a nitrile v(CN) mode at 2272 cm^{-1} (4.4 μ) and a phenyl v(CC) mode at 1613 cm^{-1} (6.2 μ) is used.

A near-infrared method using combination and overtone bands has been used[79] for carrying out the same analyses. Near-infrared spectra of four random copolymers and homopolymers are shown in Fig. 15. The bands which occur in the region 6250–4545 cm^{-1} (1.6–2.2 μ) result from overtones and combination tones which occur, respectively, in the regions 6250–5555 cm^{-1} (1.6–1.8 μ) and 5263–4545 cm^{-1} (1.9–2.2 μ).

The band near 5952 cm^{-1} (1.68 μ) is assigned as an overtone of phenyl v(CH) mode near 3000 cm^{-1} (3.33 μ) (3000 $\times$ 2 = 6000 cm^{-1} = 1.67 μ) and

TABLE 14 Log_{10} (*100R*) for polymer pyrolysates (raw samples)

Sample No.	Log_{10} (100R) at various propylene concentrations (mole%)						
	0	10	20	31	40	50	100
1	2.827	2.584	2.309	2.041	1.931	1.620	0.695
2	2.840	2.525	2.309	2.048	1.096	1.614	0.743
3	2.946	2.604	2.318	2.060	1.940	1.592	0.596
4	2.999	2.587	2.376	2.047	1.908	1.596	0.580
5	2.996	2.552	2.238	2.097	1.886	1.589	0.542
6	2.954	2.568	2.327	2.055	1.896	1.588	0.432
7	2.989	2.562	2.315	2.063	1.902	1.589	0.542
8	2.951	2.578	2.301	2.048	1.916	1.582	0.461
9	2.897	2.567	2.340	2.053	1.933	1.620	0.591
10	2.964	2.561	2.362	2.068	1.904	1.588	0.658
	2.9365	2.5687	2.3193	2.0579	1.9114	1.5979	0.5840 Av.

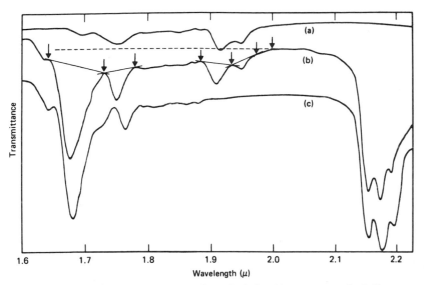

FIG. 15. Near-infrared spectra: (a) polyacrylonitrile; (b) styrene–acrylonitrile copolymer (acrylonitrile content 25.7 wt%); (c) polystyrene.

its absorbance is directly proportional to the styrene content of the
copolymer and the film thickness. The band near $5714 \, cm^{-1}$ ($1.75 \, \mu$) is
assigned as an overtone of the aliphatic $v(CH)$ mode near $2900 \, cm^{-1}$
($3.45 \, \mu$) ($2900 \times 2 = 5800 \, cm^{-1} = 1.724 \, \mu$) and both acrylonitrile and
styrene absorb at this wavelength Bands at 5235 and $5123 \, cm^{-1}$ (1.910 and
$1.952 \, \mu$) are combination tones of acrylonitrile and are assigned as,
respectively, $v(CN) + v_{asym}(CH_3)$ ($2237 + 2940 = 5177 \, cm^{-1} = 1.932 \, \mu$)
and $v(CN) + v_{asym}(CH_3)$ $2237 + 2870 = 5107 \, cm^{-1} = 1.958 \, \mu$). For calculation of the absorbance of these four characteristic bands, two different
baseline methods were used, as shown in Fig. 15. The method using the
extrapolated baseline from 2 introduces less deviation than the other
baseline method. The absorbance ratio $A_{1.675}/A_{1.910}$ proved to be the
best one for analytical measurements

Ester groups

Infrared spectroscopy has been applied to the determination of free and
combined vinyl acetate in vinylchloride–vinyl acetate copolymers.[71] This
method is based upon the quantitative measurement of the intensity of
absorption bands in the near-infrared spectral region arising from vinyl
acetate. A band at $6134 \, cm^{-1}$ ($1.63 \, \mu$) due to the vinyl group, enables the
free vinyl acetate content of the sample to be determined. A band at
$4651 \, cm^{-1}$ ($2.15 \, \mu$) is characteristic for the acetate group and arises from
both free and combined vinyl acetate. Thus, the free vinyl acetate content

may be determined by difference at 4651 cm^{-1} (2.15 μ). Polymerized vinyl-chloride does not influence either measurement.

The vinyl acetate content of films of ethylene–vinyl acetate copolymers can be determined by methods based on the measurement of absorbances at 1639 and 1389 cm^{-1} (16.1 and 13.9 μ)[72] and at 1245 cm^{-1} (8.03 μ) and 1743 cm^{-1} (5.73 μ).[73] The acrylate salt in acrylate salt–ethylene ionomers has been determined from the ratio absorbances at 1560 cm^{-1} (6.41 μ) (asymmetric vibration of the carboxylate ion) and 1380 cm^{-1} (7.25 μ).[74]

Hydroxyl groups

Infrared spectroscopy is used to determine hydroxyl groups in polymers 53–56 it has been used to determine hydroxyl groups in polymers of the following structure

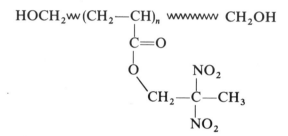

This method utilizes the strong infrared absorption band at 3450 cm^{-1} (2.90 μ). The hydroxyl concentration of approximately 30 mequiv/l is low enough that the hydroxyl groups are completely associated with the tetrahydrofuran spectroscopic solvents, and there are no apparent free or self-associated hydroxy peaks. It is essential in this method that the sample is dry, as water absorbs strongly in the 3450 cm^{-1} region of the spectrum. A further limitation of the method is that any other functional group in the sample, such as phenols, amides, amines, imines and sulphonic acid groups which absorb in the 3400–3500 cm^{-1} (2.94–2.86 μ) region, are likely to interfere in the determination of hydroxyl groups.

Unsaturation

Infrared spectroscopy has been used for the determination of unsaturation in ethylene–propylene–diene terpolymers.[28] Determination of extinction coefficients for the various terpolymers is required if quantitative work is to be done.

3.2(c)   Raman spectroscopy

This technique has found limited application to polymer analysis. It has been used for quantitative analysis of styrene–butadiene–methylmethacrylate graft copolymer, C=C stretching bands were observed for the three configurations of polybutadiene.[29]

3.2(d) Derivatization–gas chromatography

Due to their low volatility, polymers cannot be examined directly by gas chromatography. However, if the groups in the polymer, which it is required to determine, can be degraded or derivatized to produce volatiles in a preliminary chemical reaction then gas chromatography can be used as a final step in the analysis. Preliminary sample decomposition techniques that have been studied include saponification, methanolysis, silation, Ziesel reactions, alkali fusion and pyrolysis.

Saponification–gas chromatography

Ester groups occur in a wide range of polymers, e.g. polyethylene terephthalate and in copolymers such as, for example, ethylene–vinyl acetate, acrylic acid–vinyl ester, methyl acrylate–vinyl ester and polymethylacrylate. The classical chemical method for the determination of ester groups, namely saponification, can be applied to some types of polymers. For example copolymers of vinyl esters and esters of acrylic acid can be saponified in a sealed tube with 2 M sodium hydroxide. The free acids from the vinyl esters were determined by potentiometric titration or gas chromatography. The alcohols formed by the hydrolysis of the acrylate esters were determined by gas chromatography. Vinyl acetate–ethylene copolymers can be determined by saponification with 1 N ethanolic potassium hydroxide at 80°C for 3 h.[57,58] Polymethyl acrylate can be hydrolysed rapidly and completely under alkaline conditions; on the other hand, the monomer units in polymethylmethacrylate prepared and treated similarly are resistant to hydrolysis,[59] although benzoate end-groups react readily.[60] Only about 9% of the ester groups in polymethylmethacrylate reacted even during prolonged hydrolysis; hydrolysis of polymethylacrylate was complete in 0.5 h. Although only about 9% of the ester groups in methylmethacrylate homopolymers are hydrolysed by alcoholic sodium hydroxide, this proportion is increased by the introduction of comonomer units into the polymer chain. Thus, saponification techniques should be applied with caution to polymeric materials.

Methanolysis–gas chromatography

Esposito and Swann[61] published a technique involving methanolysis of a polyester resin with lithium methoxide as a catalyst: the methyl esters formed were separated from the polyols and identified by gas chromatography.

$$COOH—(CH_2)_2CO[O(CH_2)_2OCO(CH_2)_2CO]OH + 3CH_3OH \xrightarrow{LiOCH_3}$$

succininic acid–diethyl glycol polyester

$$2CH_3OOC(CH_2)_2COOCH_3 + HO(CH_2)_2OH.$$

This method was recently improved by Percival[62] by using sodium methoxide as a catalyst and injecting the reaction mixture of the transesterification directly into the gas chromatograph (without any preliminary separation).

Silation–gas chromatography

Aydin and co-workers[63] hydrolysed polyesters and converted the product acids and glycols including 1,4-cyclohexane di-methanol and isophthalic acid to the corresponding trimethylsilyl esters and ethers which were then analysed by gas chromatography.

Hydriodic acid reduction–gas chromatography

Alkoxy and ester groups have been determined in polymers and co-polymers by the Ziesel procedure involving reaction with anhydrous hydrogen iodide at 100°C.

Alkoxy groups

$$e.g.—CH_2{=}\underset{\underset{Me}{|}}{CHOR} + HI \rightarrow CH_2{=}\underset{\underset{Me}{|}}{CHI} + ROH$$

Ester groups

$$e.g.—CH_2{=}\underset{\underset{Me}{|}}{CHCOOR} + HI \rightarrow CH_2{=}\underset{\underset{Me}{|}}{CHCOOH} + RI$$

Hydrolysis using hydriodic acid has been used for the determination of the methyl, ethyl, propyl, and butyl esters of acrylates, methacrylates, or maleates,[64,65] and the determination of polyethyl esters in methyl-methacrylate copolymers.[67,68] First the total alcohol content is determined using a modified Zeisel hydriodic acid hydrolysis.[66] Secondly, the various alcohols, after being converted to the corresponding alkyl iodides, are collected in a cold trap and then separated by gas chromatography. Owing to the low volatility of the higher alkyl iodides the hydriodic acid hydrolysis technique is not suitable for the determination of alcohol groups higher than butyl alcohol. This technique has also been applied to the determination of alkoxy groups in acrylate esters.[65]

Haslam *et al.*[993] employed a procedure based on pyrolysis for the determination of polyethyl esters in methacrylate copolymers. The alkoxy groups in the polymers were reacted with hydrogen iodide and pyrolysed to their corresponding alkyl iodides which were then determined by chromatography on a dinonyl sebacate column at 75°C. Similarly, Miller[994] determined acrylate ester impurities in polymers by converting the alkoxy groups to alkyl iodides which were gas chromatographed on a di-2-ethyl hexyl sebacate column at 70°C.

Alkali fusion reaction–gas chromatography

This technique has been applied to the quantitative determination of alkyl and aryl groups in polysiloxanes[90] and of imides in aromatic polyamides and poly (amide imides).[91]

The method involves fusion of the polymer with powered potassium hydroxide, which converts alkyl and aryl groups in siloxanes into the corresponding hydrocarbons and amino and imino group to the corresponding amino or diamine;

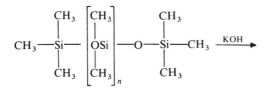

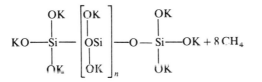

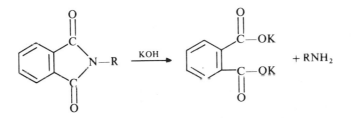

The resulting volatile products are then cold-trapped and subsequently determined by gas chromatography. It is important in these methods to establish a suitable temperature time profile when heating the samples, in order to obtain reliable results. Thus, when siloxanes are fused with alkali in a platinum boat the reaction mixture is first heated to 100°C then gradually raised to 300°C. The nitrogen compounds referred to above reacted very smoothly at 250°C. This avoided spattering of the reaction mixture and achieved quantitative clearage of methyl and phenyl groups in 5 min. Methyl contents in the range 6 to 36%, and phenyl contents in the range 10 to 81%, can be determined by this method giving results which are in good agreement with those obtained by NMR spectroscopy. Regarding nitrogen containing polymers the following polymer, upon alkali fusion, produced *m*-phenylene diamine in 100.7% yield.

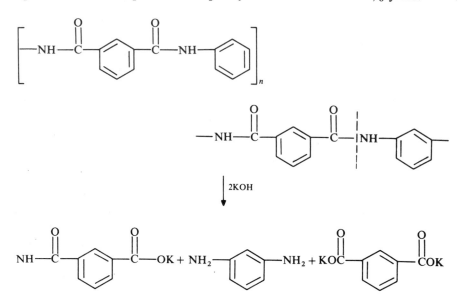

3.2(e) Pyrolysis–gas chromatography

This procedure, which is discussed in further detail in Chapter 5, involves heating the polymer to a high temperature either on a platinum filament or in a microfurnace, then sweeping the pyrolysis products with the carrier gas on to a gas chromatograph. As well as its use in microstructural studies (Chapter 5), the technique has been used for the determination of functional groups and for compositional analysis of polymers.

Composition of olefin copolymers

This technique has been applied to the gas chromatography of ethylene–butene copolymers.[88] Pyrolysis was carried out at 410°C in an evacuated gas vial and the products swept into the gas chromatograph. Under these pyrolysis conditions it is possible to analyse the pyrolysis gas components

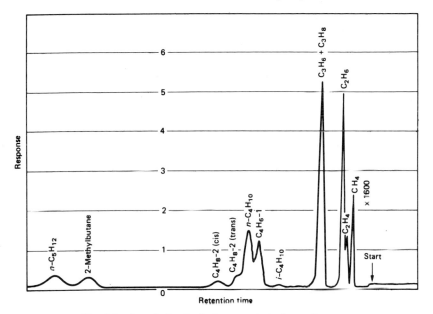

FIG. 16. A typical polyethylene pyrolysis chromatogram.

and obtain data within a range of about 10% relative. The peaks observed on the chromatogram were methane, ethylene, ethane, combined propylene and propane, isobutane, 1-butene, *trans*-2-butene, *cis*-2-butene, 2-methyl-butene and *n*-pentane. A typical pyrolysis chromatogram for polyethylene is shown in Fig. 16.

Figure 17 shows the relationship between the amount of ethylene produced on pyrolysis and the amount of butene in the ethylene–butene copolymer,

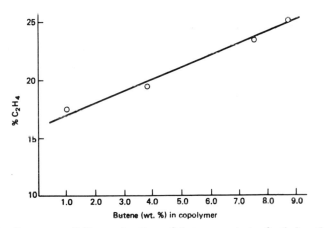

FIG. 17. Percentage C_2H_6 as function of butene content of ethylene–butene copolymers.

which was determined by an infrared analysis for ethyl branches.[89] The y intercept of 16.3% ethylene should represent that amount of ethane which would result from a purely linear polyethylene. An essentially unbranched Phillips-type polyethylene polymer yielded 14.5% ethylene, which is fairly close to the predicted 16.3%.

Alkoxy groups in ethylene oxide–propylene oxide condensates

Upon pyrolysis at 360–410°C in an evacuated vial, these polymers produce a mixture of ethylene and propylene proportional in amount to the concentrations of ethoxy and propoxy groups in the original polymers.[80,81]

$$-CH_2CH_2O-CH-CH_2O \rightarrow CH_2=CH_2 + CH_3CH=CH_2$$
$$\underset{CH_3}{|}$$

Figure 18 is a plot of the percentage ethylene as a function of the ethylene oxide content of ethylene oxide–propylene oxide condensate. As the ethoxy character of the condensate increases, so does the amount of ethylene produced. The curve is linear up to about 50% ethylene oxide, then turns sharply upward to an ethylene content of 38.6% for pure polyethylene glycol.

The relative contents of ethylene oxide and propylene oxide in polyethylene–polypropylene glycols has been determined using combined pyrolysis–gas chromatography calibrated with polyethylene glycol and polypropylene glycol standards.[82]

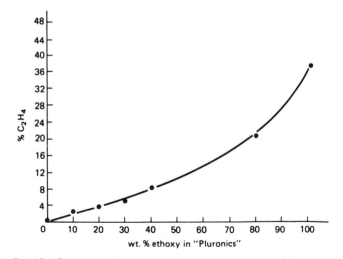

FIG. 18.   Percentage C_2H_4 as function of ethoxy content of Pluronics

Ester groups

Barrall *et al.*[69] have described a pyrolysis–gas chromatographic procedure for the analysis of polyethylene–ethyl acrylate and polyethylene–vinyl acetate copolymers and physical mixtures thereof. They used a specially constructed pyrolysis chamber as described by Porter *et al.*[70] Less than 30 s is required for the sample chamber to assume block temperature. This system has the advantages of speed of sample introduction, controlled pyrolysis temperature, and complete exclusion of air from the pyrolysis chamber. The pyrolysis chromatogram of poly(ethylene–vinyl acetate) contains two principal peaks. The first is methane and the second acetic acid.

$$-CH_2-CH_2-CH_2-CH \underset{\substack{\text{pyrolysis} \\ 300-480^\circ C}}{\xrightarrow{\hspace{1cm}}} CH_3COOH + CH_4$$
$$\overset{|}{O}OCCH_3$$

Variations from 350°C to 490°C in pyrolysis temperature produced no change in the area of the acetic acid peak, but did cause an area variation in the methane peak. The pyrolysis chromatogram of poly(ethylene–ethyl acrylate) (Fig. 19) at 475°C shows one principal peak due to ethanol. No variation in peak areas was noted in the temperature range 300°C to 480°C. Table 15 shows the analysis of 0.05 g samples of poly(ethylene–ethyl acrylate) (FEEA) and poly(ethylene–vinyl acetate) (PEVA) obtained at a pyrolysis temperature of 475°C.

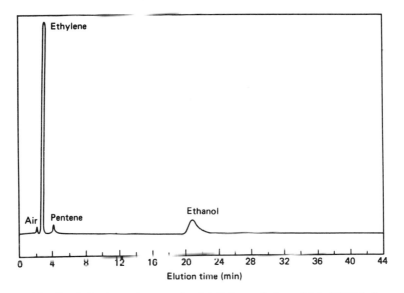

FIG. 19.   Pyrolysis chromatogram of poly(ethylene–ethyl acrylate). 475°C helium carrier, 30 cc per minute Carbowax column.

TABLE 15 *Pyrolysis results on physical mixtures of poly(ethylene–ethyl Acrylate) and poly(ethylene–vinyl acetate)*

Mixture	Acetic acid (wt%)		Ethylene (wt%)		Oxygen (wt%)	
	Found	Calculated	Found	Calculated*	Found	Calculated
50% PEEA-1 and 50% PEVA-2	9.10	9.05	2.65	2.62	7.88	8.25
33.3% PEEA-2 and 66.6% PEEA-3	12.15	12.33	0.75	0.70	7.33	7.49

*Calculated from results for acetic acid and ethylene content for individual samples on weight per cent basis.

3.2(f) Radiochemical methods

Determination of unsaturation in butyl rubber

Radiochemical methods involving the use of ^{36}Cl have been used for the measurement of higher levels of unsaturation (0.5–2 mole%) in butyl rubber.[22] This method, with slight modification, is also capable of determining unsaturation at the 0.01 to 0.1 mole% level.[23] If the specific activity of the radiochlorine in the gas phase is known, the weight of chlorine in the polymer can be found by counting. The mole% unsaturation of the polymer was calculated from the weight per cent unsaturation by assuming that one atom of chlorine enters the polymer per double bond producing a monochloro compound. With this assumption it was found that polyisobutene consumes a mean of 1.2 chlorine atoms per double bond, i.e. per molecule. This is probably an indication that side-reactions such as addition of chlorine are occurring to a small extent. Since the true unsaturation is close to one double bond per chain, it follows that the main chain-breaking reaction during the cationic polymerization of isobutene is proton transfer.

4

Fingerprinting of polymers

THE polymer analyst working in industry is frequently asked to identify a polymer, on at least to give an opinion as to the type of a polymer. This might arise as a result of an interest in the materials being used by competitors or manufacturers in components or packaging materials. Obviously, if a detailed examination of a polymer or copolymer and its additive system is required then other chapters of this book should be referred to. However, if a quick opinion is all that is required then some simple physical tests and the fingerprinting approach referred to here might suffice.

4.1 Simple physical tests

These are of limited value nowadays because the wide range of polymers now being manufactured and probably of no value in the case of copolymers. Conventional low-impact polystyrene is soluble in hot toluene, whereas high-density polyethylene or propylene have little or no solubility in this solvent. However, if the polystyrene contains some copolymerized butadiene, as occurs in the case of high-impact polystyrenes, then due to the presence of crosslinked gel the polymer would not completely dissolve in hot toluene. So even in the case of simple polymers solubility tests are of limited value, and for them to provide any useful information requires detailed knowledge. Polystyrene, on the other hand – unlike the polyolefins – due to its aromatic nature, will burn with a smoky flame, when it is held in a flame. However, if it is a non-flame grade it will not burn. PVC will, when burnt, produce an acrid odour of hydrogen chloride; so will polyvinylidene chloride and many copolymers containing vinyl chloride or vinylidene chloride. If the polymer contained fluorene, then smelling the odours produced on combustion might be a hazardous operation, as it would be in the case of acrylonitrile and polyurethane polymers, which in these circumstances are likely to produce hydrogen cyanide. Density measurement, i.e. whether the polymer sinks or floats in water, is another simple parameter that can be observed. It will distinguish polyethylene with a density of less than one from a highly chlorinated polymer with a density of greater than one. However, if the polyethylene contains, say, lead steaxate or iron oxide filler, its density may well exceed one!

Enough has been said to indicate that, mostly, simple physical tests just do not provide reliable information. Qualitative or quantitative examination for the presence of elements and the pyrolysis–gas chromatograph or infrared fingerprinting approach, as discussed below, do, however, provide more useful information and indeed in many cases will lead to a successful resolution of the problem.

Determination of elements

A cautionary note is that, in addition to the polymer itself, the polymer additive system may contain elements other than carbon, hydrogen and oxygen. The detection of an element such as nitrogen, sulphur, halogens, phosphorus, silicon or boron in a polymer is indicative that the element originates in the polymer and not the additive system if the element is present at relatively high concentrations such as several per cent. This is highlighted by the example of a high-density polyethylene which might contain 0.2–1% chlorine originating from polymerization residues and PVC homopolymer which contains more than 50% chlorine.

Useful rapid tests for elements which can be used qualitatively or quantitatively are tabulated in Table 16.

4.2 Fingerprinting by pyrolysis–gas chromatography

Pyrolysis is an analytical technique whereby complex involatile materials are broken down into smaller volatile constituent molecules by the use of very high temperatures. Polymeric materials lend themselves very readily to analysis by this technique. Providing that the pyrolysis conditions are kept constant, a sample should always degrade into the same constituent molecules. Therefore, if the degradation products are introduced into a gas chromatograph, the resulting chromatogram should always be the same and a fingerprint uniquely characteristic of the original sample should be obtained.

In one technique a pyrolyser probe is inserted into a purpose-designed adaptor which is installed in the injector unit of the gas chromatograph. Different types of adaptor are available for packed or capillary column work. Correct selection of the appropriate adaptor is a prerequisite for optimum system performance.

The choice of packed or capillary pyrolysis–gas chromatography is generally a matter of personal requirements. The type and range of samples to be analysed, the complexity of pyrogram required and the length of the analysis – all will play a significant role in decision-making.

For the general fingerprinting or analysis of a wide range of routine

TABLE 10 *Methods of determination of traces of various elements in polymer liquids*

Element	Procedure	Reference	Interferences
Sulphur up to 30 mg	Combusted in oxygen-filled flask over dilute hydrogen peroxide solution. Potentiometric titration of sulphuric acid with N/100 sodium hydroxide or photometric titration of sulphate with N/100 barium perchlorate.	92, 93	Chloride, fluoride, phosphate, nitrogen, boron, and metals, all interfere. Up to 2 mg chlorine, fluorine, nitrogen, boron and metals do not interfere in the determination of 1 mg sulphur. Phosphorus (up to 2 mg) interferes in the determination of sulphur (1 mg), but this interference can be overcome using the procedure of Colson[92-94]
Chlorine or bromine	Combustion as above. Chloride titrated potentiometrically with N/100 silver nitrate in presence of nitric acid and acetone.		
Chlorine or bromine or iodine	Combustion as above. Halide titrated with mercuric nitrate.	94	Up to 8 mg phosphorus, fluorine, sulphur, do not interfere in determination of 2 mg chlorine.
Phosphorus 0.1 g polymer	Digested with sulphuric acid/perchloric acid. Digested diluted and ammonium vandate/ammonium molybdate added. Yellow phosphovanadium-molybdate complex evaluated at 460 nm.	95	No interference by sulphur, chlorine, fluorine, nitrogen.
Nitrogen	Kjeldahl digestion carried out on 0.5.–1 g polymer (a) Kjeldahl digest made alkaline and distilled into 4% boric acid Ammonia estimated by acid titration. (b) Spectrophotometric estimation at 630 μm of phenol indophenol derivative	96	No interference by chlorine and nitrogen.
Fluorine	30 mg polymer combusted in an oxygen-filled flask over distilled water. Reacted with buffered alizarin complex/cerous nitrate, blue colour produced evaluated at 610 nm.	97	No interference by large excesses of sulphur, chlorine, phosphorus and nitrogen.
Silicon	30 mg polymer in gelatine capsule and combusted with sodium peroxidede, sucrose and benzoic acid in 22 ml capacity Parr bomb.		No interference by sulphur, halogens, phosphorus, nitrogen and boron.
Boron	0.1 g polymer is digested with concentrated nitric acid in a sealed ampoule to convert organoboron compounds to boric acid. Digest dissolved in methyl alcohol and boron estimated flame-photometrically at 5.9.5 nm	98	No interference by chlorine and nitrogen.

rubbers or plastics, packed column pyrolysis is more than adequate. However, for comparison of one batch of rubber with another batch of the same rubber, the detail obtained from capillary column pyrolysis–gas chromatography will probably give the best results because minor details can be compared.

Applications for pyrolysis are vast and include all types of man-made polymers, rubbers and plastics, as well as latexes, paints and varnishes; in fact, almost any sample that contains involatile organic material that can be contained in a tube or coated onto a platinum ribbon so that it may be pyrolysed.

The two types of pyrolyser probe available are based on a platinum ribbon or a coil. If a sample can be dissolved in a suitable solvent, it is best applied to the ribbon as a solution. This should be prepared at a concentration of about 10 mg/l (approximately 1%). About 1 to 5 μl (representing 10–50 g of sample) is applied to the centre of the ribbon using a microsyringe. The solution should be spread as evenly as possible so that the entire sample will experience the same heating profile during pyrolysis. It is essential to avoid the extreme ends of the ribbon as the points of attachment to the probe mounting provide a large heat-sink and do not reach the set pyrolysis temperature.

Before inserting the probe into the injector adaptor, the final temperature should be set to a low value (such as 100°C) with an interval of 5–10 s, so that the residual solvent is flashed off without pyrolysing the sample. For most samples the solvent can be flashed off in the atmosphere – but for extremely sensitive materials the probe should be located in the injector adaptor prior to solvent flash. After the solvent has been removed and the detector response has returned to baseline, the sample can be pyrolysed.

Insoluble but meltable samples should be applied by placing fine particles of material onto the ribbon and gradually raising the temperature with the interval set at 1–5 s until the particles adhere to the ribbon.

Samples that are insoluble and that will not melt should be pyrolysed in a quartz tube located inside the coil probe. The quartz tube must be inserted into the platinum coil probe very carefully to avoid distortion and damage.

Samples can then be inserted directly into the quartz tube and, for best results, should be between 10 and 50 μg and placed in the centre of the tube. Alternatively, the sample can be mixed with a quartz wool, and this placed in the quartz tube. Sample solutions can also be analysed in this way by injecting 1–3 μl from a microsyringe of a 1% solution of sample onto the quartz wool. The solvent should be evaporated to dryness by flashing the coil at 100°C with the interval set at 1 s. For most samples the solvent can be flashed off into the atmosphere, but for extremely sensitive material the probe should be located in the injector adaptor prior to solvent flash. After the solvent has been removed and the signal from the gas chromatograph

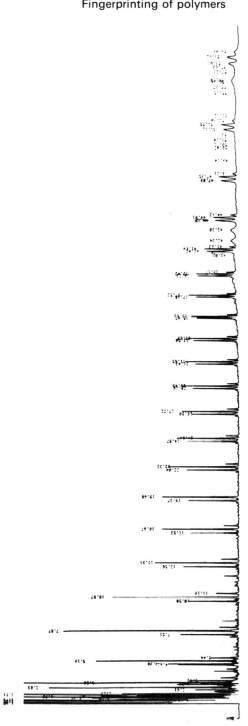

Fɪɢ. 20. Pyrolysis-gas chromatography of polyethylene. Sample: polyethylene; instrument: P-E 8310 series with split/splitless injector; column: 25 m SE30 capillary (0.25 mm i.d.); conditions: 50°C (2.0 min) 5°C min^{-1}, 230°C (after last peak); pyrolysis: 850°C for 10 s.

has returned to baseline, the sample can be pyrolysed. The coil probe generally requires a final temperature of about 150°C higher than that required by the ribbon probe.

Selection of pyrolysis temperature

An investigation to determine the lowest temperature at which pyrolysis will occur for a particular sample can be made with repeated runs on the same sample at increasing pyrolysis temperatures without removing the probe from the injector adaptor.

Alternatively, arbitrary pyrolysis conditions may be selected and the sample pyrolysed. The amount of residual organic material can be determined by repyrolysing the sample at 980°C for 1 s for the ribbon probe and 10 s for the coil probe. A large residual peak under these conditions indicates that only a small part of the sample was initially pyrolysed. A higher final pyrolysis temperature should then be used.

As mentioned earlier, packed columns, as opposed capillary columns, are often adequate for polymer fingerprinting. The chromatograms are relatively simple and the analysis time relatively short, and the technique is particularly amendable to the examination of the more volatile degradation products, such as monomers produced by thermal degradation. However, when wishing to distinguish between similar polymers such as polyethylene and an ethylene–polypropylene copolymer (Figs 20 and 21)

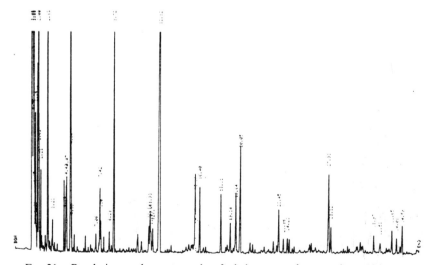

FIG. 21.   Pyrolysis–gas chromatography of ethylene–propylene copolymer. Sample: polyethylene–polypropylene copolymer; instrument: P-E 8310 series with split/splitless injector; column: 25 m SE 30 capillary (0.25 mm i.d.); conditions: 50°C (2.0 min.) 5°C min⁻¹, 230°C (after last peak); pyrolysis: 850°C for 10 s.

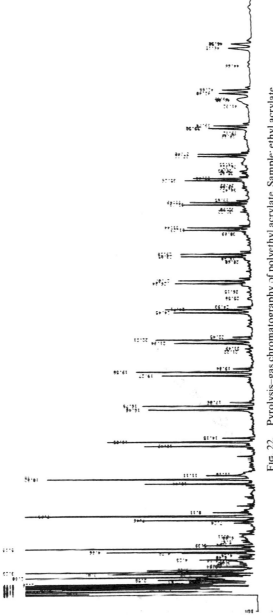

FIG. 22. Pyrolysis–gas chromatography of polyethyl acrylate. Sample: ethyl acrylate polymer; instrument: P-E 8310 series with split/splitless injector; column: 25 m SE 30 capillary (0.25 mm i.d.); conditions: 50°C (2.0 min) 5°C min⁻¹, 220°C (after last peak); pyrolysis: 850°C for 10 min.

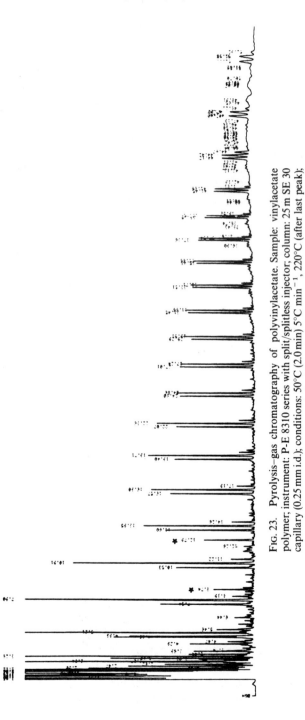

Fig. 23.　Pyrolysis–gas chromatography of polyvinylacetate. Sample: vinylacetate polymer; instrument: P-E 8310 series with split/splitless injector; column: 25 m SE 30 capillary (0.25 mm i.d.); conditions: 50°C (2.0 min) 5°C min⁻¹, 220°C (after last peak); pyrolysis: 850°C for 10 min.

it may be necessary to use capillary columns so as to produce a thermogram of more complexity, so that differences can be seen between different polymers.

Separation of all the more volatile fragments is not possible without cryogenic cooling of the oven. Under standard chromatographic conditions it is the high boiling degradation products that are readily separated, as can be seen from Figs 20 and 21. Therefore it is the latter part of the chromatogram that is used for identification and fingerprinting.

Note that almost no peak separation is achieved in the first 2.5 min, even with an initial column temperature of 50 °C. Also note that for same samples, such as polyethylene (Fig. 20), the parent polymer thermally degrades into similar fragments of varying chain length and thus gives a reiterant pattern. Other samples, such as the ethylene–propylene copolymer, thermally fragment less uniformly and such patterns are not observed.

Because it is the higher boiling degradation products that are being observed, a number of polymers give very similar fingerprints (see Figs 22 and 23). Note that, in these two examples, the pyrograms produced are almost identical. The only observable differences are the presence of two additional peaks in the vinyl acetate pyrogram (Fig. 23) at 8.74 min and 12.79 min (marked with an asterisk).

Therefore, capillary column pyrolysis–gas chromatography is slightly less useful than packed column pyrolysis as a strict identification technique. However, the complexity of the fingerprints produced allows minute details to be observed. Analysis times are two to three times longer than those for packed column pyrolysis–gas chromatography; runs are generally in the order of 60 min in length. The high split ratio and very small amounts of sample required to obtain good results with capillary column pyrolysis–gas chromatography mean that there is an obvious advantage in using this technique when only very minute amounts of sample are available for analysis.

Pyrolysis chromatograms of several common plastics are shown in Fig. 24, which illustrate the value of pyrolysis–gas chromatography in the identification of polymers by fingerprinting.

4.3   Fingerprinting by infrared spectroscopy

Infrared spectra of thin films of polymer in the region up to 4000 cm^{-1} are characteristic of the polymer. Some infrared spectra are shown in Fig. 25. Computerized retrieval from data in a library of standard polymers have been used in the pyrolysis–gas chromatography and the infrared fingerprinting techniques to facilitate polymer identification. Where doubts exist it would be advisable to attempt identification by both pyrolysis–gas chromatographic and infrared spectroscopic techniques so as to obtain confirmation.

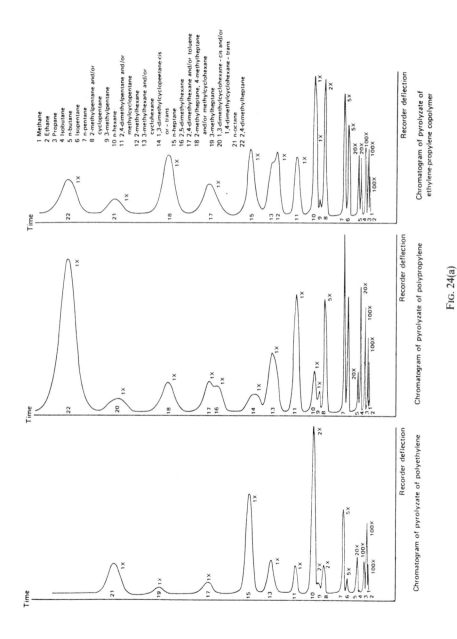

FIG. 24(a)

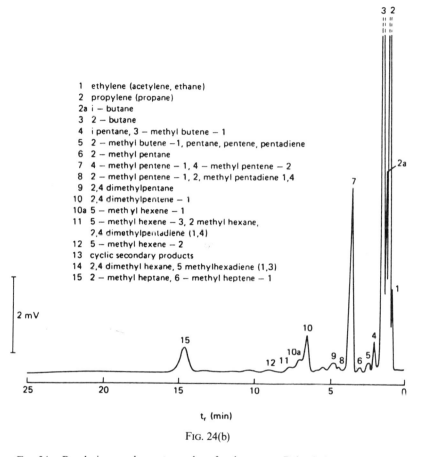

1 ethylene (acetylene, ethane)
2 propylene (propane)
2a i − butane
3 2 − butane
4 i pentane, 3 − methyl butene − 1
5 2 − methyl butene −1, pentane, pentene, pentadiene
6 2 − methyl pentane
7 4 − methyl pentene − 1, 4 − methyl pentene − 2
8 2 − methyl pentene − 1, 2, methyl pentadiene 1,4
9 2,4 dimethylpentane
10 2,4 dimethylpentene − 1
10a 5 − methyl hexene − 1
11 5 − methyl hexene − 3, 2 methyl hexane,
   2,4 dimethylpentadiene (1,4)
12 5 − methyl hexene − 2
13 cyclic secondary products
14 2,4 dimethyl hexane, 5 methylhexadiene (1,3)
15 2 − methyl heptane, 6 − methyl heptene − 1

2 mV

t, (min)

FIG. 24(b)

FIG. 24. Pyrolysis–gas chromatography of polymers. *a*, Polyethylene, polypropylene, ethylene–propylene copolymer; *b*, poly-4-methylpentene-1; *c*, polybutene-1; *d*, polybutadiene; *e*, polystyrene; *f*, polyacrylate and polymethacrylates; *g*, polymethyl methacrylate; *h*, polyvinylchloride; *i*, polyvinlidene chloride.

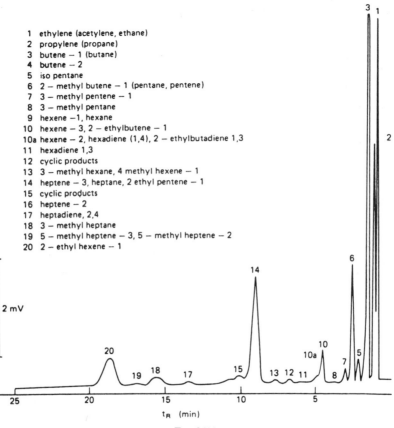

1 ethylene (acetylene, ethane)
2 propylene (propane)
3 butene — 1 (butane)
4 butene — 2
5 iso pentane
6 2 — methyl butene — 1 (pentane, pentene)
7 3 — methyl pentene — 1
8 3 — methyl pentane
9 hexene —1, hexane
10 hexene — 3, 2 — ethylbutene — 1
10a hexene — 2, hexadiene (1,4), 2 — ethylbutadiene 1,3
11 hexadiene 1,3
12 cyclic products
13 3 — methyl hexane, 4 methyl hexene — 1
14 heptene — 3, heptane, 2 ethyl pentene — 1
15 cyclic products
16 heptene — 2
17 heptadiene, 2,4
18 3 — methyl heptane
19 5 — methyl heptene — 3, 5 — methyl heptene — 2
20 2 — ethyl hexene — 1

2 mV

t_R (min)

FIG. 24(c)

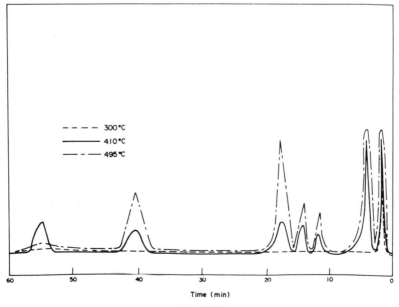

- - - - 300°C
——— 410°C
— · — 495°C

Time (min)

FIG. 24(d)

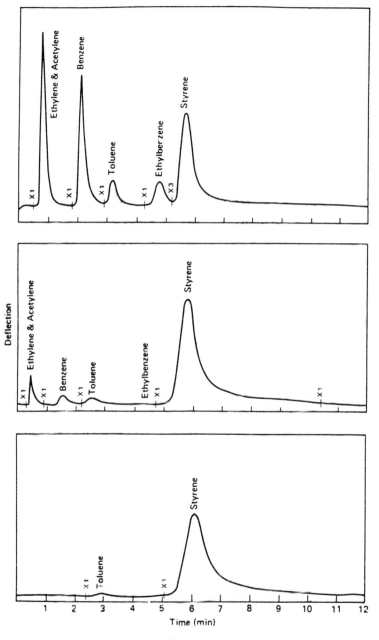

FIG. 24(e)

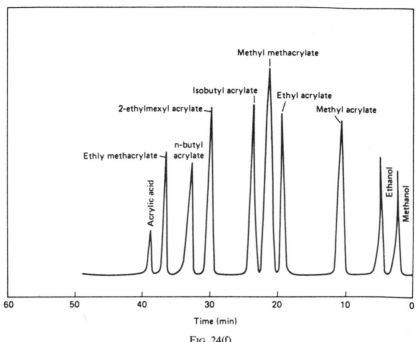

FIG. 24(f)

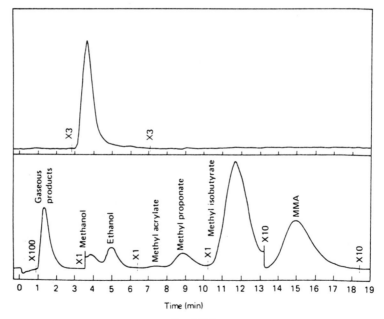

FIG. 24(g)

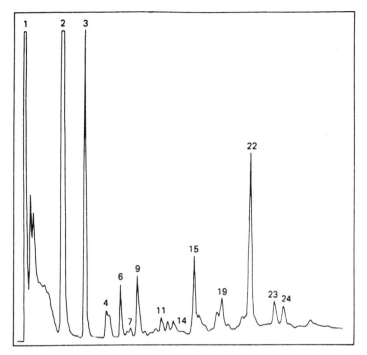

FIG. 24(h)

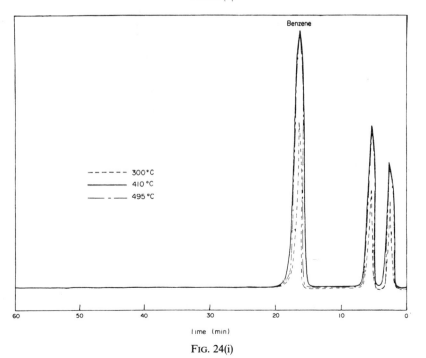

FIG. 24(i)

Functional group checks by infrared spectroscopy

Brako and Wexler[99] have described a useful technique for differentiating the presence or absence of functional groups such as hydroxyl, carboxylic acid or ester in polymers containing small percentage components of such groups. Films of lattices or polymers are subjected to chemical treatment which results in marked changes in the infrared spectrum and which can be associated with the disappearance of a functional group or its replacement by another functional group. Infrared data may be readily interpreted negatively so that one may definitely preclude the presence of hydroxyl, carbonyl, amine, amide, nitrile, ester, carboxylic, aromatic, methylene, tertiary butyl, and terminal vinyl groups if the corresponding group vibrations are absent in the infrared spectrogram. More difficult is the assignment of functional groups where multiple or several alternative possibilities exist, as in the mixture of a carboxylic and keto group or in the assignment of a band to an olefinic group.

Figure 26 shows the infrared spectra of a sodium polyacrylate film before and after exposure to hydrochloric acid vapour. Exposure to acid results in the disappearance of the broad, intense band associated with the

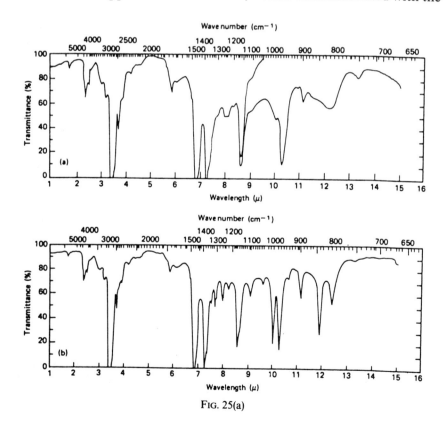

FIG. 25(a)

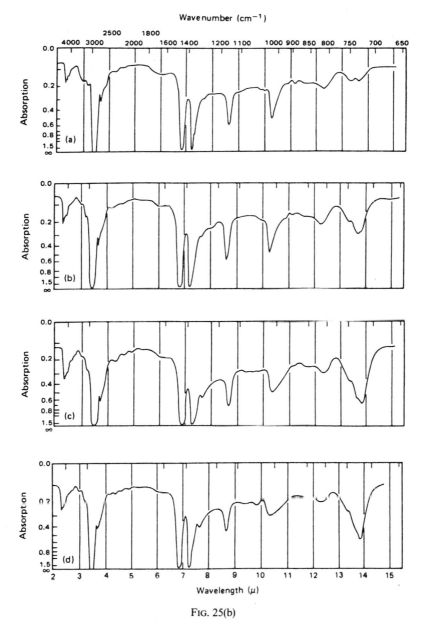

FIG. 25(b)

FIG. 25. Infrared spectra of polymers. *a*, Polypropylene; *b*, ethylene-propylene copolymers; *c*, hydrogenated polyisoprenes and hydrogenated polydimethyl butadiene; *d*, 3, 4-polyisoprene and polydimethyl butadiene; *e*, various polybutadienes; *f*, hydrogenated polybutadienes; *g*, ethylene–propylene–diene terpolymers *h*, polystyrene.

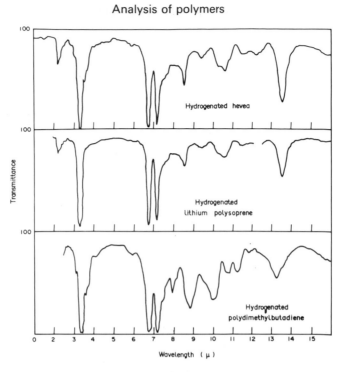

FIG. 25(c)

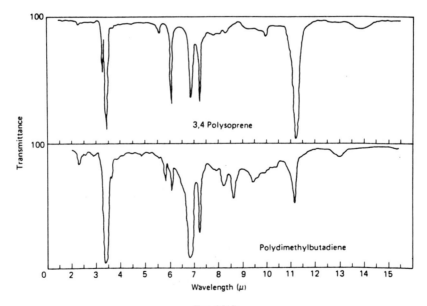

FIG. 25(d)

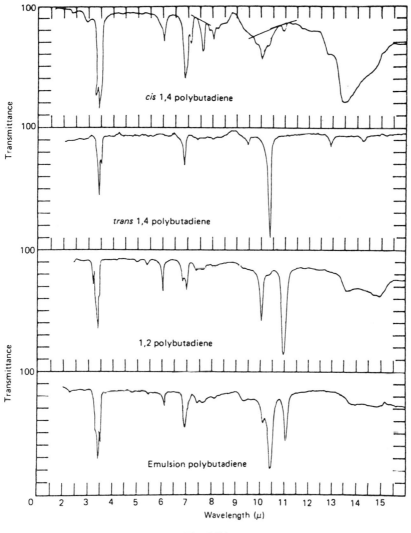

FIG. 25(e)

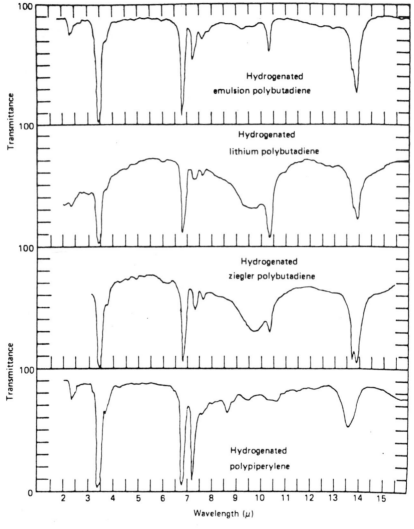

Fɪɢ. 25(f)

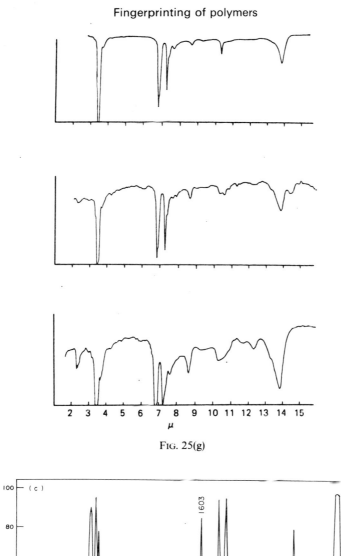

FIG. 25(g)

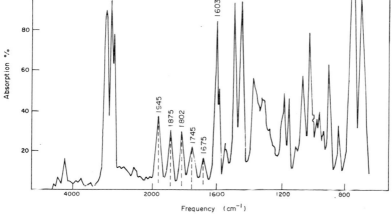

FIG. 25(h)

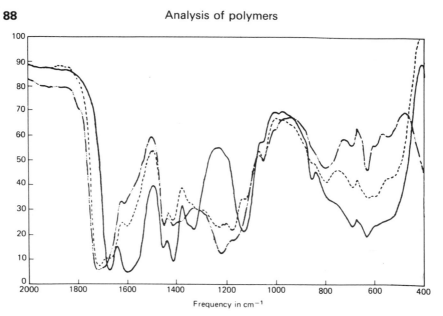

FIG. 26.   Infrared spectra of chemically treated polymers. Spectrum of sodium polyacrylate before and after exposure to hydrogen chloride vapour and the heated film of the hydrogen chloride-treated acrylate: ——, sodium polyacrylate film; – – – –, sodium polyacrylate film exposed to HCl vapour: ——— – – –, sodium polyacrylate film exposed to HCl vapour and then heated.

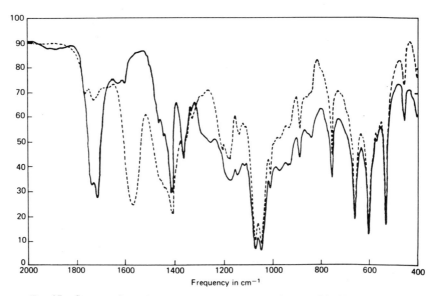

FIG. 27.   Spectra of copolymer of acrylic acid and vinylidene chloride and the film after exposure to ammonia vapour: ——, acrylic acid–vinlidene chloride copolymer film; – – – –, acrylic acid–vinylidene chloride copolymer film exposed to ammonia vapour.

carboxylate group in the polyacrylate ion at around $1600\,cm^{-1}$ ($6.25\,\mu$). Appearance of a broad intense absorption at about $1720\,cm^{-1}$ ($5.81\,\mu$) is associated with the carbonyl of the carboxylic acid group in the polymer. Significant changes are also observed in the 1100 to $1200\,cm^{-1}$ (9.09–$8.33\,\mu$) region. Heating of the acidified film resulted in minor changes in the spectrum. Figure 27 shows the changes in the infrared spectrum resulting from the exposure of acrylic acid–vinylidene chloride copolymer film to ammonia vapour. Bands associated with the carboxylic acid carbonyl stretching frequencies at 1715–$1740\,cm^{-1}$ (5.83–$5.75\,\mu$) disappear on exposure to ammonia vapour. A well-defined carboxylate band appears at $1570\,cm^{-1}$ ($6.37\,\mu$). This change is sufficient to confirm that the copolymer contains carboxylic acid groups.

5

Polymer microstructure

MICROSTRUCTURAL features are often of tremendous importance in deciding the physical, and to some extent the chemical, properties of polymers. The term microstructure does not necessarily infer that the feature of concern occurs at low concentrations. It is more concerned with the particular detail of the structure of the polymer. Thus polybutadiene unsaturation exists in various *cis* and *trans* forms, all present as constituents of the polymer. Some forms of unsaturation might exist at high concentrations, and some at low. In the case of polyethylene, however, butyl hexyl and octyl side-groups, which are of great interest from the microstructural point of view, all exist at low concentrations usually expressed as the number of such groups present per 1000 carbon atoms in the polymer chain.

In this chapter various aspects of microstructure are discussed in turn. The first section is concerned with the measurement of the various types of unsaturation in polymers. Some of the commonly occurring types of unsaturation are trans 1, 4, *cis* 1, 4-vinyl, 3, 4 addition and 1, 2 addition which occur in polymers based on butadiene and isoprene. Ozoneolysis is a very useful method for determining the amount and type of unsaturation in polymers. The ozonization of double bonds produces ozonides, which upon hydrolysis produce mixtures of carbonyl compounds which can be analysed by various chromatographic techniques. The nature and concentration of these carbonyl compounds provides useful information on the original polymer structural feature.

Branching is another aspects of polymer microstructure which is of great interest. Polyethylene, for example, can contain side-chain alkyl groups ranging from methyl to octyl or even higher, which can be identified and determined by techniques such as NMR and infrared spectroscopy. Such groups often have a profound effect on the physical properties of polyethylene and the presence of different types of alkyl side-groups accounts for the fact that many different grades of this polymer are available, each with its own particular physical properties. These techniques are also applicable to copolymers of ethylene with other olefins, e.g. ethylene–butene.

A further very important aspect of microstructure is the sequence of monomer units in a polymer. This applies whether the polymer is based on a single monomer which is capable of polymerizing in different ways, e.g. head-to-head or head-to-tail polymerization, or whether it is based on two or more different monomers when, obviously, many variants of monomer sequence are possible. Sequence distribution has an important bearing on the tacticity and other properties of polymers, as will be discussed later. Three major techniques have been used to study sequence problems in polymers, they are pyrolysis–gas chromatography, NMR and infrared spectroscopy.

As might be expected, the groups at either end of a polymer chain differ from those in the main polymer chain. Various methods are available for determining low concentrations of such end-groups in polymers.

5.1 Unsaturation

Brako and Wexler[99] have described a useful technique for testing for the presence of unsaturation in polymer films such as polybutadiene and styrene–butadiene. They expose the film to bromine vapour and record its spectrum before and after exposure (Fig. 28). This results in marked changes in the infrared spectrum. Noteworthy is the almost complete disappearance of bands at 730, 910, 965 and 1640 cm^{-1} (13.69 10.99, 0.36 and 6.10 μ) associated with unsaturation. A pronounced band possibly associated with a C–Br vibration appears at 550 cm^{-1} (18.18 μ). Bands also appear at 785, 1145 and 1250 cm^{-1} (12.73, 8.73 and 8.00 μ), which is due to exposure to bromine vapour. Exposure of butadiene–styrene copolymer (Fig. 28b) to bromine vapour results in the disappearance of bands at 910 and 965 cm^{-1} (10.99 and 10.36 μ) associated with unsaturation in the butene component of the copolymer. Some alteration of the phenyl bands at 700 and 765 cm^{-1} (14.28 and 13.07 μ) is evident. The loss of a band at 1550 μ^{-1} (6.45 μ) and the appearance of a band at 1700 cm^{-1} (5.88 μ) are probably due to the action of acidic vapours on the carboxylate surfactant of the latex.

In addition to bromination, as discussed above, chlorination has also been used to elucidate the number of double bands per molecule of a polymer. McNeill[100] used ^{36}Cl for the measurement of higher levels of rubber unsaturation 0.5–2 mol % in butyl rubbers. McGuchan and McNeill[101] developed this method further and applied it to the determination of 0.01–0.1 mole % end-group unsaturation in butyl rubbers. If the specific activity of the radiochlorine in the gas phase is known, the weight of chlorine in the polymer can be found by counting. The mole % unsaturation (UM) of the polymer was calculated from the weight % unsaturation (UW) by assuming that one atom molecule of chlorine enters the polymer per double bond producing a monochloro compound. Mole per cent unsaturation of polyisobutenes was shown to be inversely proportional to their molecular weight. Polyisobutene averaged a mean of

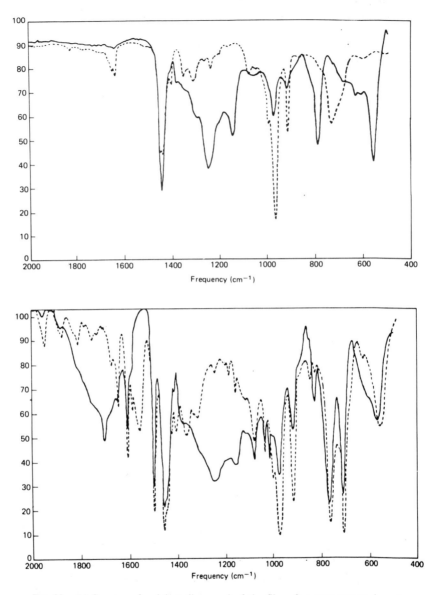

Fig. 28. (a) Spectra of polybutadiene and of the film after exposure to bromine: – – – –, polybutadiene; ———, polybutadiene after exposure to bromine vapour. (b) Spectra of a film of butadiene–styrene copolymer and of the film after exposure to bromine vapour: – – – –, copolymer of butadiene and styrene; ——— copolymer of butadiene and styrene after exposure to bromine vapour.

1.2 chlorine atoms per molecule of polymer. This is probably an indication that side-reactions, such as addition of chlorine, are occurring to a small extent. Since the true unsaturation is close to one double bond per polymer molecule, it follows that the main chain-breaking reaction during poly-merization of isobutene is proton transfer.

Infrared spectroscopy is a very useful technique for the measurement of different types of unsaturation in polymers. Polybutadiene has the following structure

$$CH_2{=}CH{-}CH{=}CH_2$$
$$1 \qquad 2 \quad 3 \qquad 4$$

Its polymers can contain the following types of unsaturation:

trans 1,4

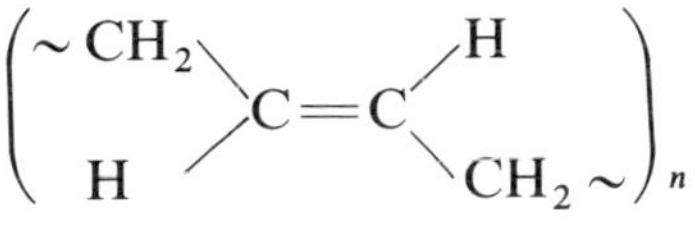

$$\left(\begin{array}{c} {\sim}CH_2 \\ \diagdown \\ H \end{array} C{=}C \begin{array}{c} H \\ \diagup \\ \diagdown CH_2{\sim} \end{array}\right)_n$$

cis 1,4

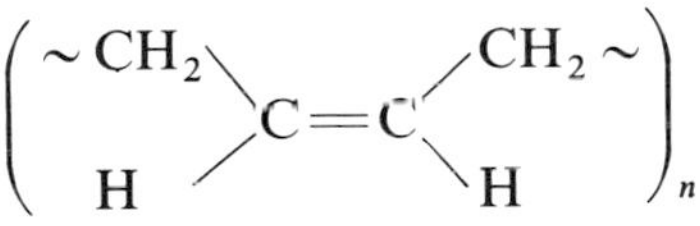

$$\left(\begin{array}{c} {\sim}CH_2 \\ \diagdown \\ H \end{array} C{=}C \begin{array}{c} CH_2{\sim} \\ \diagup \\ \diagdown H \end{array}\right)_n$$

vinyl

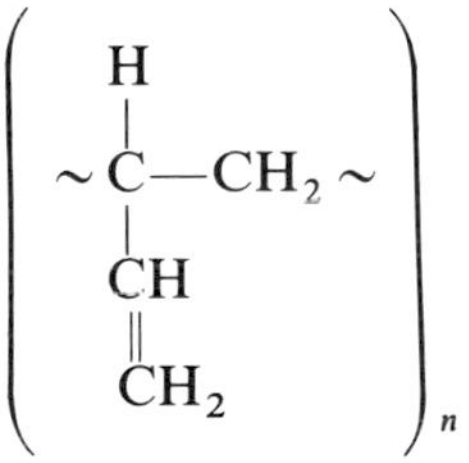

$$\left(\begin{array}{c} H \\ | \\ {\sim}C{-}CH_2{\sim} \\ | \\ CH \\ \| \\ CH_2 \end{array}\right)_n$$

Fraga[107] has developed an infrared–near–infrared method of analysis of carbon tetrachloride solutions of polybutadienes suitable for the evalua-tion of *cis*-1,4,5000–714.2 cm⁻ (2–14 μ), *trans*-1,4,9708 cm⁻¹ (10.3 μ) and vinyl, 9091 cm⁻¹ (11.0 μ) structures. Only polybutadiene is required for calibration purposes. The method is applicable to carbon tetrachloride soluble polybutadienes containing 0–97% *cis*-1,4 structure, 0–70% *trans*-1,4 structure, and 0–90% vinyl structure. Some typical spectra are shown in Fig. 29 for a high-*cis*-1,4 polybutadiene, and high-*trans*-1,4 poly-butadiene, and a high-1,2 (atactic) polybutadiene, and other samples of different *cis* and *trans* compositions.

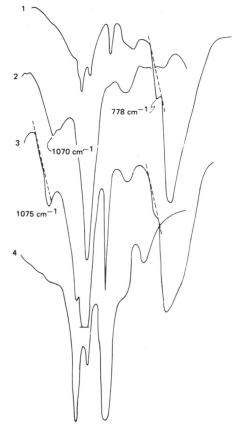

FIG. 29.　Spectra of (1) *cis*-1, 4-PBD; (2) *trans*-1, 4-PBD; (3) mixed PBD structure; (4) atactic 1, 2-PBD. Figures displayed vertically for viewing. All samples in carbon disulphide solutions; 2 mm light bath cell with NaCl window.

Figure 30 shows infrared spectra of various kinds of poly-butadienes,[103,104] which illustrate the usefulness of infrared spectroscopy for distinguishing between different types of unsaturation.

Fraga[102] has also described an infrared thin film area method for the analysis of styrene–butadiene copolymers. The integrated absorption area between 1515 and 1389 cm^{-1} (6.6 and 7.2 μ) has been found to be essentially proportional to total bound butadiene, and is independent of the isomeric-type butadiene structure present. This method can be calibrated for bound styrene contents ranging from 25 to 100%.

Infrared[105–109] and pyrolysis–infrared[110] methods have both been tried for determination of the composition of vulcanizates, but both methods have serious disadvantages. Carlson and Altenau[111] investigated NMR spectroscopy as a means of solving this problem. Various vulcani-

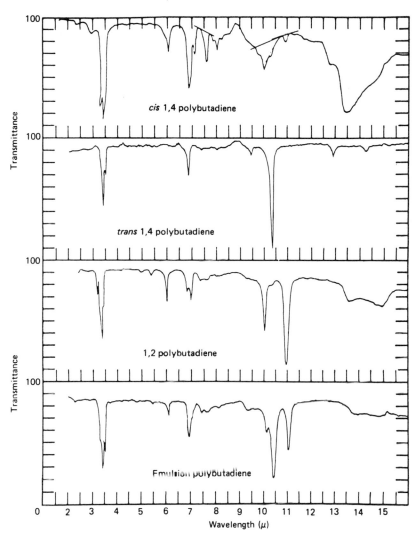

FIG. 30. Infrared spectra of various kinds of polybutadienes.

zates containing two or three of the following: natural rubber, poly-butadiene, and butadiene–styrene copolymer, were analysed in hexa-chlorobenzene solution (trimethylsilane internal standard).

Carlson and Altenau[111] used the equations to calculate polymer compositions by NMR which have been published by Mochel.[112] The quality of the spectra was comparable to that of uncompounded polymers. Good agreement was obtained between the calculated amounts of butadiene, natural rubber and styrene, and the found values; however, the method will not distinguish between butadiene from polybutadiene and

TABLE 17   *Determination of polymer composition*

Calculated compounded formula				Found compounded formula (infrared)			
		Butadiene				Butadiene	
Styrene	*cis*	*trans*	1, 2	Styrene	*cis*	*trans*	1, 2
	37	55	8		35	55	10
	14	30	56		16	27	57
15	20	53	12	16	21	51	12
21	13	52	14	23	15	49	13
26	8	53	14	28	12	48	12
17	17	43	23	22	17	39	22

butadiene–styrene copolymer. The relative amount of 1, 2-butadiene and 1, 4-butadiene were found in the cases of polybutadiene and styrene–butadiene rubber blends. The individual amounts of *cis* and *trans*-1, 4-butadiene could not be calculated because of complete overlapping of the two signals, as is also the case in uncompounded polymers.

Determination of the different microstructures in a blend of polybutadiene and/or styrene butadiene rubber with natural rubber is very difficult because the signals from the olefinic protons severly overlap one another. Carlson and Altenau[111] did not attempt to calculate the percentages of the different microstructures in these blends of rubber. Natural rubber is essentially 100% *cis*-1, 4-polyisoprene.

In further work, Carlson *et al.*[113] discuss a disadvantage of their NMR method connected with the fact that it provides only limited microstructural data. This was due to the lack of resolution of the 60-megacycle NMR. They developed alternative carbon disulphide extraction techniques to overcome these limitations and applied infrared methods as described by Binder.[114,115].

Table 17 shows infrared analyses of vulcanizates containing either polybutadiene or styrene–butadiene copolymer, or a blend of these as reported by Carlson *et al.*[113] The relative amounts and types of polybutadiene and styrene–butadiene rubber varied in these blends. These differences are reflected in the calculated percentage for the samples.

Table 18 shows infrared analyses of vulcanizates containing natural rubber, polybutadiene and styrene–butadiene rubber. The infrared results indicate that the concentration of natural rubber in the carbon disulphide solution is considerably higher than that present in the vulcanizate.

Other workers who have applied spectroscopy to the study of unsaturated polybutadiene and styrene–butadiene copolymers include Braun and Canji,[116,117] Hast and Deur Siftar,[118] Silas *et al.*,[119] Cornell and Koenig,[120] Neto and Di Lauro,[121] Binder,[122] Clark and Chen,[123] and Harwood and Richey.[124] ^{13}C NMR spectroscopy has also been used to

TABLE 18 *Determination of polymer composition*

| Calculated compounded formula | | | | | Found compounded formula (infrared) | | | | |
Styrene	cis	trans	1,2	Natural rubber	Styrene	cis	trans	1,2	Natural rubber
	19	28	4	50		13	18	3	66
	37	55	8	—		37	54	10	—
19	4	22	20	35	14	4	20	12	50
29	6	35	30	—	31	11	39	19	—
14	5	32	9	40	10	5	30	6	49
24	8	54	15	—	30	10	49	11	—
13	4	27	7	50	8	6	22	6	59
24	7	53	14	—	32	9	49	10	—

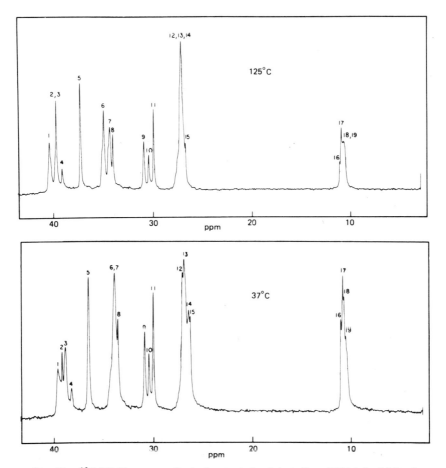

FIG. 31. ^{13}C NMR spectra of a hydrogenated polybutadiene (73% 1,2-additions) at 37 and 125°C shown with respect to a trimethylsilane internal standard.

study sequence distribution[125] and *cis*-1,4, *trans*-1,4 and 1.2 units (126, 127) in polybutadiene.

Figure 31 shows ^{13}C NMR spectra at 25.1 MHz of a hydrogenated polybutadiene that contained 73% 1,2 additions (or 183 branches per 1000 carbon atoms). The spectra were recorded at a temperature of 37 and 125°C in approximately 10 wt% polymer solutions of 1,2,4-trichlorobenzene and perdeuterobenzene. The temperature sensitivity of some of the resonances in Fig. 31 is clearly indicated by the overlap that occurs at one temperature but is resolved at the other. In all, 19 resonances, could be identified even though only 18 were visible at 37°C and only 16 at 125°C.

Isoprene has the structure

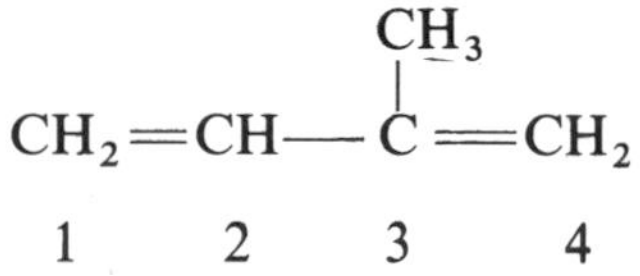

$$CH_2 = CH - \underset{\underset{4}{\overset{\overset{CH_3}{|}}{\underset{3}{C}}}}{\underset{2}{}} = CH_2$$

$$\overset{}{\underset{1}{}} \quad \underset{2}{} \quad \underset{3}{} \quad \underset{4}{}$$

Its polymers can contain the following four types of unsaturation:

trans-1,4

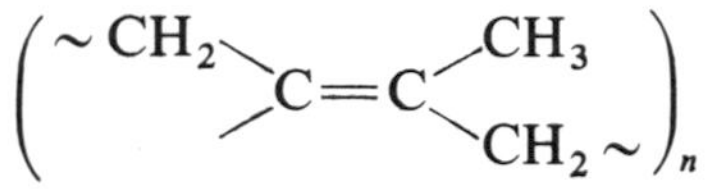

$$\left(\sim CH_2 \overset{}{\diagdown} C = C \overset{CH_3}{\diagup} \underset{CH_2 \sim}{} \right)_n$$

cis-1,4

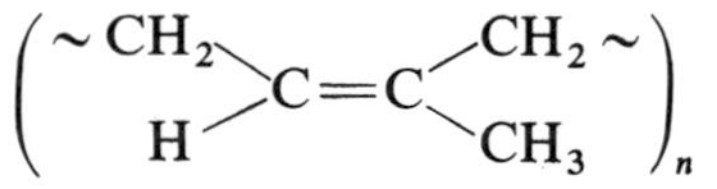

$$\left(\sim CH_2 \overset{}{\diagdown} C = C \overset{CH_2 \sim}{\diagup} \underset{CH_3}{} \right)_n$$

3,4 addition

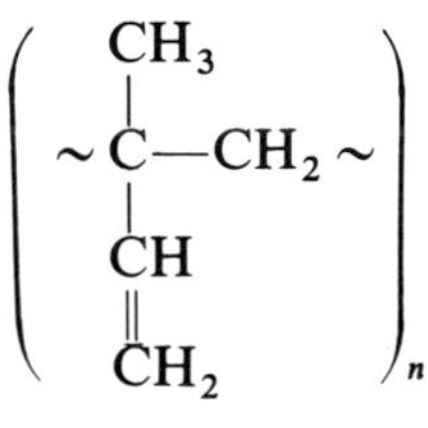

$$\left(\sim \underset{\underset{\underset{CH_2}{\parallel}}{\overset{|}{CH}}}{\overset{\overset{CH_3}{|}}{C}} - CH_2 \sim \right)_n$$

1,2 addition

$$\left(\sim CH_2 = CH \sim \underset{\underset{CH_2}{\parallel}}{\overset{|}{C}} - CH_3 \right)_n$$

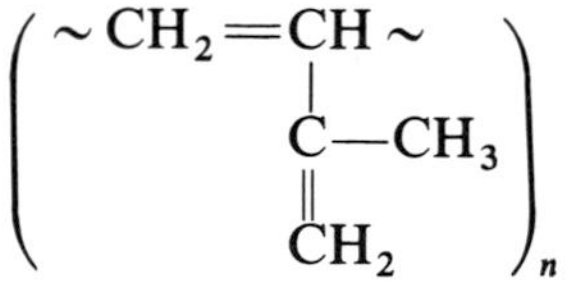

Vodchnal and Kossler[128,129] have reported an infrared method for analysis of polyisoprenes suitable for polymers with a high content of 3,4 addition and relatively small amounts of *cis*-1,4 and *trans*-1,4 structural units.

Absorptivities of the bands which are commonly used for the determination of the amount of 1, 4 structural units are about 50 times lower than the absorptivity of the band at $888 \, \text{cm}^{-1}$ ($11.26 \, \mu$) which is used for the determination of the amount of 3, 4-polyisoprene units. Therefore, in analyses of samples with high content of 3, 4-polyisoprene units it is necessary to use two concentrations or two cuvettes with different thicknesses. Application of the $1780 \, \text{cm}^{-1}$ and $3070 \, \text{cm}^{-1}$ band (5.62 and $3.26 \, \mu$) offers the possibility of using only one cuvette and one concentration. The $1780 \, \text{cm}^{-1}$ and $3070 \, \text{cm}^{-1}$ (5.62 and $3.26 \, \mu$) absorption bands do not overlap with absorption bands of other structural forms, the accuracy of analyses thus being increased. Besides exact determination of the amount of 3, 4 structural units it is possible to estimate an approximate amount of 1, 4 addition from the $840 \, \text{cm}^{-1}$, $572 \, \text{cm}^{-1}$ and $600 \, \text{cm}^{-1}$ (11.90, 17.48 and $16.66 \, \mu$) absorption bands.

Values of apparent molar absorptivities of 3, 4-polyisoprene, Hevea and balata in carbon disulphide solutions for the $572 \, \text{cm}^{-1}$, $840 \, \text{cm}^{-1}$, $888 \, \text{cm}^{-1}$, $1780 \, \text{cm}^{-1}$, and $3070 \, \text{cm}^{-1}$ (17.48, 11.90, 11.26, 5.62 and $3.26 \, \mu$) absorption bands are summarized in Table 19.

The results of measurements of the samples in carbon disulphide solutions, obtained using absorptivities from Table 19, are summarized in Table 20. The cuvettes employed were $2 \, \text{mm}$ thick, concentration of polymer varying from 0.5 to $3 \, \text{g}$ in $100 \, \text{g}$ solution. From the value of absorption at $840 \, \text{cm}^{-1}$ ($11.90 \, \mu$) the minimum amount of 1, 4 structural units was estimated assuming that all 1, 4 units are *cis*. Analysis using the

TABLE 19 *Apparent molar absorptivities k_m for CS_2 solutions (k_m; $mol^{-1} \, 1. \, cm^{-1}$)*

Sample	1, 4 units		3, 4 units		
	$572 \, \text{cm}^{-1}$ ($17.48 \, \mu$)	$840 \, \text{cm}^{-1}$ ($11.90 \, \mu$)	$888 \, \text{cm}^{-1}$ ($11.26 \, \mu$)	$1780 \, \text{cm}^{-1}$ ($5.62 \, \mu$)	$3070 \, \text{cm}^{-1}$ ($3.26 \, \mu$)
Hevea	5.7	16.6*	1.72	0	0
Balata	2.1	7.6*	0.58	0	0
3. 4-polyisoprene	6.5	0	110	3.46	30.6

TABLE 20 *Results of analyses using various absorption bands (structure %)*

3, 4 units				1, 4 units
$88 \, \text{cm}^{-1}$	$3070 \, \text{cm}^{-1}$	$1780 \, \text{cm}^{-1}$	Average	$840 \, \text{cm}^{-1}$
31.6	32.9	—	32	43
37.0	39.0	—	38	42
37.2	40.2	—	39	47
39.0	42.3	—	41	41
—	49.7	51.3	51	—
—	54.6	59.6	58	28

TABLE 21   *Results of analyses*

		3,4%		
cis-1,4 (%)	*trans*-1,4 (%)	$888\,cm^{-1}$ $(11.26\,\mu)$	$3070\,cm^{-1}$ $(3.26\,\mu)$	$1780\,cm^{-1}$ $(5.62\,\mu)$
0	10	—	57	62
60	9	15	11	—
45	17	—	20	—

$572\,cm^{-1}$ and $980\,cm^{-1}$ $(17.48$ and $10.20\,\mu)$ bands was inapplicable due to the presence of cyclic structure.

The results of analyses of the samples in potassium bromide pellets are presented in Table 21. In these analyses it was possible to utilize the $572\,cm^{-1}$ and $600\,cm^{-1}$ $(17.48$ and $16.66\,\mu)$ absorption bands only, for an approximate estimation of the relative abundance of 1, 4 structural units.

Fraga and Benson[130] have investigated a thin-film infrared method for the analysis of polyisoprene. They emphasize that clear, smooth and uniform films are necessary, and that these can be cast from a toluene solution of the polymer. Film thickness should be maintained to provide between 0.5 and 0.7 absorbance units at the peak near $1370\,cm^{-1}$ $(7.3\,\mu)$. When good quality films are used the repeatability of the method is excellent. Thinner films will give slightly lower results. Binder[131] found a direct correlation existed between the intensity of the $742\,cm^{-1}$ $(13.48\,\mu)$ and the percentage net *cis*-1,4 for various high *cis*-1,4 polyisoprenes. Synthetic *cis*-1,4 polyisoprenes prepared with Zeigler catalysts or lithium catalysts contain a small percentage of the 3,4 structure. On the other hand, naturally occurring polyisoprenes such as natural rubber (Hevea), gutta percha balata, and chicle consist exclusively of the 1,4 structure. The differences between the thermal and mechanical properties of the natural and synthetic polyisoprenes have been attributed to the amount of *cis*-1,4 units. It is reasonable to expect that the physical properties of polyisoprenes are also affected by the distribution of the isomeric structure units along the polymer chain, as well as the composition of the polymers.

Various infrared methods[140–143] have been reported for the analysis of polyisoprene with predominantly *cis*-1,4 and *trans*-1,4 structural units with low amounts of 3,4 addition.

Tanaka *et al.*[132] determined the distribution of *cis*-1,4 and *trans*-1,4 units in 1,4 polyisoprenes by using ^{13}C NMR spectroscopy, and found that *cis*-1,4 and *trans*-1,4 units are distributed almost randomly along the polymer chain in *cis–trans* isomerized polyisoprenes, and that chicle is a mixture of *cis*-1,4 and *trans*-1,4 polyisoprenes. These workers[133] have also investigated the ^{13}C NMR spectra of hydrogenated polyisoprenes and determined the distribution of 1,4 and 3,4 units along the polymer chain for *n*-butyllithium catalysed polymers, and have confirmed that these units are randomly distributed along the polymer chain. The polymers did not contain appreciable amounts of head-to-head or head-to-tail 1, 4 linkages.

Maynard and Moobel,[144] and Ferguson[145] have described infrared methods, respectively, for determining cis-1,4, trans-1,4,3,4 and 1,2 structures and 1,4 structures in polyisoprene.

Kossler and Vodchnal[146] came to the conclusion that the infrared spectra of polymers containing cis-1,4, trans-1,4 and 3,4 or cyclic structural units are not additively composed of the spectra of stereoregular polymers containing only one of these structures.

It is known that stereoregular cis-1,4 polyisoprene (Hevea) and stereoregular trans-1, 4 polyisoprene (balata) have absorption bands at 1130 and 1150 cm^{-1} (8.84 and 8.69 μ) respectively. These workers found that a polymer having a high content of 3, 4 structural units, in addition to the 1,4 structural units, has no absorption band at either 1130 or 1150 cm^{-1} (8.84 or 8.69 μ) but does have a band at 1140 cm^{-1} (8.77 μ). They attribute this band to the C—CH$_3$ vibration of the —C(CH$_3$)=CH structural unit separated by other structural units.

The appearance of the absorption band at 1140 cm^{-1} (8.77 μ) in some synthetic polyisoprenes has been mentioned several times by Binder,[147-149] with the comment that the origin of this band is not known.[149]

A similar phenomenon has been discovered by analysis of a polymer having approximately 20% trans-1,4 in addition to about 75% cis-1,4 structural units, as estimated by an analysis using the absorption bands at 572 and 980 cm^{-1} (17.48 and 10.20 μ).[150] The band at 1130 cm^{-1} (8.85 μ) was shifted towards higher values. In a mixture of Hevea and balata with the same content of 20% trans-1,4 structural units both the 1130 and 1150 cm^{-1} (8.85 and 8.69 μ) bands are quite distinct. The behaviour of the 1130 and 1150 cm^{-1} (8.85 and 8.69 μ) bands is in agreement with the finding of Golub,[150,151] who has shown that during the cis–trans isomerization of polyisoprene the 1136 cm^{-1} (8.80 μ) absorption band appear instead of the 1126 cm^{-1} (8.88 μ) band in cis-1,4 isomers or the 1149 cm^{-1} (8.70 μ) band in trans-1, 4 isomers. The statement of Maynard and Moobel[152] that the small amount of trans-1, 4 structural units may be better detected using the band pair near 1307 cm^{-1} (7.65 μ) rather than the bands at 1131 and 1152 cm^{-1} (8.84 and 8.68 μ) is also in good agreement with the findings of Kossler and Vodchnal.[146]

As a consequence of these results it is possible to conclude that only polyisoprenes having long sequences of cis-1,4 or trans-1,4 units have the absorption bands at 1130 and 1150 cm^{-1} (8.85 and 8.69 μ) respectively. It is also evident that the analysis of synthetic polyisoprenes using these absorption bands leads to distorted results. Kossler and Vodchnal[146] obtained better results using the absorption bands at 572, 980 and 888 cm^{-1} (17.48, 10.20 and 11.26 μ) for cis-1,4 and trans-1,4 and 3,4 polyisoprene structural units, respectively. The use of various combinations of different absorption bands permits one to conclude whether a polymer in question is more of the block copolymer type or a mixture of stereoregular polymers.

The diad distribution of *cis*-1,4 and *trans*-1,4 units in low molecular weight 1,4-polyisoprene has been determined from the ^{13}C NMR spectra at 350k by Morese-Seguela *et al.*[134]

Gronski and co-workers,[135] Beebe,[136] Dalinskaya *et al.*[137] and Duck and Grant,[138] used the chemical shift correction parameters for linear alkanes in the alphatic region of ^{13}C NMR spectra to determine the relative amounts of 3,4 and *cis*-1,4 units of polyisoprene.

Microstructure studies have been carried out on the cyclic content and *cis–trans* isomer distribution in polyisoprene.[139]

Much useful information regarding the types of unsaturation present in polyethylene can be gained by the application of infrared spectroscopy. Polyethylenes can contain various types of unsaturation which are of great importance from the microstructural point of view. These include vinyl, vinylidene, and *trans* olefinic end-groups.

The electron irradiation of linear and branched polyethylenes causes a number of molecular rearrangements in the chemical structure of the polymer.[153] In addition to the significant changes in the type and

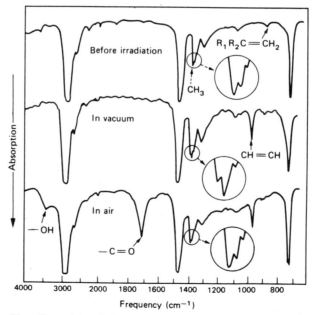

FIG. 32.   The effect of irradiation (500 Mrad) in air and vacuum on branches polyethylene.

distribution of unsaturated groups an infrared comparison of the radiation-induced chemical changes that occur in air and in a vacuum showed that the presence of oxygen has a marked influence on the structural rearrangements that occur on irradiation.

Figure 32 shows the infrared spectra of the branched polyethylene before and after irradiation in vacuum and air. Strongly absorbing *trans*-type unsaturation (CH=CH) bands at 964 cm^{-1} (10.37 μ) appear in both the vacuum and air-irradiated sample spectra. Vinylidene decay on irradiation is shown by the decrease in the $R_1R_2C=CH_2$ band at 888 cm^{-1} (11.26 μ).

Irradiation in vacuum produces a significant decrease in the methyl (—CH$_3$) content (1373 cm^{-1}) (7.28 μ), whereas in the bombardment in air there appears to be only a negligible decrease in —CH$_3$, if any. In addition a comparison of the 720–730 cm^{-1} (13.89–13.70 μ) doublet shows that only the 720 cm^{-1} (13.89 μ) component remains in the spectra of the vacuum sample, whereas there is only a slight decrease of the 730 cm^{-1} (13.69 μ) component in the air-irradiated sample. Additional evidence of structural changes is shown in the spectra of the air-irradiated sample. Here both —OH and C=O bands appear, and there is a general depression of the spectrum background from 1300 to 900 cm^{-1} (7.69 to 11.11 μ).

Figure 33 shows the unsaturation region of the spectra. The top two traces show this region for branched polyethylene before and after 6 MR irradiation. The lower traces are those of linear polyethylene before and after similar irradiation. In branched polyethylene before irradiation, most of the unsaturation is of the the external vinylidene type. After a dose of 6 MR, *trans*-unsaturation at 964 cm^{-1} (10.73 μ) increases and the vinylidene (at 888 cm^{-1}, 11.26 μ) decreases.

In linear polyethylene almost all the saturation is of the terminal vinyl type (CH=CH$_2$) as shown by the bands at 990 and 910 cm^{-1} (10.10 and 10.99 μ). Here again, after only a 6 MR dose, the *trans* groups form rapidly and the vinyl groups at 990 and 910 cm^{-1} (10.10 and 10.99 μ) decrease. Because of the rapid increase of *trans* groups during irradiation and the simultaneous decrease of the other unsaturated groups, it might appear that the *trans* groups are being formed from a reaction involving the sacrifice of the other unsaturated groups in the polymer. In order to determine the validity of this observation, Luongo and Solovay[153] exposed to similar doses of irradiation a sample of polymethylene which has no infrared detectable unsaturation or branching. In Fig. 33 (lower curves) it is seen that, in polymethylene, the *trans* unsaturation band still forms strongly after irradiation in either air or vacuum. This means that the *trans* groups come from a reaction that is independent of either unsaturation or branching. As for the vinyl and vinylidene decay, although there is no conclusive mechanism to explain their disappearance, they probably

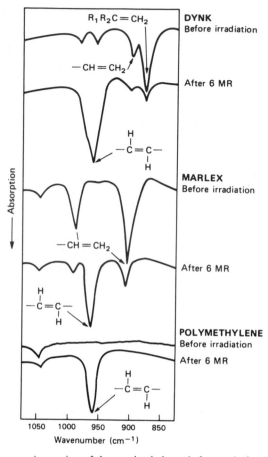

Fig. 33.   Unsaturation region of three polyethylenes before and after irradiation at 6 Mrad.

become saturated by atomic hydrogen in the system or become cross-linking sites.

Unsaturation in low-density polyethylene has been estimated to ± 0.003 C=C$/10^3$C atoms by compensating with brominated polymer of the same thickness.[154]

Rueda[155] has used infrared spectroscopy to measure vinyl, vinylidene and internal *cis* or *trans* olefinic end-groups in polyethylene.

Dankovics[156] determined the degree of unsaturation of low-density polyethylene. The total degree of unsaturation in polyethylene was determined by summing the vinyl, vinylene and vinylidene unsaturation derived from the differential infrared spectra using an unbrominated polyethylene film as the sample and a brominated film as the reference.[155]

5.2 Ozoneolysis

The oxidation of double bonds in organic compounds and polymers in a non-aqueous solvent leads to the formation of ozonides which, when acted upon by water, are hydrolysed to carbonyl compounds

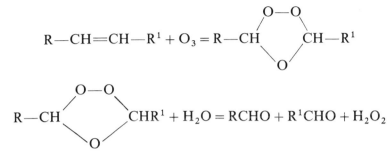

$$R-CH=CH-R^1 + O_3 = R-CH \underset{O}{\overset{O-O}{\diagup \diagdown}} CH-R^1$$

$$R-CH \underset{O}{\overset{O-O}{\diagup \diagdown}} CHR^1 + H_2O = RCHO + R^1CHO + H_2O_2$$

Triphenyl phosphine is frequently used to assist this reaction. Clearly, when applied to complex unsaturated organic molecules or polymers this reaction has great potential for the elucidation of the microstructure of the unsaturation. Further applications of the technique to various model compounds are illustrated in Table 22. Examination of the reaction products, for example by conversion of the carbonyl compounds to carboxylates then esters followed by gas chromatography, enables identifications of these products to be made.

An example of the value of the application of this technique to a polymer structural problem is the distinction between polybutadiene made up of consecutive 1,4–1,4 butadiene sequences I, and polybutadiene made up of alternating 1,4 and 1,2 butadiene sequences. 11, i.e. 1,4–1,2–1,4.

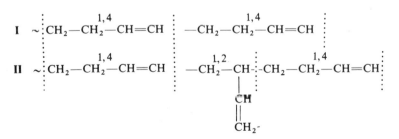

Upon ozonolysis, followed by hydrolysis, these in the case of 1,4–1,4 sequences produce succinaldehyde ($CHO-CH_2-CH_2CHO$) and in the case of 1,4–1,2–1,4 sequences produce formyl 1,6-hexane-dial

$$(CHOCH_2-CH-CH_2CH_2-CHO)$$
$$CHO$$

and formaldehyde.

TABLE 22   *Ozoneolysis of unsaturated organic compounds*

1:5 hexadiene

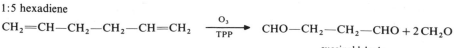

$$CH_2{=}CH{-}CH_2{-}CH_2{-}CH{=}CH_2 \xrightarrow[TPP]{O_3} CHO{-}CH_2{-}CH_2{-}CHO + 2\,CH_2O$$

succinaldehyde

4-vinyl-1 cyclohexene

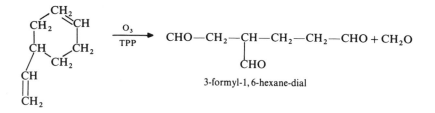

$$CHO{-}CH_2{-}CH{-}CH_2{-}CH_2{-}CHO + CH_2O$$

with CHO branch

3-formyl-1,6-hexane-dial

cis, cis-1,5 cyclooctadiene

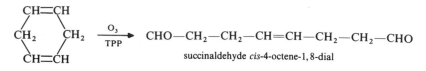

$$CHO{-}CH_2{-}CH_2{-}CH{=}CH{-}CH_2{-}CH_2{-}CHO$$

succinaldehyde *cis*-4-octene-1,8-dial

cyclooctene

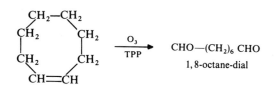

$$CHO{-}(CH_2)_6\,CHO$$

1,8-octane-dial

4-methyl-1-cyclohexene

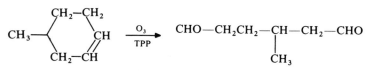

$$CHO{-}CH_2CH_2{-}CH{-}CH_2{-}CHO$$

with CH_3 branch

3-methyl-1,6-hexane-dial

cyclopentene

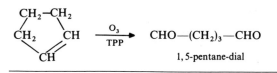

$$CHO{-}(CH_2)_3{-}CHO$$

1,5-pentane-dial

TABLE 23 *Microozonolysis of polybutadiene*

	1,4 Vinyl (*cis* + (1,2), *trans*) (%)	Area (from GC)%			
Sample		Succin-aldehyde	3-formyl 1,6-hex-anedial	4-octene 1,8-dial	1,2 units occurring in 1,4–1, 2–1,4 sequences
1	98.0	2	50	1	49	0.5
2	89.1	10.9	30	10	60	5
3	89.0	11.0	43	7	50	3
4	81.0	19.0	34	14	52	6
5	76.2	23.8	36	25	39	11
6	71.8	28.2	33	27	40	11
7	69.7	30.3	48	26	26	10
8	67.7	32.3	36	26	38	9
9	64.2	35.8	38	31	31	10
10	62.8	37.2	45	27	28	8
11	50.5	49.5	26	41	33	12
12	56.0	44.0	30	39	31	11
13	26.0	74.0	33	64	3	5

Analysis of the reaction product for concentrations of succinaldehyde and 3-formyl-1,6-hexane-dial can show whether the polymer is 1,4–1,4 or 1,4–1,2–1,4, or whether it contains both types of sequence.

Various workers[157–162] have applied this technique to the elucidation of the microstructure of polybutadiene. They found that the 3 formyl-1,6-hexane-dial content was directly proportional to the 1,2 (vinyl) content of polymers containing 1,4–1,2–1,4 butadiene sequences.

Polymers having 98% *cis*-1,4 structure, 98% *trans*-1,4 structure and a series of polymers containing from 11% to 75% 1,2 structure were ozonized (Table 23). The final products obtained from these polymers were succinaldehyde, 3-formyl-1,6 hexane-dial, and 4-octene-1,8-dial. Model compounds were ozonized and the products were compared with those from the polymers (Table 22).

Figures 34 and 35 show chromatographic separation of the ozonolysis products from polybutadienes having different amounts of 1,2 structure, as measured by infrared or NMR spectroscopy. Figure 36 shows the relationship of 1,2 content to the amount of 3-formyl-1,6 hexane-dial in the ozonolysis products.

The amount of 1,4–1,2–1,4 sequences in polybutadienes can be estimated from the amounts of the different ozonolysis products (Table 23) if one considers the amount of 1,4 structure not detected. (Since the ozonolysis technique cleaves the centre of a butadiene monomer unit, one half of a 1,4 unit remains attached to each end of a block of 1,2 units after ozonolysis; these structures do not elute from the gas chromatographic

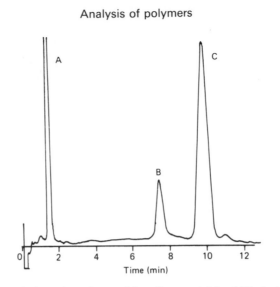

FIG. 34.   Ozonolysis products from polybutadiene containing 11% vinyl structures:
(A) succinaldehyde; (B) 3-formyl-1, 6-hexanedial; (C) 4-octene-1, 8-dial.

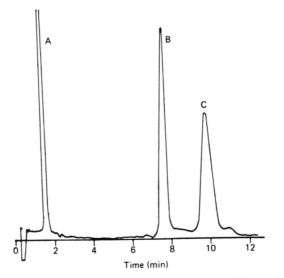

FIG. 35.   Ozonolysis products from polybutadiene containing 37.2% vinyl structure:
(A) succinaldehyde; (B) 3-formyl-1, 6-hexanedial; (C) 4-octene-1, 8-dial.

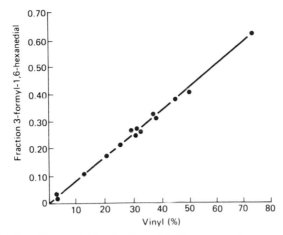

Fig. 36. Relationship of yield of 3-formyl-1,6-hexanedial from ozonolysis to percentage vinyl structures (from NMR or infrared spectra) in polybutadienes.

column.) Using random copolymer probability theory, the maximum amounts of these undetected 1,4 structures can then be calculated.

A further example concerns the application of the ozonolysis technique to butadiene–propylene copolymers.[163–165] Samples of highly alternating copolymers of butadiene and propylene yielded large amounts of 3-methyl-1,6-hexane-dial when submitted to ozonolysis. The ozonolysis product from 4-methyl-cyclohexane-1

(i.e. 3-methyl-1,6 hexane-dial, $O{=}C{-}C{-}C{-}C{-}C{-}C{=}CO$)
                                                    |
                                                    C

was used as a model compound for this structure. Ozonolysis of these polymers occurs as shown below:

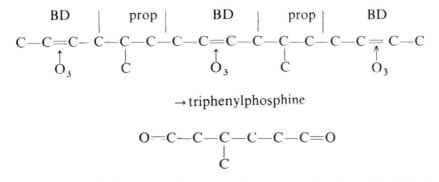

The amount of alternation in these polymers can be determined if the

TABLE 24   *Microzonolysis of butadiene–propylene copolymers*

	1,4 (%)	1,2 (%)	Propylene (mole%)	Succin-aldehyde	3-methyl −1,6- hexane-dial	3-formyl 1,6-hexane-dial	4-octene −1,8- dial	Alter-nating BD/Pr (%)
					Area (from gas chromatography) %			
A	45	5.7	49.3	5	92	1	2	77
B	47.8	2.2	50	11.5	85	0.5	3	71
C	53.1	3.2	43.7	25	61	6	8	48
D	—	—	30	49	38	1	12	33

amounts of 1,4 and 1,2 polybutadiene structure and total propylene have been determined by infrared or NMR spectroscopy. Table 24 shows results obtained for several butadiene–propylene copolymers having more or less alternating structure. Similar polymers have been analysed by Kawasaki[164] by use of conventional ozonolysis methods with esters as the final products.

The technique has been applied to various other unsaturated polymers. Thus polyisoprene, having nearly equal 1,4 and 3,4 structures, produced large amounts of levulinaldehyde, succinaldehyde and 2,5 hexanedione, indicating blocks of 1,4 structures in head–tail, tail–tail and head–head configurations.

Hill *et al.*[166] utilized ozoneolysis in their investigation of a butadiene–methyl methacrylate copolymer. The principal products were succinic acid, succindialdehyde and dicarboxylic acids containing several methyl-methacrylate residues. The percentage of butadiene (9.2%) recovered as succinic acid and succindialdehyde provided a measure of the 1,4-butadiene–1,4-butadiene linkages in the copolymers, and the percentage of methylmethacrylate units (51%) recovered as trimethyl 2-methyl-butane-1,2,4-tricarboxylate (4) $n = 1$, provided a measure of the methyl-methacrylate units in the middle of butadiene–methacrylate–butadiene triads.

Extensive ozonization followed by lithium aluminium hydride reduction of oxyalkylene groups in polypropylene glycols produces monomeric diglycols as follows:

$$\sim \!OCH_2-CH-O-CH_2-CH-O-CH-CH_2-O-CH_2-CH-O + 4O_2$$
$$\quad\quad\;\; CH_3 \quad\quad\quad\; CH_3 \quad CH_3 \quad\quad\quad\quad CH_3$$

$$\sim O\ CH_2-CO + OCH-CH-O-CH-CHO + CHO-CH-O \sim\ + 2H_2O = 4O_2$$
$$\quad\quad\; CH_3 \quad\quad\; CH_3 \quad CH_3 \quad\quad\quad\quad CH_3$$

$$CHO—CH—OCH—CHO \quad LiAlH_4 =$$
$$\qquad\quad |\qquad\ |$$
$$\qquad\quad CH_3\quad CH_3$$

$$HOCH_2 - CH - OCH—CH_2OH \qquad = \left(HOCH_2—CH \atop CH_3\right)_2 O$$
$$\qquad\qquad |\qquad\ |$$
$$\qquad\qquad CH_3\quad CH_3$$

diprimary propylene glycol

Diprimary propylene glycol is the product resulting from fusion at **b**—**b** of the head to head unit. Similarly, fusion at **c**—**c**, i.e. at a tail-to-tail unit, would produce

$$\left(HO—CH —CH_2 \atop CH_3\right)_2 O$$

(disecondary propylene glycol)

and fusion at **a**–**a**, i.e. at head-to-tail unit, would produce

$$HOCH_2—CH_3—OCH_2—CHOH$$
$$\qquad\ |\qquad\qquad\qquad |$$
$$\qquad CH_3\qquad\qquad\quad CH_3$$

(Primary–secondary propylene glycol).

Figure 37 is an example of a gas chromatogram of the ozoneolysis product of polypropylene glycol showing the presence of all three types of breakdown products mentioned above, including the two optical isomers

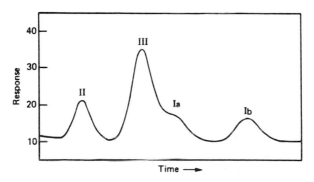

FIG. 37.   Gas–liquid chromatogram of dipropylene glycol from amorphous poly(propylene glycol), $ZnEt_2 . H_2O$ catalyst.

of diprimary propylene glycol (1a, and 1b);

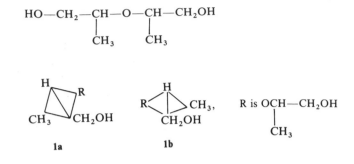

1a   1b

5.3 Alkyl group branching in polyolefins

Along with chemical composition, molecular weight, molecular weight distribution and type and amount of gel (see Chapter 7), branching is considered to be one of the fundamental parameters needed to characterize polymers fully, and this latter property, which is a microstructural feature of the polymer, has very important effects on polymer properties. Changes in branching of a given polymer such as polypropylene lead to changes in its stereochemical configuration and this, in turn, is a fundamental polymer property to formulating both polymer physical characteristics and mechanical behaviour. It is important, therefore, to be able to identify the type and amount of side-group branching in polymers.

Although molecular symmetry is well understood, until the development of proton NMR, and later ^{13}C NMR, a study of this aspect of polymer structure presented problems.

Polymer configuration in polyolefins

Technically, each methine carbon in a poly(1-olefin) is asymmetric; however, this symmetry cannot be observed because two of the attached groups are essentially equivalent for long chains. Thus a specific polymer unit configuration can be converted into its opposite configuration by a simple end-to-end rotation and subsequent translation. It is possible, however, to specify relative configurational differences, and Natta introduced the terms 'isotactic' to describe adjacent units with the same configurations, and 'syndiotactic' to describe adjacent units with opposite configurations.[167] Although originally used to describe diad configurations, 'isotactic' now describes a polymer sequence of any number of like configurations and 'syndiotactic' describes any number of alternating configurations. Diad configurations are called meso if they are alike and racemic if they are unalike.[168] Thus from a configurational standpoint, a poly(1-olefin) can be viewed as a copolymer of meso and racemic diads.

The advantages of ^{13}C NMR in measurements of polymer stereochemical configuration arise primarily from a useful chemical shift range which is approximately 20 times that obtained by proton NMR. The structural sensitivity is enhanced through an existence of well-separated resonances for different types of carbon atoms. Overlap is generally not a limiting problem. The low natural abundance (-1%) of ^{13}C nuclei is another favourably contributing factor. Spin–spin interactions among ^{13}C nuclei can be safely neglected, and proton interactions can be eliminated entirely through heteronuclear decoupling. Thus each resonance in a ^{13}C NMR spectrum represents the carbon chemical shift of a particular polymer moiety. In this respect, ^{13}C NMR resembles mass spectrometry because each signal represents some fragment of the whole polymer molecule. Finally, carbon chemical shifts are well behaved from an analytical viewpoint because each can be dissected, in a strictly additive manner, into contributions from neighbouring carbon atoms and constituents. This additive behaviour led to the Grant and Paul rules,[169] which have been carefully applied in polymer analyses, for predicting alkane carbon chemical shifts.

The advantages so clearly evident when applying ^{13}C NMR to polymer configurational analyses are not devoid of difficulties. The sensitivity of ^{13}C NMR to subtle changes in molecular structure creates a wealth of chemical shift-structural information which must be 'sorted out'. Extensive assignments are required because the chemical shifts relate to sequences from three to seven units in length. Model compounds, which are often used in ^{13}C NMR analyses, must be very close structurally to the polymer moiety reproduced. For this reason, appropriate model compounds are difficult to obtain. A model compound found useful in polypropylene configurational assignments was a heptamethylheptadecane, where the relative configurations were known.[170] To be completely accurate, the model compounds should reproduce the conformational as well as the configurational polymer structure. Thus reference polymers such as predominantly isotactic and syndiotactic polymers form the best model systems. Even when available, only two assignments are obtained from these particular polymers. Pure reference polymers can be used to generate other assignments.[171]

To obtain good quantitative ^{13}C NMR data one must understand the dynamic characteristics of the polymer under study. Fourier transform techniques, combined with signal averaging, are normally used to obtain ^{13}C NMR spectra. Equilibrium conditions must be established during signal averaging to ensure that the experimental conditions have not led to distorted spectral information. The nuclear overhauser effect (NOE), which arises from ^{1}H, ^{13}C heteronuclear decoupling during data acquisition, must also be considered.

Energy transfer, occurring between the ^{1}H and ^{13}C nuclear energy levels during spin decoupling, can lead to enhancements of the ^{13}C resonances by factors between 1 and 3. Thus the spectral relative intensities will only reflect the polymer's moiety concentrations if the differentiated Nuclear Overhauser effects (NOE) are equal or else taken into consideration. Experience has shown that polymer NOEs are generally maximal, and consequently equal, because of a polymer's restricted mobility.[172,173] To be sure, one should examine the polymer NOEs through gated decoupling or paramagnetic quenching, and thereby avoid any misinterpretation of the spectral intensity data.

The ^{13}C configurational sensitivity falls within a range from triad to pentad for most vinyl polymers. In non-crystalline polypropylenes, three distinct regions corresponding to methylene (-46 ppm), methine (-28 ppm) and methyl (-20 ppm) carbons are observed in the ^{13}C NMR spectrum. (The chemical shifts are reported with respect to an internal tetramethylsilane (TMS) standard.) The ^{13}C spectrum of a 1,2,4-trichlorbenzene solution at 125°C of a typical amorphous polypropylene is shown in Fig. 38. Although a configurational sensitivity is shown by all three spectral regions, the methyl region exhibits by far the greatest sensitivity and is consequently of the most value. At least ten resonances, assigned to the unique pentad sequences, are observed in order, mmmm, mmmr, rmmr, mmrr, mmrm, rmrr, mrmr, rrrr, rrrm and mrrm, from low to high field.[169,170,174]

The ^{13}C NMR spectrum of a crystalline polypropylene shown in Fig. 39 contains only three lines which can be identified as methylene, methine and

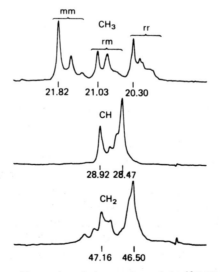

FIG. 38. Methyl, methine and methylene regions of the ^{13}C NMR spectrum of a non-crystalline polypropylene.

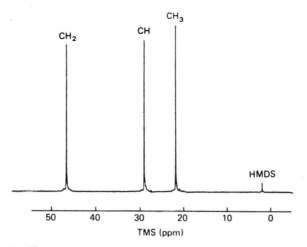

FIG. 39. ^{13}C NMR spectrum at 25.2 MHz of crystalline polypropylene.

methyl from low to high field by off-resonance decoupling. An amorphous polypropylene exhibits a ^{13}C spectrum which contains not only these three lines but additional resonances in each of the methyl, methine and methylene regions as shown in Fig. 38. The crystalline polypropylene must therefore be characterized by a single type of configurational structure. In this case the crystalline polypropylene structure is predominantly isotactic, thus the three lines in Fig. 39 must result from some particular length of meso sequences. This sequence length information is not available from the spectrum of the crystalline polymer, but can be determined from a corresponding spectrum of the amorphous polymer. To do so one must examine the structural symmetry of each carbon atom to the various possible monomer sequences. Randall[175,177] carried out a detailed study of the polypropylene methyl group in triad and pentad configurational environments.

Stehling[178] also studied polypropylene. The study of stereochemical configuration by ^{13}C NMR has not been limited to the polyolefins. Randall[179] and other workers[180,181] studied amorphous polystyrene. Randall[179] concluded that nine methylene resonances could be resolved, numbered A through L in Fig. 40. A high-field strong methine resonance is also present, but shows no apparent configurational splitting. The observation of nine resonances is in itself interesting, since a ^{13}C NMR sensitivity to just tetrad sequences would have produced six resonances, while a complete hexad sensitivity would have produced 20 resonances.

Other polymers that have been studied include ethylene–propylene copolymers, polypropylene oxide,[182–184] polyalkylvinyl esters,[185,186] PVC,[187–191] ethylene-2 chloroacrylate copolymers[192] ethylene glycidyl–acrylate copolymers[193] and poly-*trans*,1, 3-pentadiene.[194]

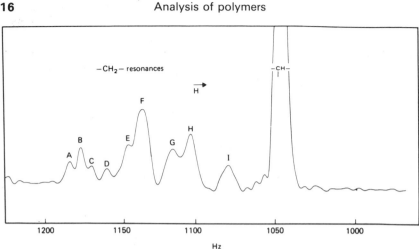

Fig. 40. The methylene and methine region of a ^{13}C NMR spectrum of an amorphous polystyrene at 25.2 MHz and 120°C. The MHz values are relative to an internal tetramethylsilane (TMS) standard.

Methyl branching in polyethylene

A regression analysis of infrared, differential thermal analysis and X-ray diffraction data by Laiber *et al.*[195] for low-pressure polyethylene showed that, as synthesis conditions varied, the number of methyl groups varied from 0 to 15 per 1000 carbon atoms and the degree of crystallinity varied from 84 to 61%.

Saturated hydrocarbons evolved during electron irradiation of polyethylene are characteristic of short side-chains in the polymer. Solovay and Pascale[196] showed that a convenient analysis is effected by programmed temperature gas chromatography. In order to minimize the relative concentrations of extraneous hydrocarbons, i.e. those not arising from selective scission of complete side-chains, it is necessary to irradiate at low temperatures and doses. Such analyses of a high-pressure polyethylene indicate that the two to three methyls per 100 carbon atoms detected in infrared absorption (low-pressure polyethylenes at least an order of magnitude lower) are probably equal amounts of ethyl and butyl branches. These arise by intramolecular chain transfer during polymerization. At a dose of 10 Mrad about 1–4% of the alkyl group are removed. Methane is the only hydrocarbon detected on irradiation of polypropylene, indicating little combination of methyl radicals to form ethane during irradiation.

The measurement of the methyl absorption at 1378 cm^{-1} (7.26 μ) in polyethylene can serve as a good estimation of branching. However, interference from the methylene absorption at 1368 cm^{-1} (7.31 μ) makes it difficult to measure the 1368 cm^{-1} (7.31 μ) band, especially in the case of relatively low methyl contents.

A method has been developed[197] which utilizes the suggestion by Neilson and Holland.[198] They associated the amorphous phase absorption of polyethylene at 1368 cm^{-1} (7.31 μ) and 1304 cm^{-1} (7.69 μ) with the *trans–trans* conformation of the polymer chain about the methylene group. Therefore the intensities of these two absorptions are proportional to one another. By placing an annealed film (ca. 10–15 mil) of high-density polyethylene in the reference beam of a double beam spectrometer and a thin, quenched film of the sample in the sample beam, most of the interference at 1368 cm^{-1} (1.31 μ) can be removed. The method has the advantage that it is not necessary to have complete compensation for the 1368 cm^{-1} (7.31 μ) band since a correction for uncompensation at 1378 cm^{-1} (7.25 μ) can be applied based on the intensity of the 1368 cm^{-1} (7.31 μ) absorption.

A calibration for the methyl absorption based on mass spectrometric studies of such gaseous products produced during electron bombardment of polyethylene has demonstrated irradiation-induced detachment of complete alkyl units.[199] In addition to saturated alkanes characteristic of the branches, small quantities of methane, other paraffins, and olefins were simultaneously evolved. It was suggested that 'extraneous' paraffins result from cleavage of the main chain.[199,200] Nerheim[201] has described a circular calibrated polymethylene wedge for the compensation of CH_2 interferences in the determination of methyl groups in polyethylene by infrared spectroscopy.

Methyl group content of low-density polyethylene has been determined with a standard deviation of 0.8% provided methylene group absorptions were compensated by polyethylene of similar structure.[202]

Nishoika et al.[203] determined the degree of chain branching in low-density polyethylene using proton NR Fourier transform NMR at 100 MHz and ^{13}C Fourier transform NMR at 25 MHz with concentrated solutions at approximately 100°C. The methyl concentrations agreed well with those of infrared based on the absorbance at 1378 cm^{-1} (7.25 μ).

Ethyl and higher alkyl groups branching in polyethylene

High-density (low-pressure) polyethylenes are usually linear, although the physical and rheological properties of some high-density polyethylenes have suggested the presence of long chain branching (butyl and higher groups) at a level one to two orders of magnitude below that found for low-density polyethylenes prepared by a high-pressure process. A measurement of long chain branching in high-density polyethylenes has been elusive because of the low concentrations involved[204] and can only be directly provided by high-field, high-sensitivity NMR spectrometers.

High-density polyethylenes prepared with a Ziegler type, titanium-based catalyst have predominantly n-alkyl or saturated end-groups. Those prepared with chromium-based catalysts have a propensity toward more olefinic end-groups. The ratio of olefinic to saturated end-groups for polyethylenes prepared with chromium-based catalysts is approximately unity. The end-group distribution is therefore another structural feature of interest in low-pressure polyethylenes, because it can be related to the catalyst employed, and possibly to the extent of long chain branching. It is possible not only to measure by ^{13}C NMR concentrations of saturated end-groups, but also the olefinic end-groups and, subsequently, an end-group distribution.

Short chain branches can be introduced deliberately in a controlled manner into polyethylenes by copolymerizing ethylene with a 1-olefin. The introduction of 1-olefins allows the density to be controlled, and butene-1 and hexene-1 are commonly used for this purpose. Once again, as in the case of high-pressure process low-density polyethylenes, ^{13}C NMR can be used to measure ethyl and butyl branch concentrations independently of the saturated end-groups. This result gives ^{13}C NMR a distinct advantage over corresponding infrared measurements because the latter technique can only detect methyl groups irrespective of whether the methyl group belongs to a butyl branch or a chain end. ^{13}C NMR also has a disadvantage in branching measurements because only branches five carbons in length and shorter can be discriminated independently of longer chain branches.[205,206] Branches six carbons in length and longer give rise to the same ^{13}C NMR spectral pattern independently of the chain length. This lack of discrimination among the longer side-chain branches is not a deterring factor, however, in the usefulness of ^{13}C NMR in a determination of long chain branching.

By far the most difficult structural measurement is long chain branching. In low-density polyethylenes the concentration of long chain branches is such (< 0.5 per 1000 carbons) that characterization through size exclusion chromatography in conjunction with either low-angle laser light scattering or intrinsic viscosity measurements becomes feasible.[204,206-209] When ^{13}C NMR measurements have been compared to results from polymer solution property measurements, good agreement has been obtained between long chain branching from solution properties with the concentration of branches six carbons long and longer.[206,207] Unfortunately, these techniques utilizing solution properties do not possess sufficient sensitivity to detect long chain branching in a range of 1 in 10,000 carbons, the level suspected in high-density polyethylenes. The availability of superconducting magnet systems has made measurements of long chain branching by ^{13}C NMR a reality because of a greatly improved sensitivity. An enhancement by factors between 20 and 30 over conventional NMR spectrometers has been achieved through a combination of higher field strengths, 20 mm probes, and the ability to examine polymer samples in essentially a melt state. The data discussed by Randall[210,211] have been

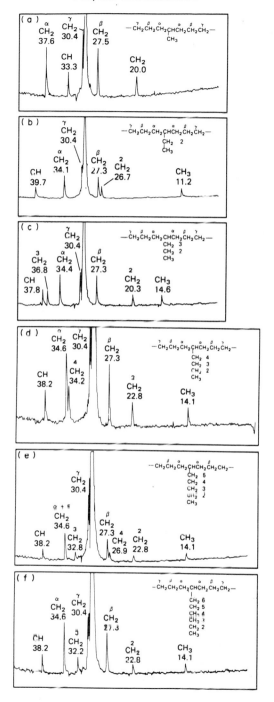

FIG. 41. ^{13}C NMR at 25.2 M Hz of (a) an ethylene-1–propene copolymer, (b) an ethylene–butene copolymer, (c) an ethylene-1-pentene copolymer, (d) an ethylene-1-hexene copolymer, (e) an ethylene-1–heptene copolymer and (f) an ethylene-1–octene copolymer.

obtained from both conventional iron magnet spectrometers with field strengths of 23.5 kG and superconducting magnets operating at 47 kG.

The ^{13}C NMR spectra from a homologous series of six linear ethylene 1-olefin copolymers beginning with 1-propene and ending with 1-octene are reproduced in Fig. 4.1 The side-chain branches are therefore linear, and progress from one to six carbons in length. Also, the respective 1-olefin concentrations are less than 3%, thus only isolated branches are produced. Unique spectral fingerprints are observed for each branch length. The chemical shifts, which can be predicted with the Grant and Paul parameters,[210,212] are given in Table 25 for this series of model ethylene–1-olefin copolymers. The nomenclature used to designate those polymer backbone and side-chain carbons discriminated by ^{13}C NMR, is as follows:

$$-CH_2-\overset{\gamma}{C}H_2-\overset{\beta}{C}H_2-\overset{\alpha}{C}H_2-\overset{1}{C}H_2-\overset{\alpha}{C}H_2-\overset{\beta}{C}H_2-\overset{\gamma}{C}H_2-CH_2-CH$$

2	CH_2
3	CH_2
4	CH_2
5	CH_2
6	CH_2

The distinguishable backbone carbons are designated by Greek symbols, while the side-chain carbons are numbered consecutively starting with the methyl group and ending with the methylene carbon bonded to the polymer backbone.[210] The identity of each resonance is indicated in Fig. 41. It should be noticed in Fig. 41 that the '6' carbon resonance for the hexyl branch is the same as α, the '5' carbon resonance is the same as β and the '4' carbon resonance is the same as γ. Resonances 1, 2 and 3, likewise, are the same as the end-group resonances observed for linear

TABLE 25   *Polyethylene backbone and side-chain ^{13}C chemical shifts in ppm from trimethylsilane (± 0.1) as a function of branch length (Carbon chemical shifts, which occur near 30.4 ppm, are not given because they are often obscured by the major 30 ppm resonance for the 'n' equivalent, recurring methylene carbons) (solvent: 1, 2, 4-trichlorobenzene; temperature: 125°C)*

Branch length	Methine			6	5	4	3	2	1
1	33.3	37.6	27.5	20.0					
2	39.7	34.1	27.3	11.2	26.7				
3	37.8	34.4	27.3	14.6	20.3	36.8			
4	38.2	34.6	27.3	14.1	23.4	—	34.2		
5	38.2	34.6	27.3	14.1	22.8	32.8	26.9	34.6	
6	38.2	34.6	27.3	14.1	22.8	32.2	30.4	27.3	34.6

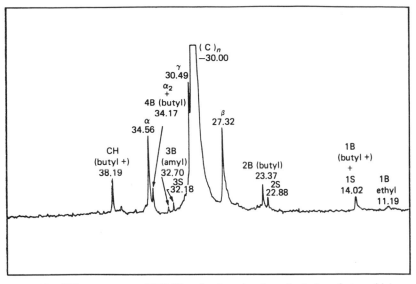

FIG. 42. ^{13}C spectrum at 25.2 MHz of a low density polyethylene from a high-pressure process.

polyethylene. Thus a six-carbon branch produces the same ^{13}C spectral pattern as any subsequent branch of greater length. Carbon-13 NMR, alone cannot therefore be used to distinguish a linear six-carbon from a branch of some intermediate length or a true long chain branch.

The capability for discerning the length of short chain branches has made ^{13}C NMR a powerful tool for characterizing low-density polyethylenes produced by free-radical, high-pressure processes. The ^{13}C NMR spectrum of such a polyethylene is shown in Fig. 42. Others are also present, and Axelson, *et al.* in a comprehensive study[213] have concluded that no unique structure can be used to characterize low-density polyethylenes. They have found nonlinear short chain branches as well as 1, 3 paired ethyl branches. Bovey *et al.*[206] compared the content of branches six carbons and longer in low-density polyethylenes with the long chain branching results obtained through a combination of gel permeation chromatography and intrinsic viscosity. An observed good agreement led to the conclusion that the principal short chain branches contained fewer than six carbons and the six and longer branching content could be related entirely to long chain branching. Others have now reported similar observations in studies where solution methods are combined with ^{13}C NMR.[207] However, as a result of the possible uncertainty of the branch lengths, associated with the resonances for branches six carbons and longer, ^{13}C NMR should be used in conjunction with independent methods to establish true long chain branching.

From the results we have seen thus far, it is easy to predict the ^{13}C NMR spectrum anticipated for essentially linear polyethylenes containing a small degree of long chain branching. An examination of a ^{13}C NMR spectrum from a completely linear polyethylene, containing both terminal olefinic and saturated end-groups, shows that only five resonances are produced. A major resonance at 30 ppm arises from equivalent, recurring methylene carbons, designated as 'm', which are four or more removed from an end-group or a branch. Resonances at 14.1, 22.9 and 32.3 ppm are from carbons 1, 2 and 3, respectively, from the saturated, linear end-group. A final resonance, which is observed at 33.9 ppm, arises from an allylic carbon, designated as 'a', from a terminal olefinic end-group. These resonances, depicted structurally below, are fundamental to the spectra of all polyethylenes.

$$\underset{1}{CH_3} - \underset{2}{CH_2} - \underset{3}{CH_2} - CH_2 - \quad \underset{\text{'n'}}{-(CH_2)_n-} \quad \underset{\text{'a'}}{-CH_2-CH=CH_2}$$

An introduction of branching, either long or short, will create additional resonances to those described above.

From the observed ^{13}C NMR spectrum of the ethylene-1-octene copolymer Randall found that the α and β and methine resonances associated with branches six carbons and longer occur at 34.56, 27.32 and 38.17 ppm respectively. Thus in high-density polyethylenes, where long chain branching is essentially the only type present, ^{13}C NMR can be used to establish unequivocally the presence of branches six carbons long and longer. If no comonomer has been used during polymerization, it is very likely that the presence of such resonances will be indicative of true long chain branching. In any event, ^{13}C NMR can be used to pinpoint the absence of long chain branching and place an upper limit upon the long chain branch concentration whenever branches six carbons and longer are detected.

It can be seen from the above considerations that ^{13}C NMR is a highly attractive method for characterizing polyethylenes. A serious drawback is not encountered even though branches six carbons in length and longer are measured collectively. The short branches are generally less than six carbons in length and truly long chain branches tend to predominate. On occasions there may be special exceptions for 'intermediate' branch lengths, so independent rheological measurements should be sought as a matter of course. Nevertheless ^{13}C NMR is a direct method, which possesses the required sensitivity to determine long chain branching in high-density polyethylenes, and provide much information of micro-structural interest.

A study of the products produced upon vacuum radiolysis of ethylene homopolymers and copolymers is another means of obtaining information

TABLE 26   *Radiolysis products of ethylene–α-olefin copolymers (^{235}U radiation source)*

Copolymers	CH_4	C_2H_6	C_3H_8	i-C_4H_{10}	n-C_4H_{10}	n-C_5H_{12}	n-C_6H_{14}
				G value $\times$ 10^2*			
Ethylene–propylene	1.7	0.1	0.1	—	0.03	—	—
Ethylene–butene-1	0.2	1.5	0.1	0.1	0.2	0.03	—
Ethylene–pentene-1	0.2	0.3	2.1	0.02	0.05	—	—
Ethylene–hexane-1	0.4	0.3	0.2	0.03	1.2	0.02	—
Ethylene–octene-1	0.2	0.3	0.1	—	0.1	0.1	1.2
Linear polyethylene	0.2	0.3	0.2	0.03	0.06	0.05	—

*G value defined as the number of molecules of the particular product produced per gram of sample per 100 e.v. of incident radiation dose. This is calculated with 10% accuracy, from the gas chromatographic analysis data of the products, weight of the polymer sample irradiated, and radiation dose to which the sample has been subjected.

on branching in these polymers.[214] If a correction is applied, to take into account the fragments arising from scission at chain ends, the remaining products can be quantitatively accounted for as entirely due to scission of side-branches introduced onto the backbone chain by the α-olefin comonomer. The cleavage of branches takes place, for all practical purposes, exclusively at the branch points at which the branches are attached to the backbone chain. The same data, together with similar radiolysis data of poly(3-methyl pentene-1) and poly(4-methyl pentene-1), further showed that all branches cleave with equal efficiency, regardless of their length. Radiolysis does, therefore, provide a reliable and convenient tool for the quantitative characterization of high-pressure polyethylene with regard to the unique short chain branching distribution that is characteristic of each.

The results obtained in a series of irradiations of ethylene- α-olefin copolymers containing about 4 mole% comonomer are shown in Table 26.

During radiolysis of polyethylene there also takes place a certain amount of random scission at chain ends in addition to the cleavage of branches. The observed extraneous hydrocarbons are simply the products derived from this fragmentation at the chain ends. It is quite probable that a portion of the extraneous hydrocarbons is derived from scission of stray branches that might have been introduced on the chains by stray impurities during polymerization, but the fact that one also observes a consistent decrease in the total amount of these extraneous hydrocarbons derived from the homopolymer with an increase in its molecular weight leads to the conclusion that the random scission at chain ends is their main cause. Obviously, then, if one makes an appropriate allowance for these radiolysis fragments derived from chain ends, then the only significant paraffin left in the radiolysis products of each copolymer is

TABLE 27 *Efficiency of radiation scission of short branches (G value* $\times 10^2$)

	CH$_3$ per 1000						Average G × 10 CH$_3$ per 1000 CH$_4$
Copolymers	CH_2	CH_4	C_2H_6	C_3H_8	n-C_4H_{10}	n-C_6H_{14}	
Ethylene–propylene	41.9	3.4					0.074
	26.4	1.7					
Ethylene–butene-1	28.7		2.0				
	21.6		1.5				0.072
	4.1		1.1				
Ethylene–pentene-1	25.6			2.1			0.072
	17.1			1.1			
Ethylene–hexene-1	23.5				1.2		0.073
	7.8				0.7		
Ethylene–octene-1	19.1					1.3	
	15.1					1.2	0.074
	11.6					0.9	

that corresponding to the branch introduced on the polyethylene backbone by the comonomer.

Since it is generally agreed that high-pressure polyethylene actually contains a variety of short branches, and not just one type of branch, it is obvious that, if one wants to translate the hydrocarbon analysis of its radiolysis products into quantitative branch-type analysis, one will also need accurate information on the relative efficiency of the scission of different branch types. To obtain this information Kamath and Barlow[214] irradiated some additional ethylene α-olefin copolymers, differing significantly in their comonomer content, and the hydrocarbons in their radiolysis products were analysed as before. Table 27 lists these results. The efficiency of scission is calculated from the known comonomer content (methyl group analysis) and the observed G values of the principal hydrocarbons after application of the appropriate correction against the small chain-end fragmentation. It is clear from the data that all branches of up to six carbon atoms or more break off with equal efficiency, and that the branch length *per se* exerts little or no effect on the ease of scission.

Table 28 presents results obtained in the radiolysis of high-pressure polyethylenes of various densities. Noteworthy is the disparity between the total number of all branches, as determined from the radiolysis data, and the methyl group content, as derived by the infrared method. This may be attributed to the fact that the usual infrared method of determining the methyl group content, based on the absorbance of 1379 cm^{-1} (7.25 μ), does not necessarily count all methyl groups. For example, if two methyl groups are attached to one carbon atom, as in polyisobutylene, the characteristic methyl absorption at 13.79 cm^{-1} (7.25 μ) splits, giving two

TABLE 28 *Distribution of short-chain branches in high pressure polyethylene (branches per 1000 methylene units)**

Resin No	Density (g/cc)	—CH₃	—C₂H₅	—C₃H₇	i-C₄H₉	—n-C₄H₉	—C₅H₁₁	Total branches per 1000 methylene	Methyl group content per 1000 methylene†
A	0.934	2.9	11.8	1.7	0.3	5.9	1.6	24.2	18.8
B	0.929	2.5	15.2	2.1	0.3	8.5	2.0	30.6	24.0
C	0.924	3.8	16.7	2.2	—	9.9	1.8	34.4	30.9

*Calculated from G values of isolated hydrocarbons assuming scission efficiency of 0.073×10^2 per 1000 carbon atoms. As short chain branches outnumber chain ends by 30:1, no correction for the fragmentation products at chain ends was made.
†Determined by infrared analysis.

bands at $1389\,\text{cm}^{-1}$ $(7.20\,\mu)$ and $1351\,\text{cm}^{-1}$ $(7.40\,\mu)$ respectively,[215] which consequently are lost in the methyl group determination procedure.

In line with the observations reported by earlier workers, the two most populous branches, according to the radiolysis method, are ethyl and n-butyl, which occur in the ratio 2:1, as observed by others.[216]

5.4 Copolymer composition

In addition to measuring the concentration of branching groups up to octyl in polyethylene ^{13}C NMR can be used to elucidate the composition and structure of olefin copolymers by measurement of the concentration of branched groups. Thus in the case of ethylene–propylene copolymers containing between 97 and 99% propylene it is possible to measure propylene groups and elucidate polymer microstructure.

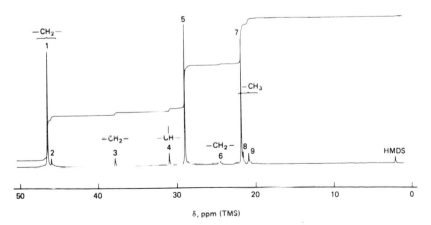

FIG. 43.   A ^{13}C NMR spectrum at 25.2 MHz and 125°C of a 97/3 propylene–ethylene copolymer in 1, 2, 4-trichlorobenzene and perdeuterobenzene. The internal standard copolymer in 1, 2, 4-trichlorobenzene and perdeuterobenzene.

As shown by a typical example in Fig. 43, each ^{13}C NMR spectrum was recorded with proton noise-decoupling to remove unwanted ^{13}C–^{1}H scalar couplings. No corrections were made for differential Nuclear Overhauser effects (NOE) since constant NOEs were assumed[217,218] in agreement with previous workers.[219-22] Constant NOEs for all major resonances in low ethylene content ethylene–propylene copolymers have been reported.[223]

The ^{13}C NMR spectrum of an ethylene–propylene copolymer, containing approximately 97% propylene in primarily isotactic sequences, is shown in Fig. 43. The major resonances are numbered consecutively from low to high field. Chemical shift data and assignments are listed in Table 29. Greek letters are used to distinguish the various methylene carbons and designate the location of the nearest methine carbons.

Paxton and Randall[224] in their method use the reference chemical shift data obtained on a predominantly isotactic polypropylene and on an ethylene–propylene copolymer (97% ethylene). They concluded that the three ethylene–propylene copolymers used in their study (97–99% propylene) contained principally isolated ethylene–ethylene linkages. Knowing the structure of their three ethylene–propylene copolymers, they used

TABLE 29   *Observed and reference ^{13}C NMR chemical shifts in ppm for ethylene–propylene copolymers and reference polypropylenes with respect to an internal trimethylsilane standard*

Line	Carbon	3/97 E/P*	E/P (ref. 222)	97/3* E/P	Sequence assignment	Reference cryst. (ref. 217)	PP amorphous (ref. 217)
1	$\alpha\alpha$-CH$_2$	46.4	46.3		PPPP	46.5	47.0–47.5 r
							46.5
2	$\alpha\alpha$-CH$_2$	46.0	45.8		PPPE		
3	$\alpha\alpha$-CH$_2$	37.8	37.8		PPEP		
4	CH	30.9	30.7		PPE		
5	CH	28.8	28.7		PPP	28.5	28.8 mmmm
							28.6 mmmr
							28.5 rmmr
							28.4 mr + rr
6	$\beta\beta$ CH$_2$	24.5	24.4		PPEPP		
7	CH$_3$	21.8	21.6	P	PPPPP	21.8	21.3–21.8 mm
							20.6–21.0 mr
							19.9–20.3 rr
8	CH$_3$	21.6	21.4		PPPE		
9	CH$_3$	20.9	20.7		PPPEP		
	CH$_3$		19.8	19.8	EPE		
	CH		33.1	33.1	EPE		
	α CH$_2$			37.4	EPE		
	β CH$_2$			27.3	EPE		
	—(CH$_2$)n$^-$		29.8	29.8	EEE		

*As measured by Paxton and Randall.[224]

the ^{13}C NMR relative intensities to determine the ethylene–propylene contents and thereby establish reference copolymers for the faster infrared method involving measurements at $732\,cm^{-1}$ (13.66 μ). After a detailed analysis of resonances Paxton and Randall[224] concluded that methine resonances 4 and 5 (Table 29) gave the best quantitative results to determine the comonomer composition. The composition of the ethylene–propylene copolymers was determined by peak heights using the the methine resonances only. In no instance was there any evidence for an inclusion of consecutive ethylene units. Thus, composition data from ^{13}C NMR could now be used to establish an infrared method based on a correlation with the $732\,cm^{-1}$ (13.66 μ) band which is attributed to a rocking mode, r, of the methylene trimer, $-(CH_a)_3-$.

Randall[225] has developed a ^{13}C NMR quantitative method for measuring ethylene–propylene mole fractions and methylene number-average sequence lengths in ethylene–propylene copolymers. He views the polymers as a succession of methylene and methyl branched methine carbons, as opposed to a succession of ethylene and propylene units. This avoids problems associated with propylene inversion and comonomer sequence assignment. He gives methylene sequence distributions from one to six and larger consecutive methylene carbons for a range of ethylene–propylene copolymers, and uses this to distinguish copolymers which have either random, blocked or alternating comonomer sequences.

5.5 Monomer unit sequence distribution in polymers

During polymerization it is possible to direct the way in which monomers join on to a growing chain. This means that side-groups (X) may be placed randomly (atactic) or symmetrically along one side of the chain (isotactic) or in a regular alternating pattern along the chain (syndiotactic) as shown below. A good example of this is polypropylene, which in the atactic form is an amorphous material of little commercial value but in the isotactic form is an extremely versatile large-tonnage plastic material.

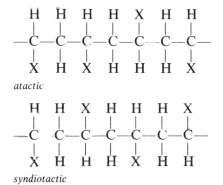

atactic

syndiotactic

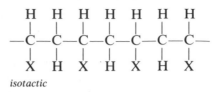

isotactic

Three-dimensionally, atactic and isotactic polypropylene may be represented as shown in Fig. 44.

Polymers can be produced by combining two or more different monomers in the polymerization process. If two monomers are used the product is called a copolymer and the second monomer is usually included in the reaction to enhance the properties of the polymer produced by the first monomer alone. It is possible to control the way in which the two monomers link up.

A good example of a polymer in which sequences of different monomer units occur is ethylene–propylene, and this will be discussed here in some detail. Processes for the manufacture of this polymer can produce several

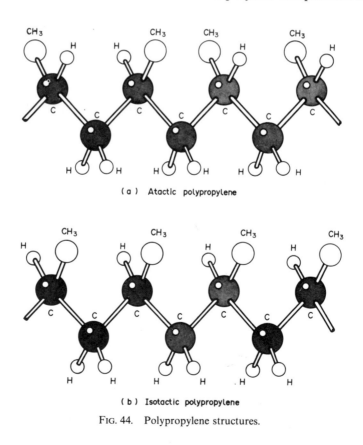

(a) Atactic polypropylene

(b) Isotactic polypropylene

Fig. 44. Polypropylene structures.

distinct types of polymer which, although they may contain similar proportions of the two monomer units, differ appreciably in their physical properties. The differences in these properties lie not only in the ratio of the two monomers present but also, and very importantly, in the detailed microstructure of the two monomer units in the polymer molecule. Ethylene–propylene copolymers might consist of mixtures of the following types of polymer:

(i) Physical mixture of ethylene homopolymer and propylene copolymer:

E–E–E–E–E–   P–P–P–P

(ii) Copolymers in which the propylene is blocked, e.g.:

E–E–P–P–P–P–P–E--E–E–E–E–E–P–P–P–

(iii) Copolymers in which the propylene is randomly distributed, e.g.:

–E–P–E–P–E--P–E–P (alternating e.g. pure *cis*-1, 4 polyisoprene)

or   –E–E–E–P–E–E–E–E–E–P–E–E–E–E

(iv) Copolymers containing random (or alternating) segments together with blocks along the chains, i.e. mixtures of (iii) and (ii) (random and block) or (iii) and (ii) (alternating and block), e.g.:

–E -P–E–P–E–P–

(v) Containing tail-to-tail propylene units in propylene blocks:

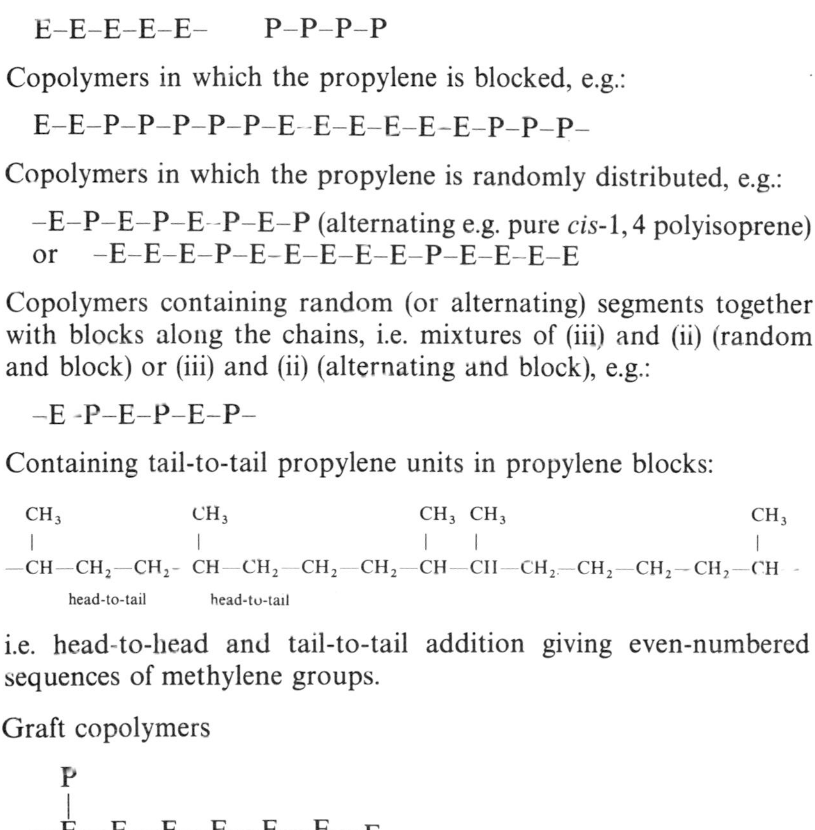

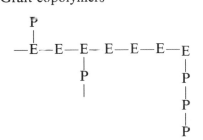

i.e. head-to-head and tail-to-tail addition giving even-numbered sequences of methylene groups.

(vi) Graft copolymers

Various methods are available for determining the percentage of ethylene and of propylene, and of the ethylene propylene ratio, in these polymers, and these methods are discussed in further detail in this section.

In fact the problem is somewhat more subtle than expressed above, and the types of measurements that are commonly required are listed below:

(a) determination of total percentage propylene in the above polymers, regardless of the manner in which the propylene is bound;
(b) determination of total percentage ethylene in the above polymers, regardless of the manner in which the ethylene is bound;
(c) Determination of proportion of total propylene content of polymer which is blocked and that which is randomly distributed along the polymer chain.

Several possible techniques are available for carrying out these analyses, and these are now discussed:

(a) Pyrolysis of sample followed by gas–liquid chromatography of pyrolysis products.
(b) Direct infrared spectroscopy of polymer, either at room temperature or at elevated temperatures.
(c) Pyrolysis of sample followed by infrared spectroscopy.
(d) Nuclear magnetic resonance spectroscopy and proton magnetic resonance spectroscopy.

5.5(a) Pyrolysis–gas chromatography

Before discussing the applications of this technique to microstructural studies of polymers it is necessary to understand the principles of the technique, and the factors which affect the results obtained. These aspects are discussed first, followed by some examples of the application of pyrolysis–gas chromatography which show its usefulness in microstructural studies.

In this technique a small quantity of the polymer is mounted on an inert metal support and either an electrical current is passed through the support (filament method) or external heat is supplied to the support (furnace method) so as to rapidly heat up and break down (i.e. pyrolyse) the polymer into a mixture of smaller molecules which, under standard pyrolysis conditions, are characteristic of the polymer being examined. The products are swept from the pyrolysis chamber by a stream of carrier gas onto a gas chromatographic column and separated into their individual components prior to passing through the detector which records their retention time (time taken, under standard conditions, to travel from pyrolysis chamber to detector) and quantity (peak height under standard conditions).

This in principle is the essence of the pyrolysis–gas chromatography technique. It is, of course, possible to then pass the separated pyrolysis

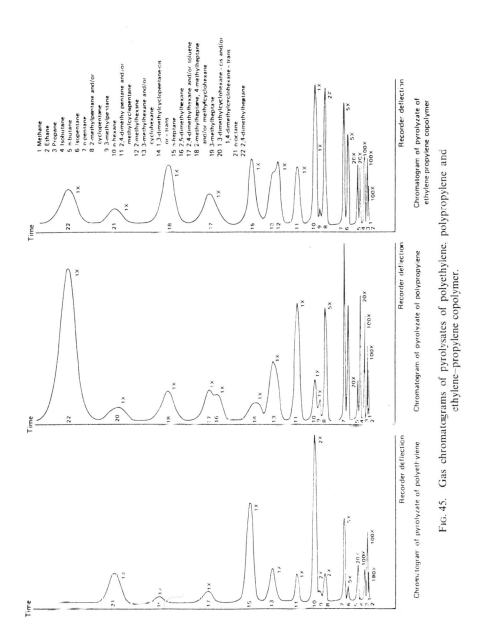

FIG. 45.   Gas chromatograms of pyrolysates of polyethylene, polypropylene and
ethylene–propylene copolymer.

products one at a time into a mass spectrometer in order to obtain definitive information regarding their precise identity, i.e. pyrolysis–mass spectrometry.

An example of the results obtainable by pyrolysis–gas chromatography is shown in Fig. 45, which compares the pyrograms of polyethylene, polypropylene and an ethylene–propylene copolymer. To obtain these results the sample (20 mg), in a platinum dish, was submitted to controlled pyrolysis in a stream of hydrogen as carrier gas. The pyrolysis products were then hydrogenated at 200°C by passing through a small hydrogenation section containing 0.75% platinum on 30/50 mesh aluminium oxide. The hydrogenated pyrolysis products are then separated on a squalane on fireback column, and the separated compounds detected by a katharometer. Under the experimental conditions used in this work only alkanes up to C_9 could be detected.

It can be seen that major differences occur between the pyrograms of these three similar polymers. Polyethylene produces major amounts of normal C_2 to C_8 alkanes and minor amounts of 2-methyl and 3-methyl compounds such as isopentane and 3-methylpentane, indicative of short chain branching on the polymer backbone. In the case of polypropylene, branched alkanes predominate, these peaks occurring in regular patterns, e.g. 2-methyl, 3-ethyl and 2, 4-dimethyl configurations. Particularly noticeable are the large peaks due to 2, 4-dimethylpentane and 2, 4-dimethylheptane, which are almost absent in the polyethylene pyrolysate. Minor components obtained from polypropylene are normal paraffins present in decreasing amounts up to normal hexane. This is to be contrasted with the pyrogram of polyethylene, where n-alkanes predominate. The ethylene–propylene copolymer, on the other hand, as might be expected, produces both normal and branched alkanes. 2, 4-dimethylpentane and 2, 4-dimethylheptane concentrations are lower than occur in polypropylene.

It is emphasized that the technique of pyrolysis–gas chromatography can be used in two ways for the examination of polymers. It can be used as a fingerprinting technique (discussed in Chapter 4), in which the pyrogram of an unknown polymer prepared under standard conditions is simply visually compared with a library of pyrograms of various known polymers prepared under the same conditions. Also, as will be discussed below, the technique can be used to obtain fine detail regarding minor, but often very important, detail concerning polymer structure such as monomer sequences, branching, crosslinking copolymer structure and the nature of end-groups, and can also be used to elucidate polymer structure between adjacent monomer units in copolymers and to determine monomer ratios in copolymers. Before the technique is discussed in further detail some important general points concerning it are discussed below. This discussion centres mainly, but not exclusively, on the polyolefins such as polyethylene, polypropylene and their copolymers, as these are the polymers which have been most extensively studied by this technique.

Design of pyrolyser unit

Heated platinum filament pyrolyser. Voigt[226-228] employed the apparatus shown in Fig. 46 attached directly to the gas inlet of the gas chromatograph for the examination of polyethylene, polypropylene and ethylene–propylene copolymers. Figure 46 too shows the Pyrex glass pyrolysis cell *b* located in the oven *a*, the carrier gas entering the cell from above, and bringing with it the pyrolysate formed on the red-hot platinum wire *c* on leaving the cell and passing through the gas inlet and a heated feed line to be gas chromatograph; *d* is the junction of chromel–alumel thermocouple, the lead wires of which are embedded in a Ni–Cr–Fe alloy tube filled with magnesium oxide. Like the platinum wire of the heating spiral, this tube has a diameter of only 0.5 mm.

Specimens weighing about 2 mg were heated for 18 s up to a maximum temperature of 550°C. During the pyrolysis the carrier gas flushes the products of pyrolysis into the separating column. The pyrolysis cell is cut out of the carrier gas flow 1 min after starting the pyrolysis. In this apparatus the pyrolysis gases were not hydrogenated, and hence a mixture of unsaturated hydrocarbons was obtained.

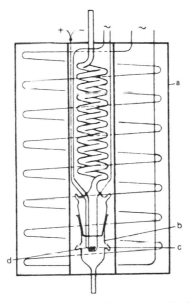

FIG. 46. Pyrolysis apparatus: (a) jacket, (b) pyrolyzer, (c) platinum filament, (d) thermoelement.

In the above technique the polymer is pyrolysed with a certain voltage applied to the platinum filament to produce a specified maximum pyrolysis temperature. It might be assumed that precise setting of the voltage provides sufficient control to reproduce pyrolysis conditions. However, it was found that it is the heating rate of the filament, rather than its maximum temperature, that is the primary factor governing reproducibility of the product distribution. Poor control of the heating rate may well explain the reported non-reproducibility of the filament, compared with the furnace pyrolyser. It is to be expected that variations in heating rate of the filament can result certainly in the case of some types of polymers, in different depths of cracking and different amounts of secondary pyrolysis reactions.

The effect of heating rate on the pyrolysis patterns is illustrated by the data given in Table 30, which were obtained by pyrolysis–hydrogenation–gas-chromatography of polyethylene. Results are compared for a filament and a surface types of pyrolyser both run at maximum temperatures between 550 and 800°C. The patterns for the linear polyethylene are compared on the basis of C_1 plus C_2, and the non-normal hydrocarbons. The filament pyrolyser results are fairly consistent for heating rates of 70°C per second and lower. Thus, below a certain heating rate one does not need to control the rate precisely. At 170°C per second and above the amounts of C_1 plus C_2 and non-normals are increased markedly.

The filament temperature programming technique has the following advantages: (1) minimizes secondary pyrolysis reactions; (2) provides a means of duplicating product distributions with different filaments; (3) offers a standard method of thermal decomposition which is suitable for polymers of widely varying thermal stabilities.

TABLE 30   *Effect of filament heating rate on pyrolysis pattern*

Sample	Heating rate (°C/s)	Maximum Temp (°C)	C_1 plus C_2	Area %* branched + cyclic C_4–C_{10}	Ratio ($1C_6/nC_6$)
Filament					
Linear polyethylene	8	550	7	9	3
(Marlex 6009)	20	550	7	7	2
	20	800	7	5	—
	70	550	9	6	3
	170	650	25	8	6
	280	800	47	9	12
Furnace					
	—	550	12	13	8
	—	650	28	21	19
	—	800	52	83	73

*Pyrogram-heated silicone column; peak areas normalized through C_{10}.

The hydrogenation technique referred to above is used to convert the unsaturated olefinic pyrolysis products to saturated alkanes. As several different olefins, upon hydrogenation, will be converted to the same alkane, hydrogenation has the effect of reducing the number of components in the pyrolysis mixture and consequently of simplifying the task of interpreting the gas chromatogram. Thus, for example, the three olefins hexene-1, hexene-2 and hexene-3, upon hydrogenation, are converted to n-hexane. Of course, if the object is to deduce structural detail of the polymer, then hydrogenation and the subsequent alterations in pyrolysis product composition might mask some important feature, and in such instances it would be advisable not to use hydrogenation in the preliminary studies. The hydrogenation technique is used principally in the pyrolysis of polyolefins and PVC, and is not used in the case of many other types of polymers such as acrylates or polystyrene, where the pyrolysis products are either not olefinic, or are fewer in number than occurs in the case of polyolefins.

The furnace pyrolyser. This type of pyrolyser, mentioned above, is different in principle from the filament type in which the filament, and consequently the polymer, is heated to a predetermined temperature by the application of a known voltage to the filament. In the furnace pyrolyser a weighed quantity of polymer is placed in a small platinum dish. A small furnace is brought to a predetermined temperature by application of an electrical current, and the platinum dish then introduced into the furnace so that the polymer is very rapidly heated up to the required temperature. The pyrolysis products are then swept into the gas chromatogram by the carrier gas.

Comparison of results obtained by filament and furnace pyrolysers

Classic work on the pyrolysis of polyethylene, polypropylene and ethylenepropylene copolymers containing between 0 and 100% propylene, was reported by Van Schooten and others.[229–232,236] In their earlier work on the furnace method the sample (20 mg) in a platinum dish was submitted to controlled pyrolysis in a stream of hydrogen as carrier gas. The pyrolysis products were then hydrogenated at 200°C by passing through a small hydrogenation section containing 0.75% platinum on 30/50 mesh aluminium oxide. The hydrogenated pyrolysis products are then separated on a squalane on fireback column and the separated compounds detected by a katharometer. Under the experimental conditions used in this work only alkanes up to C_9 could be detected.

Using this procedure, however, a large sample (20–30 mg) was required, which favoured the occurrence of consecutive and side-reactions of the primary pyrolysis products. In their improved method,[233,234] discussed below, Van Schooten and Evenhuis pyrolysed a much small polymer weight (0.4 mg) in a stream of hydrogen on an electrically heated nichrome filament at 500°C and led the pyrolysis products onto a hydrogenation catalyst to convert all products to saturated hydrocarbons. Operation with such small samples was made possible by using a sensitive flame ionization GLC detector. This and the use of programmed heating of the Apiezon N/firebrick GLC column extended the range of detectable volatile products from C_9 to iso C_{13}, with considerable improvement in resolution of the pyrograms, especially of the more volatile components.

These modifications of experimental techniques minimized the occurrence of secondary reactions, with the result that the relative amounts of lowest molecular weight fragments are considerably reduced. This is illustrated in Figs 47 and 48, where peak surface areas are given for the n-alkanes in the pyrograms of polyethylene and polypropylene as obtained by the old and the new pyrolysis apparatus. The modified technique therefore provides a much more reliable picture of the primary pyrolysis reactions, and enables some conclusions to be drawn regarding primary reaction mechanism.

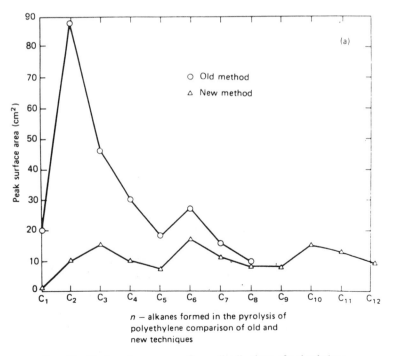

n – alkanes formed in the pyrolysis of polyethylene comparison of old and new techniques

Fig. 47. Peak surface area–n-alkane distributions of polyethylene.

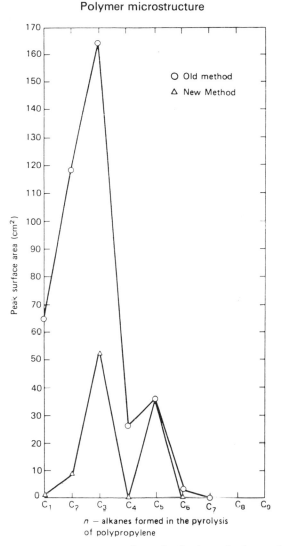

FIG. 48. Peak surface area–n-alkane distributions of polypropylene.

Effect of pyrolysis temperature on pyrogram

Polystyrene. Lehmann and Brauer[235] investigated the pyrolysis gas chromatography of polystyrene under helium at temperatures ranging from 400 to 1100°C, utilizing for the pyrolysis a silica boat surrounded by a platinum heating coil.[236] Separations were carried out on a column of Apiezon (30%) on Chlorowax 70 and dinonyl phthalate on firebrick operated at 100–140°C.

Typical chromatograms obtained on pyrolysing polystyrene are shown

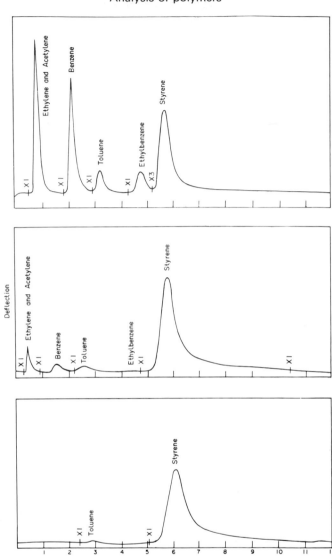

Fig. 49. Chromatograms of pyrolysis products of polystyrene. Column, Apiezon L; col. temp., 140°C; flow rate, 60 ml/min; pyrolysis temp; °C, top, 1025, middle, 825, bottom, 425. (Attenuation scale indicated by numbers in figure.)

in Fig. 49. At 425°C only styrene monomer is eluted from the column; degradation at 825°C produces a number of additional products, presumably produced by secondary reactions undergone by the styrene monomer. Reactions leading to these products are even more important at a temperature of 1025°C.

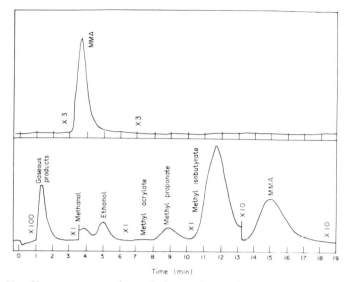

FIG. 50. Chromatograms of pyrolysis products of poly(methylmethacrylate). Column, dinoyl phthalate. Top pyrolysis temp., 425°C; col. temp., 128°C; the flow rate, 60 ml/min. Bottom pyrolysis temp., 1025°C; col. temp., 100°C; the flow rate, 20 ml/min. (MMA-methyl-methacrylate monomer.)

Methyl methacrylates. Lehmann and Brauer[235] and Brauer[237] investigated the pyrolysis–gas chromatography of polymethylmethacrylate at temperatures between 400°C and 1100°C utilizing for the pyrolysis a silica boat surrounded by a platinum heating coil.

Chromatograms obtained from pyrolysing poly(methylmethacrylate) at 425°C and 1025°C using a dinonyl phthalate column are shown in Fig. 50. Monomer is formed nearly exclusively (99.4%) at 425°C reducing to 20% at 879°C, whereas a number of additional compounds are detected at 1025°C. Non-volatile products are retained in the column. In view of these findings regarding the effect of pyrolysis temperature on product composition it is always desirable to investigate temperature effects at an early stage and, indeed, studies performed at different pyrolysis temperatures might provide additional information in polymer microstructure studies.

In the case of polyethylene, it has been shown that increasing the pyrolysis temperature to 650°C increases the yield of C_1 and C_2 and non-normal hydrocarbons due to the occurrence of secondary reactions. In the case of polystyrene and polyacrylates increase in pyrolysis temperature leads to a decrease in monomer yield and an increase of the amount of products formed in secondary reactions.

Quantitative measurements. Quantitative measurements of the amounts of various pyrolysis products can in many instances be correlated with the

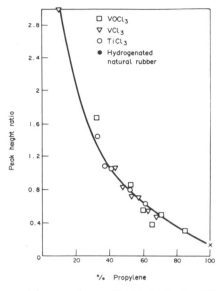

FIG. 51.   Ethylene–propylene copolymers. Peak height ratio $nMC_7/(2MC_7 + 4MC_7)$
as a function of propylene content.

percentage composition of a copolymer, or with the concentration of a
particular microconstituent in the polymer. Thus, Van Schooten and
Evenhuis[229,230] applied their pyrolysis–hydrogenation–gas chromatography technique to the quantitative determination of copolymer
composition of ethylene–propylene copolymers, an analysis which presents difficulties in solvent solution–infrared methods, especially with
samples that are only partly soluble in suitable solvents such as carbon
tetrachloride. Since the hydrogenation pyrogram of polyethylene consists
almost exclusively of normal alkanes and that of polypropylene of iso-
alkanes, the ratio of the peak heights of a n-alkane to an iso-alkane is a
good measure of the copolymer composition. The ratio n-C$_7$ (2-methyl C$_7$
+ 4-methyl C$_7$) was a good measure of ethylene–propylene ratio in
copolymers (Fig. 51).

Newmann and Nadeau[238] applied pyrolysis–gas chromatography to
the examination of ethylene–butene copolymers. Pyrolyses were carried
out at 410°C in an evacuated gas vial and the products swept into the gas
chromatograph. Under these pyrolysis conditions it is possible to analyse
the pyrolysis gas components and obtain data within a range of about
10% relative. The peaks observed on the chromatogram were methane,
ethylene, ethane, combined propylene and propane, isobutane, 1-butene,
n-butene, $trans$-2-butene, cis-2-butene, 2-methylbutane, and n-pentane.

Figure 52 shows the relationship between the amount of ethylene
produced on pyrolysis and the amount of butene in the ethylene–butene

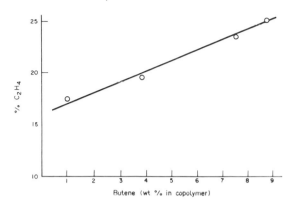

FIG. 52.   Percentage C_2H_6 as function of butene content of ethylene–butene copolymers.

copolymers, which was determined by an infrared analysis for ethyl branches. The y intercept of 16.3 ethylene represents the amount of ethane which would result from a purely linear polyethylene.

As one would expect, ethyl branches will increase with the amount of butene copolymerized with the ethylene. This is in agreement with the work of Madorsky and Straus,[239] who found that the thermal stability and breakdown products obtained on pyrolysis can be related to the strength of the C—C bonds in the polymer chain, i.e. secondary > tertiary > quaternary.

Neumann and Nadeau[238] studied the pyrolysis of ethylene oxide, propylene oxide and condensates containing various proportions of ethylene oxide. There were no significant differences in the pyrolysis chromatograms of samples heated for 0.5 to 2 h. The only effect of time of pyrolysis is the total amount of gas produced. The relative concentrations of components are not significantly changed. The temperature of pyrolysis, however, does play a large part in both the amount and type of components in the volatile gases. Between 390°C and 410°C there was no noticeable change in products.

Figure 53 is a plot of the percentage ethylene as a function of the ethylene oxide content of the condensate. As the ethylene oxide content of the copolymer increases, so does the amount of ethylene produced. The curve is linar up to about 50% ethylene oxide, then turns sharply upward to an ethylene content of 38.6% for pure polyethylene glycol. The relative contents of ethylene oxide and propylene oxide in polyethylene–polypropylene glycols can be determined using combined pyrolysis–gas chromatography calibrated with polyethylene glycol and polypropylene glycol standards.

The consensus of opinion is that, certainly in the cases of those polymers such as the polyolefins where complex pyrograms are produced, filament

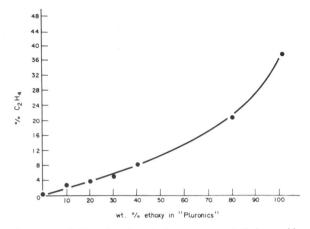

Fig. 53. Percentage C_2H_4 as function of ethoxy content of ethylene oxide–propylene oxide condensates.

pyrolysis is the preferred method. For the purposes of fundamental studies, pyrolysis at a variety of temperatures and heating rates is preferred. Smaller sample weights of the order of 1–2 mg are preferred as these prevent or reduce the occurrence of secondary side reactions in the pyrolysis which might confuse the interpretation of the pyrogram when carrying out polymer structural studies. Certainly, sample weights of the above 3 mg should be avoided. The use of small sample sizes necessitates the use of the more sensitive types of gas chromatograph detectors such as flame ionization. In particular circumstances where the occurrence of a microstructural feature is being studied, failure to use a sensitive enough detector could result in the pyrolysis product being missed. In such studies, to improve sensitivity, a limited increase in sample weight, say from 1–2 mg to 4, might be permitted.

In-line hydrogenation is a useful innovation for simplifying the pyrogram obtained for those polymers which produce complicated mixtures, but should be used with caution in fundamental studies. More information might be obtained by carrying out studies with and without in-line hydrogenation. Finally, the type of gas chromatograph separation column used should be the subject of close scrutiny. The information gained in pyrolysis studies is only as good as the degree and type of separation achieved on the column and, certainly in the early stages of investigation work, a variety of columns should be studied.

Polymer pyrolysis mechanisms. Under pyrolysis conditions, different types of polymers break down in different ways. Two important different methods of decomposition are: (i) those involving the formation of free radicals with consequent hydrogen atom transfer followed by subsequent reactions and (ii) the unzipping mechanism. Of course, both mechanisms can occur simultaneously, and the representations below are oversimplified.

Free radical formation:

$$\sim 2CH_2-CH_2-CH_2-CH_2 \sim \; \rightarrow \; \sim CH_2-CH_2^{\cdot} + {}^{\cdot}CH_2-CH_2 \sim$$

(depolymerization)

Polyacrylic acid

Unzipping mechanism:

$$\left(\begin{array}{c} -CH_2-CH- \\ | \\ COOH) \end{array} \right)_n \rightarrow nCH_2 = \begin{array}{c} CH \\ | \\ COOH \end{array}$$

It is important to gain an insight into these reaction mechanisms if progress is to be made in applying the pyrolysis–gas chromatographic technique to studies of the microstructure of polymers. This aspect is discussed below.

Polyethylene. Voigt[226] states that the formation of decomposition products of polyethylene is a result of a primary step in a statistical thermal breaking down of a chain:

$$\sim CH_2-CH_2-CH_2-CH_2 \sim \; \rightarrow \; \sim CH_2-CH_2^{\cdot} + {}^{\cdot}CH_2-CH \sim$$

in which the free radicals formed can react further either by depolymerization (unzipping) or by the transfer of hydrogen atoms. The first type of reaction leads to the corresponding monomers, while the latter competes with the continuing decomposition of the chains to form a spectrum of hydrocarbons with different chain lengths. This spectrum contains, apart from saturated hydrocarbons and diolefins, olefins which, in their structure and quantitative distribution, are dependent in a characteristic manner on the nature of the initial polymers, as might well be expected. The very marked depolymerization which takes place with other polymers – such as polymethacrylates, polystyrene, and the like – does not appear to play any major role in the cases of polyethylene or other polyolefins, since the corresponding monomers do not exist in any considerable excess as compared with the other decomposition products. As is to be expected, only unbranched fractions arising from statistical decomposition arise in the case of polyethylene, which breaks down with reactions of the following type:

$$\sim CH_2-CH_2-CH_2^{\cdot} + {}^{\cdot}CH_2-CH_2 \sim \; \rightarrow$$
$$\sim CH_2-CH=CH_2 + CH_3-CH_2 \sim$$

or

$$\sim CH_2—CH_2—CH_2 \sim + \cdot CH_2—CH_2 \sim \rightarrow$$
$$\sim CH_2—CH—CH_2 \sim + CH_3—CH_2 \sim$$

$$\sim CH_2—CH_2—CH_2—CH_2—CH_2 \sim \rightarrow$$
$$\sim CH_2—CH{=}CH_2 + \cdot CH_2—CH_2 \sim$$

The very small amounts of branched molecules which are present even in low-pressure polyethylenes do not make themselves noticeable on the gas chromatogram under the conditions employed by Voigt.[226] Branched molecules are, however, found under other more searching gas chromatographic conditions (see later). In this connection, it should be pointed out that high-pressure polyethylenes produced completely identical gas chromatograms.

According to Wall *et al.*[240] the pyrolysis of polyethylene proceeds by a radical chain mechanism. The products formed result from the process of random-chain cleavage, followed by intermolecular or intramolecular hydrogen abstraction. Hydrogen abstraction occurs preferentially at tertiary carbon atoms, and produce formation results from homolysis of the carbon–carbon bond β to the radical site. The major products formed are the *n*-alkanes and α, ω-diolefins. The peaks between the triplets result from chain branching.

The effect of chain branching on pyrolysis of polyethylene may be viewed simply as promoting cleavage of the main polymer chain at carbon atoms α and β to the branch site.[241–243] Cleavage of the branch from the main chain becomes more important with increasing branch length. The elegant work of Seeger, Exner and Cantow[242,243] with labelled ethylene–propylene copolymers on the relative amounts of α and β cleavage compared to statistical chain cleavage provides a foundation for discussion of the products arising from these reactions. They concluded that the probability of backbone cleavage at the branch sites is much greater than statistical chain cleavage, and that α and β cleavage occur with equal frequency.

Since there are a number of different olefin products possible from cleavage at a branch site, the determination of the type of chain branch is best accomplished by pyrolysis–hydrogenation–gas chromatography, On-line catalytic hydrogenation of the pyrolysis products in the injection port of the chromatograph converts all of the olefins to saturated hydrocarbons, therefore reducing the number of possible products. Cleavage of the carbon–carbon bond in the polymer backbone at the branch site (α cleavage) results in the formation of *n*-alkanes. Backbone cleavage at the bond β to the branch site gives methylalkanes. In the case of methyl branch sited, β cleavages gives 2-methylalkanes, and chain cleavage β to ethyl and butyl branches gives 3-methyl and

5-methylalkanes, respectively.[171] Statistical cleavage will produce a mixture of branched alkanes.

Additional proof of the above has been provided from studies on model ethylene-1–butene copolymers and ethylene–1-hexene copolymers.[233] Pyrolysis–hydrogenation of these model polymers shows that the areas of 3-methylalkanes are increased relative to the areas of the corresponding 2-methylalkanes in the pyrograms of ethylene-1–butene copolymers, and that 5-methylalkanes are indicative of butyl branching.[233,244] Other indications of C_4 branches are given by an increase in the relative amounts of 2-methylhexane (2-MC$_6$) and 3-methylheptane (3-MC$_7$) and an increase in the relative amount of the n-C$_4$ peak. However, the areas of 2-MC$_6$ and 3-MC$_7$ are equal in the case of C_4 branch, while for ethyl branches the area of 3-MC$_7$ is enhanced relative to 2-MC$_6$. Thus not only must individual peaks be interpreted for the determination of the type of branch, but the overall pyrolysis pattern must be considered as well.

From an examination of the products formed from the pyrolysis of high-density polyethylene and polymethylene, Tsuchiya and Sumi[245,246] proposed that the major hydrogen-abstraction reaction is due to an intramolecular cyclization. They proposed that, following initial radical formation at C_1, successive intramolecular hydrogen abstractions occur along the chain resulting in the formation of new radicals at C_5, C_9 and C_{13} as shown in equation 1–4.

$$R\!-\!R \rightarrow 2R \qquad (1)$$

$$R\!\cdot\!=3CH_2 \quad \underset{\substack{| \\ 2CH_2 \\ \diagdown 1CH_2}}{\overset{\substack{H_2 \\ |}}{\underset{}{\diagup C \diagdown}}} \quad \underset{\substack{| \\ H}}{C\dot{H}} \quad \leadsto \quad CH_3(CH_2)_3\!-\!\overset{\cdot 5}{C}H\!-\!CH_2\!-\!R \qquad (2)$$

$$CH_3\!-\!(CH_2)_3\!-\!\overset{5}{CH}\!-\!CH_2\!-\!R \rightarrow$$
$$CH_3\!-\!(CH_2)_6\!-\!CH_2\!-\!\overset{9}{C}H\!-\!CH_2\!-\!R \qquad (3)$$

$$CH_3\!-\!(CH_2)_6\!-\!CH_2\!-\!\overset{9}{C}H\!-\!CH_2\!-\!R \rightarrow$$
$$CH_3\!-\!(CH_2)_{11}\!-\!\overset{13}{C}H\!-\!CH_2\!-\!R \qquad (4)$$

Cleavage of the carbon carbon bonds β to these macroradicals results in the formation of increased amounts of C_6, C_{10} and C_{14} α-olefins and C_3, C_7, and C_{11} n-alkanes over that which could be predicted from statistical-chain cleavage.

Ahlstrom *et al.*[247] propose that cyclic intramolecular hydrogen abstraction should proceed further along the chain for less-branched high-density polyethylene than for the more highly branched low-density polyethylene. Thus, at longer chain lengths, the relative contribution of intramolecular hydrogen abstraction to the products formed as a result of this reaction should be greater for high-density polyethylene than for low-density polyethylene.

A careful examination of the data on pyrolysis in helium shows that this difference in C_{14}, C_{15} peak maxima is due to a relatively greater increase in the amount of C_{15} *n*-alkane formed from high-density polyethylene than to an increase in C_{15} α-olefin concentration. They showed that for both high-density polyethylene and low-density polyethylene the C_{14} α-olefin is the favoured product from decomposition of the C_{13} macroradical. However, due to a relatively greater contribution of products from C_{17} macroradical in high-density polyethylene the increase in the amount of C_{15} *n*-alkane and C_{18} α-olefin formed is favoured relative to that in low-density polyethylene.

From this, these workers conclude that the cyclic intramolecular hydrogen abstraction is more favourable at longer chain lengths for high-density polyethylene, and this explains the observed differences in peak maxima in the pyrolysis of low-density polyethylene and high-density polyethylene.

Polypropylene. The studies of Wall and Straus[250] indicate that pyrolysis of polypropylene proceeds, principally by random cleavage of the polymer chain. Voigt[226] observed no straight-chain decomposition products of polypropylene with a chain length greater than C_5, a fact that is in agreement with the prediction readily deduced from the head-to-tail structures of the polymer. The larger decomposition products all contain methyl branches, and the multiple-branched chains the expected 2, 4-branching.

Van Schooten and Evenhuis[233,234] found identical pyrolysis patterns for various polypropylene samples. The surface areas of the main peaks are shown in Fig. 54. Depolymerization is much more important for polypropylene, as shown by the large propane peak. Two other very large peaks in the polypropylene pyrogram can be interpreted as originating from intramolecular hydrogen transfer with the fifth carbon atom of the secondary radical:

$$\bar{1}, n\text{-pentane} + \bar{1}\bar{1}, 2, 4\text{-dimethylheptane}$$

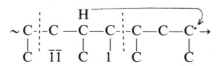

The primary radical gives it this way:

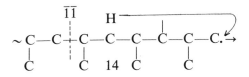

1, isobutane + II, 2, 4-dimethylheptane + III CH_3.

The isobutane and especially the methane peaks are rather small, indicating that this reaction is not very frequent for the primary radical. For this radical, hydrogen transfer with the sixth carbon atom might be more important, as the 2-methylpentane peak could be explained in this way. Another possibility is that this peak has to be ascribed to intra-molecular hydrogen transfer reactions (i.e. formation of 2-methylpentadiene). Table 31 lists the products expected from intra-molecular hydrogen transfer during the pyrolysis of polypropylene. These can be compared with the products found in the pyrogram in Fig. 54.

Ethylene–propylene copolymers and polyisoprene. Van Schooten and Evenhuis[249] prepared gas chromatograms of hydrogenated pyrolysis products of polyethylene, polypropylene, ethylene–propylene copolymers and polyisoprene (hydrogenated natural rubber). These indicate a high degree of alternation in the ethylene–propylene copolymers. They identified most peaks in the chromatograms and ascribed them to a single component, some to two or three different iso-alkanes or cyclo-alkanes.

A survey of the relative concentrations of the pyrolysis products is presented in Figs 55 and 56, where peak areas have been given for the

TABLE 31 *Products expected from intramolecular hydrogen transfer during pyrolysis of polypropylene (main peaks found in pyrogram in Fig. 55 are underlined)*

	Number of the hydrogen-donating carbon atom				
	5	6	7	8	9
Secondary radical	$\underline{nC_5}$	CH_4	$4MC_7$	CH_4	$4.6MC_9$
~C—C—C—C—C—C—C	$\underline{2.4MC_7}$	$2MC_5$	$2.4.6MC_9$	$2.4MC_7$	$2.4.6.8MC_{11}$
│ │ │ │			$4MC_7^*$		$4.6MC_9^*$
C C C C			$4.6MC_9$		$4.6.8MC_{11}$
Primary radical	$1C_4$	$\underline{2MC_5}$	CH_4	$2.4MC_7$	CH_4
~C—C—C—C—C—C—C—	$2MC_5^*$	$2.4.6MC_7$	$2.4MC_5$	$2.4.6.8MC_9$	$2.4.6MC_7$
│ │ │	$\underline{2.4MC_7}$		$2.4MC_7^*$		$2.4.6MC_9^*$
C C C	CH_4		$2.4.6MC_9$		$2.4.6.8MC_{11}$

*Formed by hydrogen exchange with a methyl group.

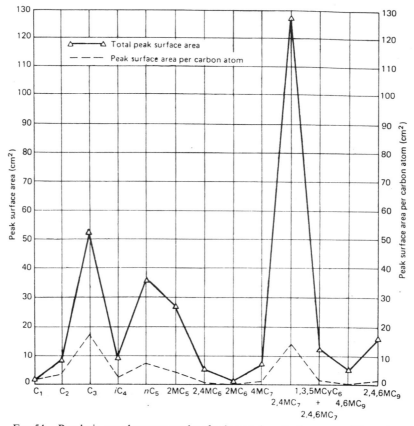

F$_{IG}$. 54. Pyrolysis–gas chromatography of polypropylene. Surface areas of the main peaks.

pyrolysis products of polyethylene, polypropylene, polyisoprene and for a copolymer containing about 50% m methylene. A cursory examination of these figures shows that, apart from a few minor differences, the pyrolysis spectra of ethylene–propylene rubber of 50% m C_2 and polyisoprene are very similar. This would indicate a high degree of alternation in the ethylene–propylene copolymer, because presumably the spectrum of a C_2/C_3 copolymer, having poor alternation, resembles that of a 1:1 mixture of polyethylene and polypropylene. The same conclusion was drawn from the size of the 2,4-dimethylheptane peak.

The main feature of the pyrolysis spectra of polyethylene, polypropylene and isoprene can be interpreted on the basis of simultaneous breakage of carbon–carbon bonds in the main polymer chain. For a more detailed interpretation of the pyrolysis spectrum it is necessary to assume a series of radical-chain reactions in which intramolecular hydrogen abstraction plays an important role.

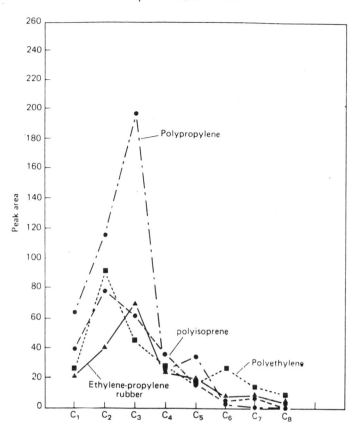

FIG. 55. Peak area of *n*-alkanes obtained after pyrolysis of some polymers.

Polyisoprene. Polyisoprene (hydrogenated natural rubber) is a completely alternating ethylene propylene copolymer (i.e. has no ethylene or propylene blocking) and is therefore an interesting substance for pyrolysis studies. The surface area of the main peaks up to C_{13} obtained by Van Schooten and Evenhuis[234] indicate that the unzipping reaction, which would yield equal amounts of ethylene and propylene in the hydrogenated pyrolysate, evidently takes place to some extent, but is less important than the hydrogen transfer reactions. The large number of peaks produced reflects the many possible transfer reactions for this polymer. Hydrogen transfer reactions from the fifth carbon atom predominate, as indicated by the large butane, 3-methylhexane, 2-methylheptane peaks, followed by transfer from the ninth carbon atom, which is shown by the size of the 3-methyloctane, 3,7-methyldecane (C_{10}) and 2,6-dimethylundecane (C_{11}) peaks.

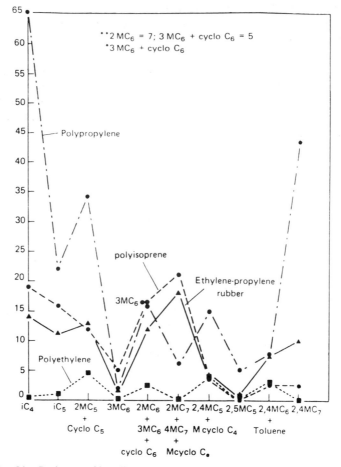

FIG. 56.   Peak area of iso-alkanes obtained after pyrolysis of some polymers.

Polybutene 1.   Whereas in the case of polyethylene and polypropylene the splitting up of the main chain is virtually the only result of thermal reaction, the stripping off of side-chains in polybutene-1 is clearly indicated by the nature of the pyrolysis products.

In the case of polybutene-1 a mechanism involving stripping off of side-chains is involved in the formation of ethylene:

$$\sim CH_2\!-\!\underset{\substack{|\\CH_2\\|\\CH_2}}{CH}\ \sim\ \rightarrow\ \sim CH_2\!-\!CH\sim\ +\ \underset{\substack{\|\\CH_2}}{CH_2}$$

In actual fact a considerable yield of ethylene is obtained from polybutene-1, whereas the production of propylene is much smaller because this can only emanate from the principal chain. The complicated structure of polybutene-1, as compared with the structures of polyethylene and polypropylene, gives rise to the formation of a markedly larger number of structurally isomeric degradation products. Theoretically with a maximum molecular size of C_8, 37 different hydrocarbon decomposition products should be obtainable from polybutene-1 and, in fact, Voigt[226] was able to identify, or indicate the probable existence of, 33 compounds for polybutene-1.

Polyisobutylene. Wall and Straus[250] indicate that pyrolysis of polyisobutylene proceeds principally by random cleavage of the polymer chain. However, the exceptionally high yield of monomer (20%) from polyisobutylene reported in their studies suggests a non-statistical distribution of pyrolysis products. Barrall *et al.*[248] investigated the structure and composition of the homopolymers and copolymers of isobutylene. These workers state that each polymer and copolymer exhibits a specific pyrolysis temperature for maximum isobutylene yield.

Polyvinylchloride. Poly(vinylchloride) upon pyrolysis in a combustion furnace undergoes non-chain scission reactions.[251] Non-chain scission

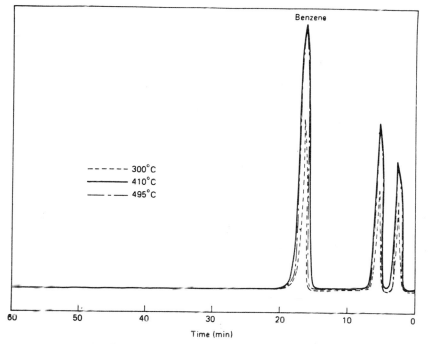

FIG. 57. Gas chromatogram of pyrolysate of poly(vinylchloride).

reactions are common to short chain esters or halogenated molecules which have a hydrogen atom attached to the carbon atom beta to the substituted group. These molecules dissociate into an olefinic structure and an acid, the main pyrolysis products of polyvinyl chloride are volatile products, including hydrochloric acid and benzene (Fig. 57). The benzene is formed as the result of a cyclization of the unsaturated chain ends.

Ohtani and Ishikawa[252] pyrolyzed polyvinychloride at 425°C in a nitrogen atmosphere and studied the resulting degradation products by infrared and ultraviolet spectroscopy. Their findings showed that: (1) the pyrolysis products (other than hydrogen chloride) consisted of aliphatic and aromatic hydrocarbons, and (2) the types of pyrolysis products from polyvinylchloride were a function of the stereoregularity (tacticity) in the ungraded polymer. Additional qualitative or quantitative measurements were not made.

O'Mara[253] carried out pyrolysis of polyvinylchloride (Geon 103, CI = 57.3%) by two general techniques. The first method involved heating the resin in the heated (325°C) inlet of a mass spectrometer in order to obtain a mass spectrum of the total pyrolzate. The second, more detailed, method consisted of degrading the resin in a pyrolysis–gas chromatograph interfaced with a mass spectrometer through a molecule enricher.[254] Samples of polyvinylchloride and plastisols (10–20 mg) were pyrolyzed at 600°C in a helium carrier gas flow. Since a stoichiometric amount of hydrogen chloride is released (58.3%) from polyvinylchloride when heated at 600°C, over half of the degradation products, by weight, is hydrogen chloride.

A typical pyrogram of a polyvinylchloride resin obtained by this method using an SE32 column is shown in Fig. 58. The major components resulting from the pyrolysis of polyvinylchloride are hydrogen chloride, benzene, toluene and napthalene. In addition to these major products, a homologous series of aliphatic and olefinic hydrocarbons ranging from C_1 to C_4 are formed. O'Mara[253] obtained a linear correlation between weight of polyvinylchloride pyrolyzed and weight of hydrogen chloride obtained by gas chromatography.

The thermal decomposition of polyvinylchloride has usually been represented as involving the dehydrochlorination of the polymer backbone. On the other hand Stromberg *et al.*[255] assumed the release of chlorine molecules in their kinetics of the decomposition of polyvinylchloride. The isolation of small amounts of hydrogen and chlorine was reported by Tsuchiya and Sumi[256] on pyrolysis of polyvinylchloride in an inert gas. Ohta[257] suggested that a part of the hydrogen chloride released from polyvinylchloride may recombine with the double bonds along the chain introduced by the dehydrochlorination. After pyrolysis of polyvinylchloride many kinds of hydrocarbons consisting of aliphatics

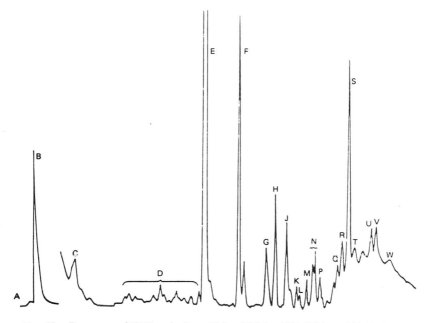

FIG. 58. Pyrogram of PVC resin from 20 ft × 3/16 in. 10% SE32 on 80/100 CRW chromatographic column. Lettered peaks refer to identifications below:

Peak identification	Components
A	CH_4 * CH, * CO_2 *, C_2H_4, * C_2H_6 *
B	HCl, C_3H_6, * C_3H_8 *;
C	Butane[a], butene[a], butadiene[a], diacetylene[a]
D	C_5 and C_6 aliphatic and olefinic hydrocarbons
E	Benzene
F	Toluene
G	Chlorobenzene
H	Xylene
J	Allylbenzene
K	C_9H_{12}
L	C_9H_{12}
M	Indane
N	Indene, ethyltoluene
P	Methylindane
R	Methylindanes
S	Naphthalene
T	Dimethylindane
U	Methylnaphthalene
V	Methylnaphthalene, acenaphthalene
W	Dimethylnaphthalene

*Separated and identified on an 8ft. Poropak QS. HCl ∼ 58.3%, ash 3–4%.

and aromatics, in addition to hydrogen chloride, were detected by Stromberg and other workers.[255,258-261]

Application of pyrolysis–gas chromatography to microstructure determination

Elucidation of short chain branching in polyethylene. Van Schooten and Evenhuis[233,262] applied their pyrolysis (at 500°C) hydrogenation–gas chromatographic technique to the measurement of short chain branching and structural details of three commercial polyethylene samples, a linear polyethylene, a Ziegler low-pressure polyethylene of density 0.945 and a high-pressure low-density polyethylene of density 0.92. Details of the pyrograms are given in Tables 32 and 33. It was observed that the sizes of the iso-alkane peaks increase strongly with increasing amount of short chain branching, i.e. increase as we go from linear polyethylene to high-pressure polyethylene. None of the *n*-alkane peaks in the Ziegler low-pressure polyethylene pyrogram is significantly greater than the corresponding peak in the pyrogram of a linear polyethylene.

The pattern of the increased iso-alkane peaks in the Ziegler polyethylene pyrogram strongly suggests that these peaks are mainly due to ethyl side-groups. This is in good agreement with the results of electron irradiation experiments. For the high-pressure polyethylene Van Schooten and Evenhuis[233,262] found that the *n*-butane peak of the

TABLE 32   *Relative sizes of n-alkane peaks in pyrograms of various polyethylenes*

Peak ratio	Linear polyethylene	Ziegler polyethylene	High-pressure polyethylene
$n\text{-}C_4\text{-}n\text{-}C_7$	0.71	0.74	1.38
$n\text{-}C_5\text{-}n\text{-}C_7$	0.59	0.57	0.75
$n\text{-}C_6\text{-}n\text{-}C_7$	1.52	1.25	1.28
$n\text{-}C_8\text{-}n\text{-}C_7$	0.63	0.65	0.68
$n\text{-}C_9\text{-}n\text{-}C_7$	0.67	0.64	0.72
$n\text{-}C_{10}\text{-}n\text{-}C_7$	1.02	0.91	0.88
$n\text{-}C_{11}\text{-}n\text{-}C_7$	1.00	0.71	0.75

TABLE 33   *Iso-alkane peaks in polyethylene pyrograms*

Sample	Iso-alkane peak
Ziegler polyethylene	i-C_5, 3MC_5, (3MC_6), 3MC_7, 3MC_8, (3MC_9),
High-pressure polyethylene	iC_4, 2C_5, 2MC_5, 3MC_5, 2MC_6, 3MC_6 2MC_7–4MC_7, 3MC_7, 2MC_8–4MC_8, 3MC_8, 4MC_9–5MC_9, 2MC_9, 3MC_9, 4MC_{10}–5MC_{10}, 2MC_{10}–4MC_{10}, 3MC_{10}

pyrogram showed a clear increase in size, and the n-pentane peak a smaller, although probably significant, increase. The increases in the n-butane and n-pentane peaks are probably due to n-butyl and n-pentyl side-groups, respectively, pentyl groups being much less numerous than butyl groups. However, from the pyrograms of the ethylene–butene, ethylene–hexane-I and ethylene–octene-I copolymers it is known than n-butyl side groups give a 2-methyl C_6 peak which is at least equal to the 3-methyl C_7 peak. In the high-pressure polyethylene pyrogram, however, the 3-methyl C_7 peak is more than three times as large. The main part of the 3-methyl C_7 peak, therefore, is probably due to ethyl side-groups. This is in agreement with the sizes of the other 3-methylalkane peaks. It may be concluded that in high-pressure polyethylene the short-chain branches are mainly ethyl and, for a smaller part, n-butyl groups, while some n-pentyl groups may also be present.

Van Schooten and Evenhuis[233,262] concluded that the results obtained by pyrolysis–hydrogenation–gas chromatography appear to be in good agreement with those obtained by infrared and electron irradiation studies. They showed that the pyrograms of a linear polyethylene contained only very small peaks for branched and cyclic alkanes and very large n-alkane peaks. The largest peaks in the pyrogram are those for propane, n-hexane, n-heptane, n-decane and n-undecane, indicating important hydrogen exchange reactions followed by scission with the fifth (C_3 and $n\text{-}C_6$), ninth ($n\text{-}C_7$ and $n\text{-}C_{10}$) and thirteenth ($n\text{-}C_{11}$ and $n\text{-}C_{14}$) carbon atoms. Hydrogen transfer with the sixth carbon atom would

TABLE 34   *Branching frequency of polyethylenes estimated from pyrogram (branches/1000 carbon atoms, from peak surface area ratios)*

Sample	n-Butane	Iso-pentane	3-Methyl pentane	3-Methyl heptane
Marlex 50 ex. Phillips (very little branching – 1 methyl group/1000 carbon atoms)	(I)	(I)	(I)	(I)
Marlex 5003 ex. Phillips (poly-ethylene containing a little copolymerised butane-1)	0.5	4	2	2
Ziegler low-pressure poly-ethylene (about 3–6 branches/1000 carbon atoms)	0.3	7	5	5
Alkathene 2 ex. ICI high-pressure polyethylene (20–30 branches/1000 carbon atoms)	21	19	17	19
Lupolen H ex. BASF high-pressure polyethylene (20–30 branches/1000 carbon atoms)	(24)	(24)	(24)	(24)

account for the rather large $n\text{-}C_4$ peak ($n\text{-}C_4$ and $n\text{-}C_7$), but this peak could also be due to intermolecular chain transfer reactions.

Pyrograms were also prepared for a range of polyethylene containing different amounts of short chain branching (see Table 34 for details of samples). Previous work by high-energy electron irradiation and mass spectrometry has shown that the short branches in high-pressure polyethylenes are mainly ethyl and n-butyl groups, but other short branches have also been supposed to be present.

The pyrograms obtained on these various polymers are shown in Figs 60 and 61 ($n\text{-}C_6$ peak taken as reference). These pyrograms show marked differences which can be attributed to differences in short chain branching. The small amount of branching in Marlex 5003 and Ziegler polyethylenes is reflected only in the somewhat larger iso-alkane peaks, whereas the n-alkane pattern is practically the same as found for Marlex 50 (low branching). The Alkathene 2 and Lupolen H high-pressure polyethylenes show, on the other hand, larger n-butane and n-pentane peaks (Fig. 59). The iso-alkane peaks that show the largest increase (for Alkathane 2 and Lupolen H) are the iso-pentane and 3-methyl alkane peaks (Fig. 60). The results in Figs. 54 and 60 clearly show that the highest amount of branching is present in Lupolen H, the lowest in Marlex 50.

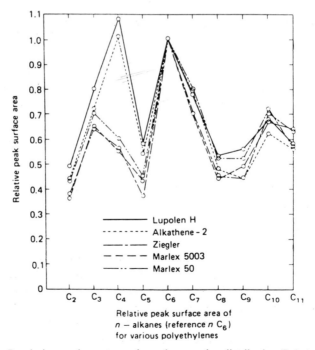

FIG. 59. Pyrolysis–gas chromatography carbon number distribution (Relative peak surface areas of n-alkanes, reference $n\text{-}C_6$) of various polyethylenes.

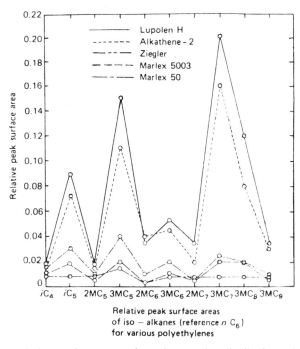

Relative peak surface areas
of iso − alkanes (reference n C_6)
for various polyethylenes

FIG. 60. Pyrolysis–gas chromatography carbon number distribution (relative peak surface areas of iso-alkanes, references n-C_6) of various polyethylenes.

Assuming arbitrarily that these polymers, respectively, contain 24 and 1 short side chains/1000 carbon atoms, and that the relative increase of the n-butane and the iso-alkane peaks is linearly related with the amount of branching, then the branching frequency of the other three samples can be obtained by interpolation. These values are in good agreement with those found in the literature. From the large increase in the n-butane peak and the relatively small increase in the ethane peak it is concluded that the two high-pressure polyethylenes (Alkathene 2 and Lupolen H) contain mainly n-butyl side-chains

Microstructure of ethylene propylene copolymers. High-vacuum pyrolysis at 40°C of ethylene–propylene copolymers carried out by Van Schooten et al.,[229] followed by trapping of released volatiles in dry ice–acetone, produced a mixture of olefins and α-olefins. The most volatile fractions, collected in dry ice–acetone, were hydrogenated to saturated hydrocarbons, which were analysed by gas–liquid chromatography. For samples of copolymer prepared using either titanium trichloride or vanadium oxychloride catalysts the chromatogram of this fraction showed peaks of 2,4-dimethylheptane, 2-methylheptane, 4-methylheptane, 2,4-

dimethylhexane, 3-methylhexane and 2-methylhexane, but only in the chromatogram of the volatile fraction from the copolymer produced using vanadium was a peak of 2, 5-dimethylhexane found. This is an indication that polymers prepared with a catalyst containing vanadium oxychloride contain methylene sequences of two units between branches. Van Schooten et al.[229] conclude that ethylene–propylene copolymers prepared with vanadium-containing catalysts, especially those with $VOCl_3$ or $VO(OR)_3$, have methylene sequences of two and four units.

1, 4/3, 4 Sequence distribution in polyisoprene.   Pyrolysis–gas chromatography has been applied to the investigation of the sequence distribution of the 1, 4 and 3, 4 units in polyisoprenes.[263,264] This method is based on the structural relationship between the isoprene dimers and the diad sequences of 1, 4 and 3, 4 units. It has been pointed out by Tanaka et al.[265] that it is difficult to deduce accurately the polymer structure by this technique, because of two major problems. First, the dimer fraction is only a minor pyrolytic product (about 30%), the monomer being the major product and other products constituting a small percentage. Second, during the pyrolysis the chain-scission reaction is often complicated by side-reactions which can alter the structure of the products.

Short chain branching in ethylene–higher olefin copolymers.   Van Schooten and Evenhuis[233] reported on the application of their pyrolysis (at 500°C)–hydrogenation–gas chromatographic technique to the measurement of short chain branching and structural details in modified polyethylene which contain small amounts of comonomer (about 10% by weight of butene-1, hexane-1, and octene-1). A survey of the isoalkane peaks of the pyrograms of these polymers with their probable assignment is given in Table 35. The effect of the comonomer on the relative sizes of the n-alkane peaks is given in Table 36. The peaks stemming from the total side group appear to be somewhat increased in intensity (n-C_4 peak in the polyethylene–hexane pyrogram, n-C_6 peak in polyethylene–octene pyrogram). The polyethylene–octene pyrogram also shows that the peak stemming from the n-alkane with one carbon atom less is somewhat enlarged.

This suggests that chain scission may occur both at the α and at the β C—C bond. Only the latter type of scission will produce methyl-substituted iso-alkanes, e.g. after intramolecular hydrogen transfer. A list of iso-alkanes that may be expected to be formed in this way is compared in Table 37 with a list of those that have been found to be significantly increased in size in the copolymer pyrograms.

Seeger and Barrall[266] have investigated the applicability of pyrolysis–hydrogenation–gas chromatography to the elucidation of side-chain

TABLE 35 *Areas of the main iso-alkane peaks in the pyrograms of linear polyethylene, ethylene–butene-1, ethylene–hexene-1 and ethylene–octene-1 copolymers*

Component	Peak area in arbitrary units			
	Linear PE	Ethylene–butene-I	Ethylene–hexene-I	Ethylene–octene-I
i-C$_4$	10	20	_37_	_30_
i-C$_5$	18	_47_	62	_46_
2MC	9	14	18	20
3MC$_5$–CyC$_5$	31	_54_	46	76
2MC$_6$	14	13	74	22
3MC$_6$	13	_33_	24	26
2MC$_7$–4MC$_7$	16	17	25	27
3MC$_7$–3EC$_6$	16	_67_	_64_	32
2MC$_8$–4MC$_8$–ECyC$_6$	50	44	38	_119_
3MC$_8$	9	_39_	12	17
4MC$_9$–5MC$_9$–4EC$_8$	11	21	_51_	36
i-PCyC$_6$–BuCyC$_5$			_94_	
2MC$_9$ n-PCyC$_6$	40	35	40	65
3MC$_9$	8	_27_	6	_45_
4MC$_{10}$–5MC$_{10}$ sec BuCyC$_6$	25	23	_67_	32
2MC$_{10}$–4MC$_{10}$–n-BuCyC$_6$	33	30	47	_102_
3MC$_{10}$	15	_32_	14	34

TABLE 36 *Relative sizes of n-alkane peaks for polyethylene and for ethylene–butene-1, ethylene–hexene-1 and ethylene–octene-1 copolymers*

Peak ratio	Marlex 50	Ethylene–butene-1	Ethylene–hexene-1	Ethylene–octene-1
n-C$_4$–n-C$_7$	0.71	0.86	_1.15_	0.86
n-C$_6$–n-C$_7$	0.59	0.69	0.61	_0.91_
n-C$_6$–n-C$_7$	0.52	1.66	1.62	_1.88_
n-C$_8$–n-C$_7$	0.63	0.58	0.66	0.79
n-C$_9$–n-C$_7$	0.67	0.63	0.62	0.75
n-C$_{10}$–n-C$_7$	1.02	0.93	0.99	1.05
n-C$_{11}$–n-C$_7$	1.00	0.82	0.90	0.92

TABLE 37 *Expected and observed increased iso-alkane peaks in copolymer pyrograms*

Ethylene–butene-1		Ethylene–hexene-1		Ethylene–octene		Number of backbone C atoms in fragment
Expected	Observed	Expected	Observed	Expected	Observed	
—	—	—	i-C$_4$	—	i-C$_4$	—
—	—	—	i-C$_5$	—	i-C$_5$	—
i-C$_5$	i-C$_5$	2MC$_6$	2MC$_6$	2MC$_8$	2MC$_8$	3
3MC$_5$	3MC$_5$	3MC$_7$	3MC$_7$	3MC$_9$	3MC$_9$	4
3MC$_6$	3MC$_6$	4MC$_8$	—	4MC$_{10}$	4MC$_{10}$	5
3MC$_7$	3MC$_7$	5MC$_9$	5MC$_9$	5MC$_{11}$	—	6
3MC$_8$	3MC$_8$	5MC$_{10}$	5MC$_{10}$	6MC$_{12}$	—	7
3MC$_9$	3MC$_9$	5MC$_{11}$	—	7MC$_{13}$	—	8

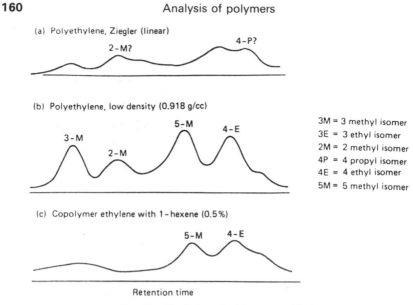

(a) Polyethylene, Ziegler (linear)

2 - M?

4 - P?

(b) Polyethylene, low density (0.918 g/cc)

3 - M

2 - M

5 - M

4 - E

3M = 3 methyl isomer
3E = 3 ethyl isomer
2M = 2 methyl isomer
4P = 4 propyl isomer
4E = 4 ethyl isomer
5M = 5 methyl isomer

(c) Copolymer ethylene with 1 - hexene (0.5 %)

5 - M

4 - E

Retention time

FIG. 61.   Single-branched fragments (C_{10}).

branching in ethylene–hexene-1 copolymers. Figure 61 shows the single-branched fragment pattern up to ten carbon atoms obtained for this copolymer compared with those obtained for Ziegler (linear) polyethylene and low-density (0.918 g/cc) polyethylene.

It is possible not only to measure the kind of branching present, but also to identify the polymer with this simple series. The distribution of isomers varies significantly with the type of branching. In the copolymer with 1-hexene (butyl branches) the 5-methyl and 4-methyl isomers are dominant. The low-density polyethylene (0.918 g/cc) shows a high 3-methyl peak as well as a high 5-methyl and 4-methyl peak yield. This indicates that both ethyl and butyl branches are present in the polyethylene material. Such evidence confirms former suggestions, mainly on the basis of infrared measurements, that low-density polyethylene has branches which are the result of certain intramolecular transfer reactions during high-pressure polymerization.[267]

Diene content of ethylene–propylene–diene terpolymers.   Pyrolysis–gas chromatography has been used to determine the overall composition of ethylene–propylene–diene terpolymers.[268] In attempting to determine the third component in ethylene–propylene–diene terpolymers, difficulties might be anticipated, since this component is normally present in amounts around 5 wt%. However, dicyclopentadiene was identified in ethylene–propylene–diene terpolymers even when the amount incorporated was very low.

Van Schooten and Evenhuis[233,234] applied their pyrolysis (500°C)–hydrogenation–gas chromatographic technique to unsaturated ethylene–propylene copolymers, i.e. ethylene–propylene–dicyclopentadiene and ethylene–propylene–norbornene terpolymers. The pyrograms show that very large cyclic peaks are obtained from the unsaturated ring: methyl cyclopentane is found when methylnorbornadiene is incorporated; cyclopentane when dicyclopentadiene is incorporated; methylcyclohexane and 1, 2-methylcyclohexane when the addition compounds of norbornadiene with, respectively, isoprene and dimethylbutadiene are incorporated; and methylcyclopentane when the dimer of methylcyclopentadiene is incorporated. The saturated cyclopentane rings present in the same ring system in equal concentrations, however, give rise to peaks which are an order of magnitude smaller. Obviously, therefore, the peaks which stem from the termonomer could be used to determine its content if a suitable calibration procedure could be found.

Van Schooten and Evenhuis[233,234] subjected a number of terpolymers containing dicyclopentadiene, and having different amounts of unsaturation, to pyrolysis–gas chromotographic analysis, and plotted the height of the characteristic peaks (or the ratio of the heights of these peaks to the height of the n-C peak) against unsaturation measured by ozone absorption.[269] In Fig. 62 these relationships are given for the cyclopentane peaks. Similar curves were found for the methylcyclopentane or ethylcylopentane peaks.

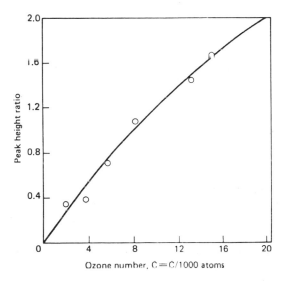

Ozone number, $C = C/1000$ atoms

FIG. 62. Ethylene–propylene–dicyclopentadiene rubbers. Peak height ratio CyC_5/n-C_4 versus ozone number.

Microstructure of polyvinylchloride

It has been recognized since the work of Cotman[270] that by studying the polyolefin obtained from the reduction of PVC with lithium aluminium hydride, valuable structural information can be gained concerning the starting molecule. This reduction reaction has been investigated and refined[271] to the point where conditions have been established so that the chlorine may be efficiently removed from the polymer without degradation. The reduced polymer is similar to high-density polyethylene in almost all respects.[272] Thus, studies that have been applied to polyethylene may also be applied to reduced PVC. Indeed better qualitative agreement has resulted when γ-ray radiolysis followed by identification of the gaseous hydrocarbons by mass spectrometry used in conjunction with infrared measurements is applied to reduced PVC than when this technique is applied to conventional polyethylene. It was concluded from the large yields of methane relative to butane and ethane that the predominant side-chains along the PVC backbone are mainly one carbon long. It has been demonstrated by ^{13}C NMR that most of the short branches in PVC are pendent chloromethyl groups.[273] This information was obtained from PVC samples reduced with lithium aluminium hydride and lithium aluminium denteride respectively.

Ahlstrom *et al.*[271] applied the techniques of pyrolysis–gas chromatography and pyrolysis–hydrogenation–gas chromatography to the determination of short chain branches in PVC and reduced PVC. Their attempts to determine the short chain branches in PVC by pyrolysis–gas chromatography were complicated by an inability to separate all of the parameters affecting the degradation of the polymer. Not only does degree of branching change the pyrolysis pattern, but so do tacticity[274] and crosslinking.[275]

In order to eliminate some of the above parameters, and improve on the then existing methods of polymer characterization, these workers examined the pyrolysis of polyethylene and studied several PVCs and lithium aluminium hydride-reduced PVCs differing in the amount of branch content, as obtained by infrared spectroscopy and ^{13}C NMR, but not in tacticity.

For the pyrolysis of PVC, a ribbon probe was used. On-line hydrogenation of the pyrolysis products was accomplished by using hydrogen as the carrier gas with 1% palladium on Chromosorb-P catalyst inserted in the injection port liner. Maximum triplet formation occurred at C_{14} for low-density polyethylene and for reduced PVC, and at C_{15} for high-density polyethylene. The occurrence of the peak maxima at C_{14} for reduced PVC indicates that the total branch content is higher than that of high-density polyethylene. But aside from the C_{14}, C_{15} peak maxima difference, the pyrolysis pattern for even the most highly branched

TABLE 38 *Relative total branch content of high-density polyethylene, LAH-reduced PVC and low-density polyethylene*

Sample	Branched products (%)
High-density polyethylene	12.0
Reduced PVC*	19.0
Low-density polyethylene	26.0

*Average value.

PVCs resembles high-density polyethylene more than low-density polyethylene. These data indicate that the type of short-chain branch in PVC is qualitatively more like that of high-density polyethylene, but that the sequence length between branch sites is shorter in low-density polyethylene and PVC. Since low-density polyethylene contains a large amount of ethyl and butyl branches, and PVC and high-density polyethylene contain mainly methyl and some ethyl branches, this qualitative resemblance would be expected. A relative measure of the total amount of short-chain branches for these polymers can be obtained by calculating the percentage of branched products formed (Table 38).

More information about the specific type of short-chain branch in PVC can be found from an examination of the C_1—C_{11} hydrocarbons (Fig. 63). Here quantitative differences between the reduced PVCs become more apparent. The most obvious differences occur in the amounts of iso-C_7 and iso-C_8 products formed, which indicate differences in the total branch content. As the amount of short-chain branching (C_{10}%) in the reduced PVCs increases, there is a decrease in the amount of iso-alkanes formed (Table 39). The data in Table 39 show small but distinguishable differences in the short-chain branch content of the reduced PVCs.

5.5(b) Infrared spectroscopy

The infrared absorption of ethylene copolymers in the 700–850 cm^{-1} (14.28–11.76 μ) region can provide information about their sequence distributions. Studies on model hydrocarbons by Van Schooten and others[276-279] have shown that the absorption of methylene groups in this region is dependent on the size of methylene sequences in the compounds. Methylene absorptions observed in this region, and their relation to structures occurring in ethylene copolymers, are shown in Table 40.

The absorptions at 724 and 731 cm^{-1} (13.81 and 13.68 μ) of several ethylene–propylene copolymers have been assigned[280] to structures (6) and (4), respectively (Table 40). The relative intensities observed were qualitatively consistent with sequence distributions calculated for the

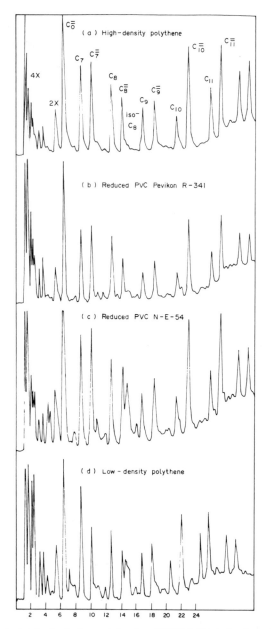

Fig. 63.   Pyrolysis of polyethylenes and reduced PVC; (a) high-density polyethylene; (b) reduced PVC Pevikron R-341; (c) reduced PVC N-E-54; (d) low-density polyethylene. Durapak column fragments C_1–C_{11}.

TABLE 39   *Short-chain branch content in LAH-reduced PVC*

Sample	iso-$C_8^=$/n-$C_8^=$	$C_{10}^=$ (%)
High-density polyethylene	0.1	8.6
Pevikron R-341	0.52	7.7
Nordforsk E-80	0.71	7.2
Nordforsk S-80	1.20	7.0
Nordforsk S-54	1.59	6.8
Nordforsk E-54	1.63	6.5
Ravinil R100/650	1.63	6.5
Shin-Etsu TK1000	1.87	6.3
Low-density polyethylene	1.83*	5.2

*This ratio does not include C_4 branch content.

TABLE 40   *Methylene absorption and copolymer structure responsible for absorption in ethylene copolymers*

—$(CH_2)_n$—	cm^{-1}	μ	Sequence	
—$(CH_2)_1$	815	11.76	~ $CH_2CHRCH_2CHRCH_2CHR$ ~	(2)
$(CH_2)_2$	751	13.31	~ $CH_2CHRCH_2CH_2CHRCH_2$ ~	(3a)
			or	
			~ $CHRCHRCH_2CH_2CHRCHR$ ~	(3b)
$(CH_2)_3$	733	13.64	~ $CH_2CHRCH_2CH_2CH_2CHR$ ~	(4)
$(CH_2)_4$	726	13.77	~ $CH_2CHR(CH_2CH_2)_2CHRCH_2$ ~	(5)
			or	
			~ $CHRCHR(CH_2CH_2)_2CHRCHR$ ~	(5b)
$(CH_2)_4$	722	13.85	~ $CH_2CHR(CH_2CH_2)_nCH_2CHR$ ~	(6)

copolymers, assuming reactivity ratios of 7.08 and 0.088 for ethylene and propylene, respectively. Veerkamp and Veermans[281] developed a technique to measure the intensities of these absorptions accurately. By assuming similar extinction coefficients for the two bands, the ratio of methylene units in $(CH_2)_3$ and larger methylene sequences in a number of copolymers was determined. The results agreed reasonably well with theoretical values. The presence of $(CH_2)_2$ sequences in ethylene-propylene copolymers has been considered in studies of their structure.[282] Such sequences could result from the tail–head incorporation of propylene units into the copolymers (structure (3a), Table 40).

Bucci and Simonazzi[283] also investigated the assignment of infrared bands of ethylene–propylene copolymers in the spectral region 900–650 cm^{-1} (11.11–15.38 μ). They calculated absorbances at various frequencies and attempted a numerical evaluation of the distribution of monomeric units in the copolymer. The spectra of ethylene–propylene copolymers in this frequency range show peaks or shoulders at 815, 752, 733 and 722 cm^{-1} (12.27, 13.30, 13.64 and 13.85 μ), whose intensity changes with composition (Fig. 64). For the assignment of these bands they

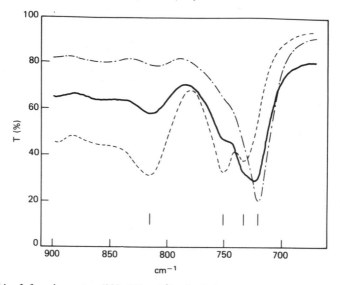

FIG. 64. Infrared spectra (900–670 cm⁻¹) of ethylene–propylene copolymers at various compositions: (–·–·25% (3 mol.)); (———50% (3 mol.)); (– – – –75% (3 mol.)).

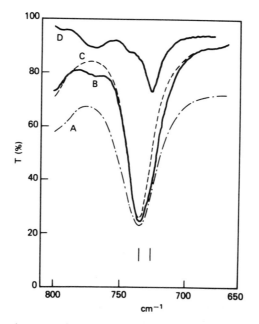

FIG. 65. Infrared spectra of model compounds: (A) hydrogenated natural rubber; (B) squalane; (C) 2, 6, 10, 14-tetramethylpentadecane; (D) squalane compensated with 2, 6, 10, 14-tetramethylpentadecane.

TABLE 41   *Assignment of infrared bands of ethylene–propylene copolymers*
(frequency, cm⁻¹)

Sequences and compounds	Bucci and Simonazzi[283]	McMurray and Thornton[284]	Van Schooten et al.[282]	Natta et al.[280,285]
$(-CH_2-)_1$	815	785–770		
	$(12.27\,\mu)$	$(12.74–12.99\,\mu)$		
Atactic polypropylene	815			
	$(12.27\,\mu)$			
$(-CH_2-)_2$	752	743–734		
	$(13.30\,\mu)$	$(13.86–13.62\,\mu)$		
2,5-Dimethylhexane			753	
			$(13.28\,\mu)$	
Ethylene–*cis*-butene-2 copolymer	752			752
	$(13.30\,\mu)$			$(13.30\,\mu)$
Hydrogenated poly-2,3-dimethyl-butadiene			752	
			$(13.30\,\mu)$	
$(-CH_2-)_3$	733–735	729–726		
	$(13.64–13.60\,\mu)$	$(13.72–13.77\,\mu)$		
2,6,10,14-Tetramethylpentadecane	735			
	$(13.60\,\mu)$			
Squalane	735			
	$(13.60\,\mu)$			
Hydrogenated natural	735		739	
	$(13.60\,\mu)$		$(13.53\,\mu)$	
$(-CH_2-)_4$	726	726–721	728	
	$(13.77\,\mu)$	$(13.77–13.86\,\mu)$	$(13.73\,\mu)$	
Squalane compensated with 2,6,10,14-tetramethylpentadecane	722–721	724–722	722	
$(-CH_2-)_5$ or more	$(13.85–13.86\,\mu)$	$(13.81–13.85\,\mu)$	$(13.85\,\mu)$	
n-Heptane	722			
	$(13.85\,\mu)$			
n-Decane	721			
	$(13.86\,\mu)$			
n-Nonadecane	721			
	$(13.56\,\mu)$			

TABLE 42   *Assignment of infrared bands of ethylene–propylene copolymers*

Sequences	Absorption frequency		Model compounds	Absorptivities for $CH_2(Kvi)$ $(cm^{-1}\,mol^{-1}\,cm^{-1}\,ml)$
	cm⁻¹	μ		
$(-CH_2-)_1$	815	12.27	Atactic polypropylene with head-to-tail linking	$0.6\,0.10 \times 10^4$
$(-CH_2-)_2$	752	13.30	Atactic polypropylene with a definite amount of head-to-head linking*	$1.0\,0.15 \times 10^4$
$(-CH_2-)_3$	733	13.34	Hydrogenated natural rubber, squalane, 2,6,10,14-tetra-methylpentadecane	$1.2\,0.20 \times 10^4$
$(-CH_2-)_5$	722	13.85	Linear C_{10}—C_{19} hydrocarbons	$1.2\,0.25 \times 10^4$

examined spectra of several model compounds which contain $(-CH_2-)_n$ sequences with different values of n (Table 41). In the same table they compare their results with those given in the literature[284,277,282,255]. On the basis of this comparison they assign the bands at 815, 752, 733 and 722 cm^{-1} to the $(-CH_2-)$ sequences with $n = 1, 2, 3$, and 5 or more, respectively (see Table 42).

This assignment does not agree with that proposed by Van Schooten and co-workers, who assign bands at approximately these wavelengths to $(-CH_2-)_n$ sequences where $n = 1, 2, 3$, and 4.

Bucci and Simonazzi[283] think that the difference lies in the fact that these authors assigned the band at 733 cm^{-1} to the sequence $(-CH_2-)_4$. Bucci and Simonazzi were not able to detect in ethylene–propylene copolymer any absorption at 725 cm^{-1} (13.79 μ) where the sequence $(-CH_2)_4$ should occur. They recorded the spectrum of squalane compensated with 2, 6, 10, 14-tetramethylpentadecane, and did find a band due to $(-CH_2-)_4$ at 726 cm^{-1} (13.77 μ) (Fig. 65).

Natta et al.[280] determined the degree of alternation of ethylene and propylene units in ethylene–propylene copolymers from the infrared spectrum, using peaks at 750, 731 and 724 cm^{-1} (13.35, 13.70 and 13.83 μ), the one at 731 cm^{-1} being attributed to a sequence of three methylene groups between branch points, presumably due to the insertion of one ethylene between two similarly oriented propylene molecules.

In order to check the correctness of this interpretation Van Schooten et al.[277] examined the infrared absorption spectra between 700 and 770 cm^{-1} (13 and 14 μ) of ethylene–propylene copolymers prepared with various catalyst systems, and compared them with the spectra of some model compounds, namely 2, 5-dimethylhexane, 2, 7-dimethyloctane, 4-methylpentadecane, 4-n-propyltridecane, polypropylene, polyethylene, polybutene-1 and hydrogenated polyisoprene, the last being considered as an ideally alternating ethylene–propylene copolymer.

The infrared absorption bands between 700 and 770 cm^{-1} of the various samples are given in Fig. 66. These bands are due to rocking modes of the CH_2 groups, and their frequency depends on the length of the CH_2 sequences. In 2, 5-dimethylhexane, hydrogenated polyisoprene and 2, 7-dimethyloctane there are two, three and four CH_2 groups between branches, respectively, corresponding to peaks in the infrared spectra at 754 cm^{-1} (13.28 μ) (ε spec. = 0.0231/g.cm), 740 cm^{-1} (13.53 μ) (ε spec. = 0.058 1/g. cm) and 729 cm^{-1} (13.75μ) (ε spec. = 0.042 1/g.cm) Amorphous and crystalline polypropylene do not show any absorption at all in this region, whereas crystalline polyethylene shows two peaks at 734 and 723 cm^{-1} (13.65 and 13.85 μ). The latter peak is also found in liquid low molecular weight hydrocarbons,[286] and in amorphous polymers containing long sequences of CH_2 groups;[286] the 734 cm^{-1} peak is a crystalline band, attributed to the CH_2 rocking in the polyethylene crystal.[286-288] From the spectra of an amorphous polybutene-1 it was concluded that ethyl side-groups give rise to a peak at around 765 cm^{-1} (13.1 μ) and from the spectra of 4-methylpentadecane and 4-n-propyltridecane that n-propyl groups absorb at 740 cm^{-1} (13.53 μ), proving that the shoulder at 752 cm^{-1} (13.3 μ) found in some polymers is not due to ethyl or n-propyl end-groups. A survey of the various peaks and shoulders present in the spectra is given in Table 43.

In the spectra of all copolymers, except sample 6, they found a sharp peak at 725–722 cm^{-1} (13.80–13.85 μ), which must be ascribed to

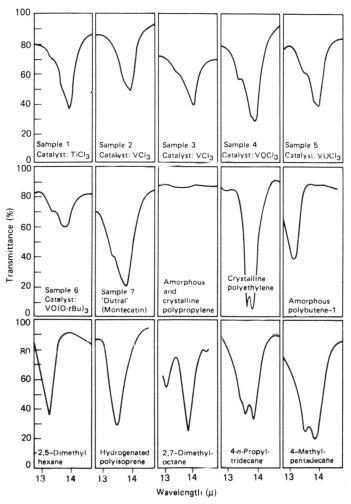

FIG. 66. Infrared spectra between 700 and 770 cm^{-1} (13–14 μ) of C$_2$–C$_2$ copolymers and model compounds.

TABLE 43   *Peaks and shoulders between 770 and 770 cm⁻¹ (13–14 μ) in infrared spectra of C₂—C₃ copolymers*

Sample	Frequency (cm⁻¹) wavelength				
	13.8–13.9 μ (725–719 cm⁻¹)	13.7–13.8 μ (730–725 cm⁻¹)	13.6–13.7 μ (735–730 cm⁻¹)	13.50–13.60 μ (741–731 cm⁻¹)	13.0–13.35 μ (769–749 cm⁻¹)
2,5-Dimethylhexane	—	—	—	—	peak at 13.28 μ (753 cm⁻¹)
Hydrogenated polyisoprene	—	—	—	peak at 13.53 μ (739 cm⁻¹)	—
2,7-Dimethyloctane	—	peak at 13.73 μ (728 cm⁻¹)	—	—	—
Amorphous butene-1 polymer	—	—	—	—	peak at 13.12 μ (762 cm⁻¹)
4-n-Propyltridecane	peak at 13.87 μ (721 cm⁻¹)	—	—	peak at 13.57 μ (737 cm⁻¹)	—
4-Methylpentadecane	peak at 13.85 μ (722 cm⁻¹)	—	—	peak at 13.53 μ (739 cm⁻¹)	—
Cryst. polypropylene					
Cryst. polyethylene	peak at 13.85 μ (722 cm⁻¹)	—	peak at 13.65 μ (733 cm⁻¹)	—	—
Sample 1 TiCl₃ catalyst	peak at 13.86 μ (721 cm⁻¹)	—	—	—	vague shoulder
Sample 2 VCl₃ catalyst	peak at 13.85 μ (722 cm⁻¹)	—	—	vague shoulder around 13.60 μ (735 cm⁻¹)	—

Sample 3 VCl₃ catalyst	peak at 13.86 μ (721 cm⁻¹)	—	—	—	shoulder at 13.3 μ (752 cm⁻¹)
Sample 4 VOCl₃ catalyst	peak at 13.35 μ (722 cm⁻¹)	vague shoulder at 13.7 μ (730 cm⁻¹)	—	—	pronounced shoulder at 13.3 μ (752 cm⁻¹)
Sample 5 VOCl₃ catalyst	peak at 13.85 μ (721 cm⁻¹)	shoulder at 13.75 μ (727 cm⁻¹)	—	—	shoulder at 13.35 μ (749 cm⁻¹)
Sample 6 VO(0-t-Bu)₃ catalyst	—	broad peak at 13.75 μ (727 cm⁻¹)	—	—	peak at 13.35 μ (749 cm⁻¹)
Sample 7 Dutral	peak at 13.82 μ (723 cm⁻¹)	—	—	—	pronounced shoulder at 13.3 μ (752 cm⁻¹)

sequences of CH_2 groups longer than four, i.e. sequences of more than two ethylene units.

Significant differences are, however, observed between the various spectra in the wavelength region from 769 to 73 cm^{-1} (13.0 to 13.7 μ). None of the copolymer spectra shows a clear shoulder at 739 cm^{-1} (13.53 μ), where hydrogenated polyisoprene shows maximum absorption.

This means that in all the samples the content of $(CH_2)_3$ sequences is low. There are, however, several samples showing a pronounced shoulder at around 752 cm^{-1} (13.3 μ), viz. samples 3, 4, 5, 6 and 7 (Table 43). This shoulder should probably be assigned to sequences of two CH_2 groups between branch points (cf. spectrum of 2, 5-dimethylhexane). Of these samples two have been prepared with a $VOCl_3$-containing catalyst and one with a $VO(0\text{-}t\text{-Bu})_3$-containing catalyst. Only a few of the samples prepared with a catalyst containing VCl_3 (No. 3), showed the shoulder at 752 cm^{-1} (13.3 μ). This shoulder was always observed in samples which had been prepared with a catalyst obtained from $VOCl_3$ or $VO(OR)_3$. It might well be that this band arises from the presence of $(CH_2)_4$ sequences. This holds for samples 4, 5 and 6, which were prepared with catalysts containing $VO(0\text{-}t\text{-Bu})_3$ or $VOCl_3$ and which, as we have seen, also display absorption bands that indicate the presence of C_2 sequences. None of the copolymer spectra obtained thus far indicates the presence of the structure.

which would give rise to absorption at 1125 cm^{-1} (8.9 μ). This means that head-to-head orientation of propylene occurs only after addition of an ethylene unit.

Van Schooten et al.[277] have shown that ethylene–propylene copolymers prepared with $VOCl_3$- or $VO(OR)_3$-containing catalysts not only contain odd-numbered methylene sequences, as was expected, but also sequences of two units. Some indications for the presence of methylene sequences of four units were found. They[282] also investigated copolymers with much lower ethylene contents and pure polypropylenes prepared with catalysts consisting of $VOCl_3$ or $VO(OR)_3$ and alkylaluminium sesquichloride.

The infrared spectra of the ethylene–propylene copolymers containing 15 to 30 mole% of ethylene show little or no absorption above 730 cm^{-1} (13.70 μ) (Fig. 67). This means that very few methylene sequences of four or more units are present. The absorption peaks at 752 cm^{-1} and 735 cm^{-1} (13.30 and 13.60 μ) reveal the presence of sequences of two and

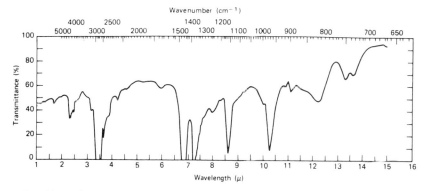

FIG. 67.   Infrared spectrum of ethylene–propylene copolymer containing 85 mode%
propylene.

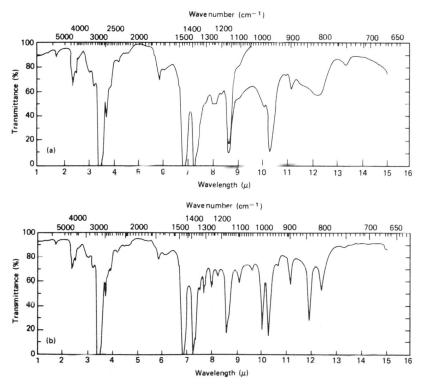

FIG. 68.   Infrared spectrum of polypropylene prepared with vanadyl-based catalyst:
(a) amorphous part, (b) crystalline part.

three units respectively. Thus these copolymers contained nearly all of the ethylene in isolated monomer units.

The earlier work[277] indicated that polypropylenes obtained with vanadyl catalysts possessed the normal head-to-tail structure. More detailed examination, however, showed that the amorphous fractions isolated from these polypropylenes do show absorption at $752\,cm^{-1}$ $(13.3\,\mu)$ pointing to the presence of methylene sequences of two units, which can only mean that tail-to-tail arrangement of propylene units does occur (Fig. 68). A very small absorption peak at $752\,cm^{-1}$ $(13.3\,\mu)$ was also found in the spectrum of the crystalline fractions (Fig. 68). Polypropylenes prepared with catalysts based on VCl_3 show only the normal head-to-tail arrangement in both amorphous and crystalline fractions, as do polymers prepared with $TiCl_3$ catalysts.

The amount of propylene units coupled tail-to-tail was estimated to range from about 5 to 15% for the amorphous and from about 1 to 5% for the crystalline fractions. The amount of propylene units in tail-to-tail arrangement was calculated from spectra of thin films by comparing the ratio of the absorbances at $752\,cm^{-1}$ and $1156\ cm^{-1}$ $(13.30\ \text{and}\ 8.65\,\mu)$ (methyl band) in the spectra of the polypropylenes with the ratio of the absorbances at $752\,cm^{-1}$ $(13.60\,\mu)$ and $1156\ cm^{-1}$ $(8.65\,\mu)$ in the spectrum of hydrogenated natural rubber. This implies the assumption that the absorbance per CH_2 group is the same at $752\,cm^{-1}$ $(13.30\,\mu)$ for $(CH_2)_2$ sequences as at $735\,cm^{-1}$ $(13.60\,\mu)$ for $(CH_2)_3$ sequences.

The differences in amount of tail-to-tail coupled units between crystalline and amorphous fractions are to be expected, since every head-to-head and tail-to-tail configuration disturbs the regularity of the isotactic chain. In the polypropylenes every tail-to-tail configuration must necessarily be accompanied by a head-to-head coupling. This would be expected to show up in an absorption peak at $1137\,cm^{-1}$ $(8.8\,\mu)$ to $1111\,cm^{-1}$ $(9.0\,\mu)$, characteristic of the structure

which is also found in hydrogenated poly-2, 3-dimethylbutadiene, used as a model compound and in alternating copolymers of ethylene and butene-2.[285] In the polypropylenes examined by Van Schooten and Mostert,[282] and in the copolymers, they did indeed find an absorption band near $1111\,cm^{-1}$ $(9.0\,\mu)$, although, unlike Van Schooten *et al.*[282] (see above) it is much less sharp than in the model compound.

All spectra containing the $752\,cm^{-1}$ $(13.3\,\mu)$ peak show a further small band at $917\,cm^{-1}$ $(10.9\,\mu)$ which is also found in the spectrum of poly-2, 3-dimethylbutadiene. The spectrum of the amorphous alternating copolymer of ethylene and butene-2, published by Natta,[285] shows a band at about $726\,cm^{-1}$ $(10.8\,\mu)$. The fact that the polypropylenes prepared with $VOCl_{3-}$ or $VO(OR)_{3-}$ containing catalysts show tail-to-tail arrangement means that tail-to-tail coupling of propylene units may also occur in the ethylene–propylene copolymers. However, because the content of $(CH_2)_2$ sequences in the copolymers is much higher than in the polypropylenes prepared with the same catalysts, a large part of these sequences most likely stems from isolated ethylene units between two head-to-head oriented propylene units, their relative amount depending on the ratio of reaction rates of formation of the sequences.

$$
\begin{array}{c}
\mid \quad\;\; \mid \quad\;\; \mid \quad\;\; \mid \\
+\,C\!-\!C\,+\,C\!-\!C\,+\,C\!-\!C\,+ \\
\mid \quad\;\; \mid \quad\;\;\;\;\; \mid \quad\;\; \mid \\
\;\;\;\;\; C \quad\quad\quad\;\; C \\
\;\;\;\;\; \mid \quad\quad\quad\;\; \mid \\
\mid \quad\;\; \mid \quad\;\; \mid \quad\;\; \mid \\
+\,C\!-\!C\,+\,C\!-\!C\,+ \\
\mid \quad\;\; \mid \;\; \mid \quad\;\; \mid \\
\;\;\;\;\; C \;\; C
\end{array}
$$

Veerkamp and Veermans[281] have reported on the differential measurement of $(CH_2)_2$ and $(CH_2)_3$ units at $745\,cm^{-1}$ $(13.42\,\mu)$ and $630\,cm^{-1}$ $(15.87\,\mu)$, respectively, in ethylene–propylene copolymers. The band at $720\,cm^{-1}$ $(13.89\,\mu)$ was assigned to units of 5 or more CH_2 units. Thus, the absorbance ratio $A_{745cm^{-1}}/A_{720cm^{-1}}$ provides a measure of the number of isolated ethylene units, and when corrected for the effect of composition (by multiplying by the ratio mole% C_3/mole% C_2) is related to the inverse of the ratio $A_{968cm^{-1}}/A_{1150cm^{-1}}$ times the ratio mole% C_2/mole% C_3. The more random the copolymer, as indicated by the absorbance ratio $A_{745cm^{-1}}/A_{720cm^{-1}}$ for the relative number of isolated ethylene units, the more diffuse the $968\,cm^{-1}$ band becomes.

In summary, the appearance of contiguous vs. isolated monomer units may be studied by the spectral characteristics of the methyl and methylene group rocking and wagging bands. The C—H stretching frequencies are useful for measurement and copolymer composition when spectrophotometers of sufficient resolution are available. A typical infrared spectrum of a cast film of ethylene–propylene copolymer, showing the base line, is reproduced in Fig. 69. The ratio between the CH_3 and CH_2 asymmetrical C—H stretching band was not used because of the large difference in relative intensities. The ratio between the asymmetrical CH_3 and the symmetrical CH_2 bands, which have nearly the same relative intensities, produced more reliable results. Resolution of the CH_3 from the CH_3 asymmetrical as well as the symmetrical bands must be satisfactory, as no curvature of the calibration curve was seen. If the CH_3

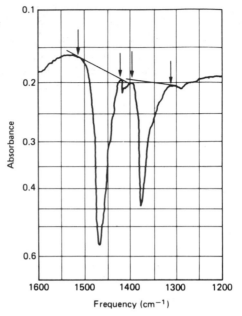

Fig. 69. Typical infrared spectrum of an ethylene–propylene copolymer in the C–H bending region of the spectrum. Base line points shown by arrows. Grating to prism changeover at 1416 cm^{-1} (7.06 μ).

symmetrical band was not resolved from the CH$_2$ band, a change in the slope of the curve at high propylene contents would have been observed.

Liang et al.,[289] on the basis of frequencies, relative intensities, polarization properties, and effects on deuteration of polypropylene, have tentatively assigned the 968 cm^{-1} (10.33 μ) band in polypropylene to the methyl rocking mode mixed with CH$_2$ and CH rocking vibrations. From this assignment, perhaps it is a change in the magnitude of the mixing with the CH$_2$ and CH modes in the case of the isolated propylene units in the copolymer which gives rise to a more diffuse absorption band at 968 cm^{-1} (10.33 μ). In this connection, Liang and Watt[290] prepared non-stereospecific polypropylene in which the 968 cm^{-1} (10.33 μ) and 1150-cm^{-1} (8.69 μ) bands are either missing or very weak. Also, the occurrence of head-to-head arrangements of contiguous propylene units would be expected to influence the behaviour of the methyl rocking mode and any mixing which may be involved.

For the purpose of polymer characterization the ratio of absorbances at 968 and 1150 cm^{-1} should provide a measure of the degree of randomness with respect to the introduction of propylene units into the copolymer. On the basis of the initial observations, the ratio $A_{968\,cm^{-1}}/A_{1150\,cm^{-1}}$ should decrease as the randomness increases. In addition, an increase in ethylene content is expected to increase the probability of producing propylene units isolated by ethylene units. For comparison of the 968 cm^{-1} to 1150 cm^{-1} absorbance ratio with other data, correction for the effect of composition is made by multiplying by the ratio mole% C_2/mole% C_3, the molar ratio of ethylene to propylene in the copolymer. Relationships have been found between this modified absorbance ratio and other established physical or spectral properties indicative of randomness (or conversely, block polymer).

Determination of ethylene and propylene contents of ethylene–propylene copolymers

Several methods have been reported for determining propylene in ethylene–propylene copolymers. The basis for calibration of many of these methods is the work published by Natta *et al.*,[291] which involves measuring the infrared absorption of the polymer in carbon tetrachloride solutions. The absorption at 1379 cm^{-1} (7.25 μ) presumably due to methyl vibrations, is related to the propylene concentration in the copolymer. In some cases the dissolution of copolymers with low propylene content or some particular structures is difficult.[292,293] Moreover, Natta's solution

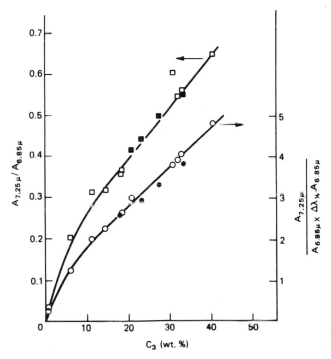

Fig. 70. Calibration curves for the determination of C_3 in high-C_2 copolymers. Full points refer to model hydrogenated poly-3 methyl alkenamers.

method was calibrated against his radiochemical method,[291] for which the precision of the method was not stated; a considerable amount of scatter is evident in the data presented. Typical methods that have used Natta's solution procedure[291] for calibration are described in publications by Wei[294] and Gossl.[295] These infrared methods avoid the solution problem by employing intensity measurements made on pressed films. The ratio of the absorption at $717 \, cm^{-1}$ $(13.95 \, \mu)$ to that at $1149 \, cm^{-1}$ $(8.70 \, \mu)$ is related to the propylene content of the copolymer. Some objections[296] to the use of solid films have been raised, because of the effect of crystallinity on the absorption spectra in copolymers with low propylene content. These film methods are reliable only over the range of 30–50 mole% propylene.

Tosi and Simonazzi[297] have described an infrared method for the evaluation of the propylene content of ethylene-rich ethylene–propylene copolymers. This is based on the ratio between absorbances of the $1379 \, cm^{-1}$ $(7.25 \, \mu)$ band and the product of the absorbances by the half-width of the $1459 \, cm^{-1}$ $(6.85 \, \mu)$ obtained on diecast polymer film at $160°C$.

The calibration curve, based on a series of standard copolymers prepared with ^{14}C-labelled either ethylene or propylene, as shown in Fig. 70, is obtained by plotting the $A_{7.25}/A_{6.85}$ ratio against the C_3 weight fraction. To improve precision Tosi and Simonazzi[297] ratioed $A_{7.25}\mu$ to the product of the absorbance of the $6.85 \, \mu$ band by its half-width. For the sake of simplicity the latter quantity was expressed in wavelengths

TABLE 44 *Calibration points for the composition analysis of C_2—C_3 copolymers (spectra recorded at 160°C)*

Sample	C_3 (wt%)*	$A_{7.25}/A_{6.85}$	(%)†	$\dfrac{A_{7.25}}{A_{6.85}}$	(%)†
3137–25	40.5	0.652 ± 0.011	1.7	4.80 ± 0.20	4.1
3629–48	32.9	0.561 ± 0.053	9.5	4.07 ± 0.06	1.5
3629–44	32.0	0.548 ± 0.032	5.8	3.92 ± 0.06	1.5
3137–38	30.7	$0.606 \pm 0.024^{\ddagger}$	4.0	3.80 ± 0.07	1.9
3137–31	20.7	0.416 ± 0.018	4.2	3.02 ± 0.05	1.7
3297–55	18.2	0.365 ± 0.037	10.0	2.62 ± 0.09	3.4
3297–59	18.0	0.357 ± 0.015	4.3	2.59 ± 0.07	2.7
3274–53	14.2	0.318 ± 0.015	4.8	2.25 ± 0.04	1.8
Blend No. 1	11.0	0.315 ± 0.022	6.8	2.00 ± 0.13	6.5
Blend No. 2	6.0	0.204 ± 0.013	6.4	1.25 ± 0.06	5.0
Polyethylene	0.7	0.036 ± 0.007	18.5	0.28 ± 0.05	20.3

*Radiochemical analysis.
† The standard deviation has been multiplied by the Student's $t_{95\%}$ coefficient to the number of replications (on the average five) for each calibration point.

instead of wavenumbers: $\Delta\lambda_{\frac{1}{2}}A_{6.85}$ is approximately $0.10\,\mu$ at room temperature and 0.14 at 160°C.

Table 44 lists the calibration data of Tosi and Simonazzi.[297] Tosi and Simonazzi[297] compared the upper curve of Fig. 70 to calibration curves obtained by other workers for C_2—C_3 copolymers based on the $A_{7.25}/A_{6.85}$ ratio. This comparison, when the C_3 content is not too low, should hold irrespective of the recording temperature: in fact they found only minor differences in the $A_{7.25}/A_{6.85}$ ratio at room temperature and at 160°C.

Ciampelli et al.[298] have developed two methods based on infrared spectroscopy of carbon tetrachloride solutions of the polymer at $1379\,\text{cm}^{-1}$ $(7.25\,\mu)$, $1156\,\text{cm}^{-1}$ $(8.65\,\mu)$ and $4310\,\text{cm}^{-1}$ $(2.32\,\mu)$ for the analysis of ethylene–propylene copolymers containing greater than 30% propylene. One method can be applied to copolymers soluble in solvents suitable for infrared analysis; the other can be applied to solvent-insoluble polymer films. The absorption band at $1379\,\text{cm}^{-1}$ $(7.25\,\mu)$ due to methyl groups is used in the former case, whereas the ratio of the band at $1162\,\text{cm}^{-1}$ $(8.6\,\mu)$ to the band at $4310\,\text{cm}^{-1}$ $(2.32\,\mu)$ is used in the latter. Infrared spectra of polymers containing between 55.5 and 85.5% ethylene are shown in Fig. 71. This approach has also been discussed by Bucci and Simonazzi[299]. This method is based on measurement of absorption of solutions of the polymer at $5910\,\text{cm}^{-1}$ $(1.692\,\mu)$ $(CH_3$ groups $5668\,\text{cm}^{-1}$ $(1.764\,\mu)\,(CH_2$ groups), and is applicable to polymers containing as high as 52 mole% propylene. Lomonte and Tirpak[300] have developed a method for the determination of percent age ethylene incorporation in ethylene–propylene block copolymers. Standardization is done from mixtures of the homopolymers. Both standards and samples are scanned at 180°C in a spring-loaded demountable cell. The standardization was confirmed by the analysis of copolymers of known ethylene content prepared with ^{14}C-labelled ethylene. By comparison of the infrared results from the analyses performed at 180°C and also at room temperature, the presence of ethylene homopolymer can be detected. These workers derived an equation for the quantitative estimation of percentage ethylene present as copolymer blocks. The method distinguishes between true copolymers and physical mixtures of copolymers, and makes use of a characteristic infrared rocking vibration due to sequences of consecutive methylene groups. Such sequences are found in polyethylene and in the segments of ethylene blocks in ethylene–propylene copolymers. This makes it possible to detect them at $730\,\text{cm}^{-1}$ $(13.70\,\mu)$ and $720\,\text{cm}^{-1}$ $(13.89\,\mu)$. There are bands at both these locations in the infrared spectrum of the crystalline phase but only at $720\,\text{cm}^{-1}$ $(13.89\,\mu)$ in the amorphous phase. The ratio of these two bands in the infrared spectrum of a polymer film at room temperature is a rough measure of crystallinity. As seen by this ratio, the infrared spectra of the copolymers show varying degrees of polyethylene-type crystallinity,

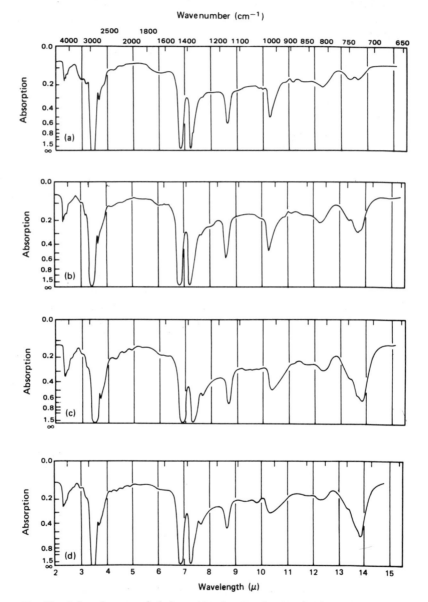

FIG. 71.   Infrared spectra of ethylene–propylene copolymers of various compositions:
(A) 85.5% polyethylene; (B) 74.0% polyethylene; (C) 65.9% polyethylene; (D) 55.5%
polyethylene.

dependent on ethylene concentration and method of incorporation. It is this varying degree of crystallinity which allows the qualitative detection of ethylene homopolymer in these materials. A calibration curve of absorbance at $720 \, \mathrm{cm}^{-1}$ ($13.89 \, \mu$) versus ethylene was made from known mixtures for both hot and cold runs. Both plots resulted in straight lines from which the following equations were calculated.

$$\text{Percentage ethylene at } 180°C = A/(0.55b)$$

where A = absorbance measured at $720 \, \mathrm{cm}^{-1}$ and b = thickness of wire spacer in centimetres and

$$\text{Percentage ethylene at room temperature} = A/(3.0b)$$

A series of ethylene–propylene block copolymers prepared with ^{14}C-labelled ethylene was analysed for percentage ethylene incorporation by radiochemical methods. These samples, when scanned at 180°C, gave results which checked with the radiochemical assay quite well. However, when the cooled samples were scanned the results from the cold calibration were low in comparison with the known ethylene content. These data are shown in Table 45.

A pair of samples were prepared in which the active sites on the growing propylene polymer were eliminated by hydrogen before addition of ethylene. Practically identical values for percentage ethylene incorporation were calculated for both the hot and the cold scans. These data are shown in Table 46. Lomonte and Tirpak[300] discuss the implications of these findings.

Developments in the field of applying infrared spectroscopy to the elucidation of sequence distribution in polymers are continuing. As has

TABLE 45 ^{14}C labelled ethylene-propylene copolymers (ethylene, %)

Sample No.	Radiochemistry	Hot infrared scan	Cold infrared scan
3401	2.4	2.9	0.9
3402	4.0	3.65	1.3
3403A	22.4	20.7	14.7
3403B	24.5	22.2	15.3
3404	12.4	13.0	7.1
3405	14.0	14.1	7.5

TABLE 46 Samples with hydrogen-reduced active sites (ethylene, %)

Sample No.	Hot infrared scan	Cold infrared scan
1487	5.0	5.3
1553	7.1	6.9

been seen, dispute still exists in some cases in the assignment of absorptions at particular wavelengths to microstructural features in the polymer molecule. The examples discussed above will give some idea of the complexity of this work, and the results that can be achieved by careful planning of experimental work.

A final example of the application of infrared spectroscopy is concerned with the measurement of tacticity. Tacticity is defined as the ratio present in the polymer of the syndiotactic to the isotactic structure. Thus, PVC can contain two configurations:

(a) Containing isotactic diads head to head

$$-CH-CH_2-CH_2-CH-$$
$$\quad | \qquad\qquad\qquad | $$
$$\quad Cl \qquad\qquad\qquad Cl$$

(b) Containing syndiotactic diads head to tail

$$-CH_2-CH-CH-CH_2-$$
$$\qquad\quad | \quad\; | $$
$$\qquad\quad Cl \; Cl$$

Schneider et al.[301] showed that the absorption around $690\,cm^{-1}$ ($14.49\,\mu$) was proportional to the number of isotactic diads, and in the region of $600–640\,cm^{-1}$ ($16.66–15.62\,\mu$) to syndiotactic diads. Based on this finding they proposed a method for determining the tacticity of amorphous samples of PVC. As some samples cannot be easily transformed into an amorphous state Schneider *et al.*[302] devised an infrared method of tacticity determination which is independent of the crystallinity of the samples. From the temperature dependence of infrared spectra of poly(vinylchloride) samples prepared by different methods, the intensity of the band at $690\,cm^{-1}$ ($14.40\,\mu$) (proportional to the number of isotactic diads in the sample), as well as that of the tacticity-independent C—H stretching band, was found to be independent of the crystallinity of the sample. These lines were applied for the tacticity determination in poly(vinylchloride), measured in the form of potassium bromide pellets. The numerical tacticity value was obtained from the known values of absorbance coefficients of S_{CH} and S_{HH} type C—Cl stretching bands in solution, and from the shape of the spectrum.

To determine the isotactic content of polypropylene Peraldo[303] carried out a normal vibrational analysis. He considered the primary unit as an isolated three-fold helix. From this work and a number of subsequent publications,[304–307] it was suggested that the absorptions at $1167\,cm^{-1}$ ($8.57\,\mu$) $997\,cm^{-1}$ ($10.02\,\mu$) and $841\,cm^{-1}$ ($11.9\,\mu$) were indicative of the helical conformation of the isotactic form. Measurements of the isotactic

TABLE 47 *Sequence measurement in polymers*

Polythene	Sequence measured	Reference
Ethylene–propylene	—(CH_2)—	310–313
1-Hexene–4-methyl-1-pentene	block structure	314
C_1–C_2O alkylethylene	methylene sequences	35
Ethylene–2-butene	—$(CH_2$—$CH_2)$ units	316
Acrylonitrile–2-substituted propene	—CH_2—C CH_3—	317
Acrylonitrile–styrene	sequences	
Styrene–acrylonitrile	sequence distributions of styrene	318, 319
Styrene–vinyl acetate	sequence distributions	320
α-Methylstyrene–methylene block copolymers	α-methyl styrene sequences	321
Styrene–methacrolein	sequence length of styrene units	322
Styrene–maleic anhydride	styrene sequences	323
Styrene–methylmethacrylate	styrene sequences	324
Vinyl chloride–vinylidene-chloride	methylene sequences	325, 326
Chlorinated polyethylene	methylene sequences	327
Vinyl chloride–isobutylene	head-to-tail structure	328
Polypropylene oxide	sequence distribution	329

contents of a series of polypropylene fractions based on these three bands were made,[307] and compared with results from Flory's melting point theory.[308] Melting points were determined as the point of disappearance of the birefringence on highly annealed samples. All three bands give qualitative agreement with the melting point data; however, only the method based on the 1167 cm^{-1} (8.57 μ) band gives quantitative agreement. Therefore, the method based on the 1167 cm^{-1} band appears to give a good measure of the isotactic content in polypropylene, at least in the 60–100% range.

The approximate degree of isotacticity of crystalline poly(3-methyl-1-butene) has been determined[309] from the ratio of the absorbance at 778 cm^{-1} (12.85 μ) to 1180 cm^{-1} (7.47 μ).

Further work on the application of infrared spectroscopy to sequence measurement in polymers is summarized in Table 47.

5.5(c) Nuclear magnetic resonance spectroscopy and proton magnetic resonance spectroscopy

Tacticity of polypropylene

As discussed previously, polypropylene can exist in three stereochemical forms – syndiotactic, isotactic and atactic:

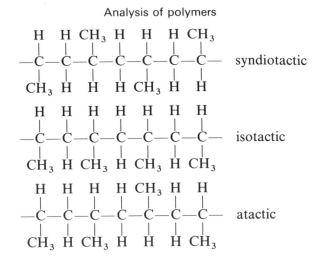

Reilly[330] has investigated an NMR method for determining the syndiotactic content of polypropylene. The results are not entirely consistent with the syndiotactic crystallinity as determined by alternative methods such as the density method. A possible explanation for the lack of consistency is that the NMR method does not require as long a block in the chain in order for the syndiotactic placement of methyl groups to be detected. He attempted to determine the syndiotactic content of some experimental polypropylenes. NMR spectra of samples dissolved in o-dichlorobenzene were obtained at 170°C at 100 Mc/s. Spectra of syndiotactic and isotactic samples gave the following NMR parameters:

	Syndiotactic	*Isotactic*
δCH_3	0.835 ppm	0.895 ppm
δCH_2	1.075 ppm	1.895 and 1.310 ppm
δCH	1.570 ppm	1.615 ppm
J_{CH_3-CH}	6.0 cps	6.0 cps
J_{CH-CH_2}	6.0 cps	6.0 cps
$J_{H-H(geminal)}$	Indeterminate	− 14.0 cps

The reference for the chemical shift measurements was hexamethyl disiloxane. The methylene hydrogens in the isotactic material are nonequivalent − as expected on geometrical grounds. Spectra calculated with the above parameters agreed reasonably well with the observed spectra.[331] High-resolution proton magnetic resonance spectroscopy has been applied to an examination of very highly isotactic, very highly syndiotactic, and stereoblock polypropylenes in o-dichlorobenzene solutions. They discuss the effects of stereoregulation on proton shieldings and some of the complexities of the methylene proton resonances and determine tactic placement contents for several polymers by a method based on the methylene proton resonances. Tactic pair contents were determined for two stereoblock fractions by a method based on the methyl proton resonances. Their results revealed much higher stereoblock characters than those determined for the same fractions from melting data. All of the PMR results were in very good accord with the results obtained on several polymers by infrared, X-ray diffraction, and differential thermal analysis.

Mitani[331] found that the diad and triad content of isotactic polypropylene, syndiotactic polypropylene, and atactic polypropylene, as determined from 100-MHz NMR spectra, were in agreement with the values determined from the 220-MHz NMR and ^{13}C NMR spectra.

Inoue et al.[332], Zambelli et al.[333] and Randall[334] have shown that ^{13}NMR is an informative technique for measuring stereochemical sequence distributions in polypropylene. These workers reported chemical shift sensitivities to configurational tetrad, pentad and hexad placements for this polymer.

The sequence lengths of stereochemical additions in amorphous and semicrystalline polypropylene were accurately measured using ^{13}C NMR[335]. The method has some limitations for addition polymers having predominantly isotactic sequences.

Randall[336] has described work on the application of quantitative ^{13}C NMR to measurements of average sequence length of like stereochemical additions in polypropylene. He describes sequence lengths of stereochemical additions in vinyl polymers in terms of the number-average lengths of like configurational placements. Under these circumstances a pure syndiotactic polymer has a number-average sequence length of 1.0; a polymer with 50:50 meso:racemic additions has a number-average sequence length of 2.0 and polymers with more meso than racemic additions have number-average sequence lengths greater than 2. Amorphous and crystaline polypropylenes were examined using ^{13}C NMR as examples of the applicability of the average sequence length method. The results appear to be accurate for amorphous and semi-crystalline polymers but limitations are present when this method is applied to highly stereoregular vinyl polymers containing predominantly isotactic sequences.[337] Randall has measured the ^{13}C NMR spin lattice relaxation times of isotactic and syndiotactic sequences in amorphous polypropylene. Spin-lattice relaxation times for methyl, methylene, and methine carbons in an amorphous polypropylene were measured as a function of temperature from 46 to 138°C. The carbons from isotactic sequences characteristically exhibited the longest spin relaxation times of those observed. The spin relaxation time differences increased with temperature, with the largest difference occurring for methine carbons, where a 32% difference was observed. Randall determined activation energies for the motional processes affecting spin relaxation times for isotactic and syndiotactic sequences. Essentially no dependence upon configuration was noted. High-resolution NMR spectra of isotactic and syndiotactic polypropylene have been used by Cavelli[338] to provide conformational information. Brosio *et al.*[339] reported on ^{13}C NMR spectra for measurements of tacticity, terminal conformation and configuration of polypropylene. Multiple sequences of syndiotactic or isotactic units can exist in polypropylene. Thus diad and triad of isotactic polypropylene would have the structures:

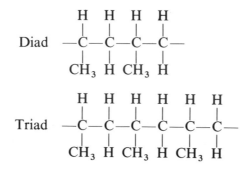

Similarly a pentad and hexad of syndiotactic polypropylene would have the structures:

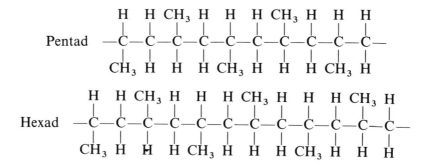

As well as the above sequences ^{13}C NMR spectroscopy can be used to determine isolated head-to-head and tail-to-tail units in polypropylene.[340]

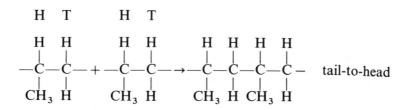

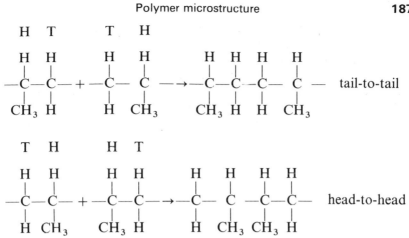

Figure 72 shows the PMR spectrum of hot *n*-heptane solution of atactic polypropylene containing up to 40% syndiotactic placement, and which by Natta's definition may be called stereoblock[341]. The spectrum is inherently complex, as a first-order theoretical calculation, and this indicates the possibility of at least 15 peaks with considerable overlap between peaks,

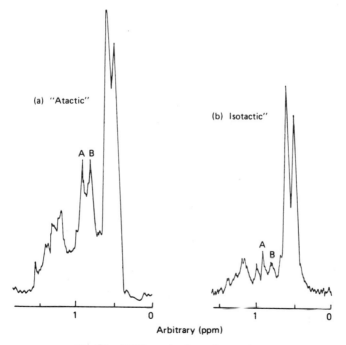

FIG. 72. PMR spectra for polypropylenes.

because differences in chemical shifts are about the same magnitude as the splitting due to spin–spin coupling.

The largest peak, at high field in Fig. 72, represents pendant methyls in propylene units. It is characteristically split by the tertiary hydrogen. By area integration, about 20% of the nominal methyl proton peak is due to overlap of absorption from chain methylenes. This overlap is consistent with a reported syndiotactic triplet, two peaks of which are close to A and B in Fig. 72 and a third peak which falls with the low field branch of the methyl split.[342] The absence of a strong singlet peak in the methylene range indicates the virtual absences of 'amorphous' polymer in the atactic polypropylene shown in Fig. 72,[341,342] which could possibly be due to head-to-head and tail-to-tail units. The low field peak represents the partial resolution of tertiary protons which are opposite the methyls on the hydrocarbon chain.

Figure 72 also shows a spectrum for an isotactic polypropylene, > 95% isotactic by solubility (< 5% soluble in boiling heptane). This spectrum has the same general character as the atactic polypropylene. The important difference is a marked decrease of peak intensity in the chain methylene region. This decrease is caused by extensive splitting, and the difference in chemical shifts for the non-equivalent methylene hydrogens in isotactic environments. This is in accord with the study of Stehling[343] on deuterated polypropylenes, which indicates that much of isotactic methylene absorption is 'buried' beneath the methyl and tertiary hydrogen peaks. The fractional area in the nominal methylene region of the spectrum is thus sensitive to the number of isotactic and syndiotactic diads, and therefore may be used as a measure of polypropylene tacticity. Figure 73 shows the spectrum of a physical mixture of a polypropylene and linear polyethylene. The low field absorption in Fig. 73, characteristic of aromatic hydrogens, is due to the polymer solvent, diphenyl ether, which was used throughout. Polymer concentrations in solution can be readily calculated from the ratios of peak areas adjusted to the same sensitivity. The superpositions of spectra that were obtained separately on the homopolymers show the same pattern, with different intensities, as spectra on physical mixtures. The polyethylene absorption falls on the peak marked A of the chain methylene complex in polypropylene. Peak B, also due to chain methylenes in polypropylene, is resolved in both spectra in Fig. 73, which also gives the spectrum of an ethylene–propylene block copolymer. The ethylene contribution again falls on peak A.

Various workers have developed analyses for physical mixtures and block copolymers based on the ratio of the incremental methylene area to the total polymer proton absorption. This concept has been tested by Barrall et al.[344] using PMR analyses on a series of physical mixtures and block copolymers synthesized with ^{14}C-labelled propylene and others with ^{14}C-labelled ethylene. A most important feature of this analysis is that

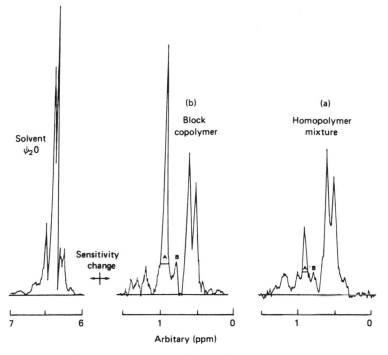

FIG. 73.   PMR spectra for polymers of ethylene and propylene.

the methylene peaks A and B have virtually the same relative heights in polypropylenes with a variety of tacticities (note Fig. 72). This is also true for PMR spectra given by Satoh and others for a tactic series of polypropylenes.[345] This suggests that PMR analyses for ethylene are independent of tacticity, since the area increment of peak A above peak B has been used for analysis

Qualitative polypropylene tacticities can be estimated by PMR not only for homopolymers (Fig. 72), but also in the presence of polyethylene and ethylene copolymer blocks. The relative heights of the peaks for secondary and for tertiary hydrogen in Fig. 73 indicate that the polypropylene in the copolymer and the physical mixture is dominantly isotactic.

TABLE 48   *Analyses of linear polyethylene polypropylene physical mixtures* (*polyethylene, wt%*)

Sample	Made up	PMR	Tracer
6	5.0	4.6	4.2
7	9.0	8.6	7.6
5	17.0	16.7	16.3
8	34.0	31.1	31.0

Table 48 gives analyses carried out by Porter[341] for a series of physical mixtures made up with tagged polypropylene standardized by radiocounting. The three sets of values are in good agreement.

Ethylene–propylene copolymers

Ethylene–propylene copolymers can contain up to four types of sequence distribution of monomeric units. These are propylene to propylene (head-to-tail and head-to-head), ethylene–propylene and ethylene to ethylene. These four types of sequence, and their average sequence lengths of both monomer units, can be measured by the Tanaka and Hatada[346] method. Measurements were made at 15.1 MHz. Assignments of the signals were carried out by using the method of Grant and Paul,[347] and also by comparing the spectra with those of squalane, hydrogenated natural rubber, polyethylene and atactic polypropylene. The accuracy and the precision of intensity measurements, that is the deviation from the theoretical values and the scatter of the measurements, respectively, were checked for the spectra of squalane and hydrogenated natural rubber, and were shown to be at most 12% for most of the signals.

Tanaka and Hatada[346] studied the accuracy and precision of such intensity measurements in ^{13}C NMR spectra and elucidated the distribution of monomeric units in ethylene–propylene copolymers. These analyses are particularly straightforward if one of the monomer units is present at a level of 95% or greater, because the other monomer will then occur primarily as an isolated unit.

The investigation of the chain structure of copolymers by ^{13}C NMR has several advantages because of the large chemical shift involved. Crain *et al.*[348] and Cannon and Wilkes[349] have demonstrated this for the determination of the sequence distribution in ethylene–propylene copolymers. In these studies the assignments of the signals were deduced by using model compounds, and the empirical equation derived by Grant and Paul for low molecular weight of linear and branched alkanes.

Ray *et al.*[350] determined comonomer content in isotactic ethylene–propylene copolymers, complete diad and triad content as well as partial tetrad and pentad distributions using ^{13}C NMR.

Proton magnetic block copolymers have been used to characterize ethylene–propylene block copolymers, containing from zero to 40% ethylene, either as homopolymer or copolymer blocks.[341] The test is independent of tacticity and provides qualitative information on copolymer sequencing and propylene chain structure. The analysis was developed using a series of standard reference polymers synthesized to contain various ratios of ^{14}C-tagged ethylene and propylene. All measurements were made on 10% polymer solutions in diphenyl ether. Analyses are accurate to about $\pm 10\%$ at higher ethylene concentrations. The method is

TABLE 49   *Application of ^{13}C NMR spectroscopy to sequence measurement in polymers*

Polymer	Measurement	Reference
Isotactic-I-butene-propylene	dia, triad and tetrad sequences	351
Butadiene–propylene	measurement of monomer sequence distribution	352
Polybutene	measurement of isotactic content	353
1, 4-poly(2, 3 dimethyl-1, 3 butadiene)	sequence distribution	354
Poly(1, 4-butadiene)	measurement methylene peaks	355, 356
2-Chlorobutadiene–1, 3-2, 3 dichlorobutadiene-1, 3.	measurement of sequences	357
Butadiene–acrylonitrile	measurement of acrylonitrile methine and methylene protons	358, 359
Butadiene vinylchloride	measurement of block length	360
Polychloroprene	measurement of diad and triad sequences	361
Chlorinated PVC	measurement of CH_2, CHCl and CCl_2 groups	362, 363, 364
PVC	diad tacticities	365
Polyvinylidene chloride	measurement of head-to-head and tail-to-tail sequences	366–369
PVC	measuring of distribution of stereochemical sequences tetrad, pentad and hexad placements	370, 371, 372–376
Ethylene–vinylchloride	measurement of diads	377
Ethylene glycol–sebacic acid-terephthalic acid	measurement of distribution of monomer units	378
Chlorinated polyethylene	measurement of methylene sequences	379, 380
Chlorinated polypropylene	measurement of chlorine distribution	381
Chlorosulphonated polyethylene	measurement of different types of methylene groups	382
Poly α-methylstyrene	measurement of tacticity	383
Polystyrene	measurement of: stereochemical sequence distribution tetrad, pentad and hexad placements	384–386
	measurements of average sequence length of like stereochemical configurations	387
Styrene-1–chloro 1, 3 butadiene	measurement of sequence distribution	388
Styrene–maleic anhydride	measurement of distribution of styrene monomer units	389
Styrene–acrylic acid	measurement of isotacticity	390
α-Methyl styrene–methacrylonitrile	measurement of syndiotacticity	391
Styrene–acrylonitrile	measurement of diad and triad distribution	392
Styrene–isobutene	measurement of randomness of structure	393

TABLE 49   Contd.

Polymer	Measurement	Reference
Poly(*p*-fluoro-α-methyl styrene)	measurement of syndiotactic and heterotactic triads	394
Polyvinyl acetate	measurement of diad tacticities, triad and pentad contents	395, 396
Ethylene–vinyl acetate	measurement of tacticity– measurement of sequence distribution	396–399
Polymethylmethacrylate	measurement of isotactic triads, tetrad, pentad and hexad resonances	400 401
	conformation	402
	tacticity	403
Polymethylacrylate	measurement of tetrad configurations of triad, tetrad and pentad configurations	405
Polyalkylmethacrylates	measurement of tacticity	406
Polyallylmethacrylate	measurement of tacticity and sequence length	407
Methacrylic acid– methylmethacrylate	measurement of triad and pentad configurations	408, 409
Methylacrylate–methylmethacrylate	measurement of sequence distribution	410
Methylacrylate–butadiene	measurement of isotactic configuration	411
Methylmethacrylate–butadiene	measurement of triad and pentad configurations	412
Styrene–methylmethacrylate	measurement of diad, triad and pentad configurations	413
Styrene methacrylic and	sequence distribution characterization	414
Ethylene–methylacrylate	measurement of sequence distribution	415
Ethylene–2-chloro-acrylate and ethylene–glycidyl acrylate	measurement of sequence distribution	416
Methacrylphenone– methylmethacrylate copolymers	measurement of sequence distribution	417
Methyl α-chloro acrylate– methylmethacrylate	measurement of diad configurations and mean sequence length	418
Chloroprene-methylmethacrylate	measurement of sequence distribution	419
Polymethyl(vinylether)	measurement of diad, triad and pentad sequences	420
Ethylene oxide–propylene oxide	measurement of diad and triad sequences	421
Ethylene–carbon monoxide	measurement of comonomer sequencing	422

sensitive, with less precision, to below 1% ethylene either as blocks or homopolymer. Porter and co-workers[341,342] found that resolution improves with increasing temperature and with decreasing polymer concentration.

Nuclear magnetic resonance spectroscopy has been used in many other applications related to the measurement of sequences in polymers. Some examples are given in Table 49.

5.6 Oxygen-containing monostructural groups

Extensive infrared studies have been carried out on oxidized polyethylene.[423][429] Infrared absorption spectrometry has been widely used to determine the oxidation products and the rate of formation of these products during the thermal or photo-oxidation of polyethylene.[430,431,423] Acids, ketones and aldehydes, the end-products reported from these oxidations, have similar spectra in the 1819–1666 cm^{-1} region. It is only in this carbonyl stretching region that the products have suitable absorptivity to give quantitative data. The absorption band of the acid (1712 cm^{-1}, 5.84 μ) ketone (1721 cm^{-1}, 5.81 μ) and aldehyde (1733 cm^{-1}, 5.77 μ) groups present in oxidized polyethylene are so overlapped as to give only a broad band on relatively low-resolution infrared spectrometers. Interpretation of these data, based on the increase in total carbonyl rather than on a single chemical moiety, could lead to incorrect conclusions because of the large differences in the absorptivity of the various oxidation products. Acid absorptivity has been reported[433] to be 2.4 times greater than that of ketones and 3.1 times greater than that of aldehydes. Rugg *et al.*,[424] using a grating spectrometer for increased resolution, have demonstrated that the carbonyl groups formed by heat oxidation are mainly ketonic, while in highly photo-oxidized polyethylene the amounts of aldehyde, ketone and acid are approximately equal. This procedure is adequate for qualitative, but suitable for accurate quantitative, data.

An infrared study of oxidative crystallization of polyethylene was made from examination of the 1894 cm^{-1} (5.28 μ) 'crystallinity' band and 1303 cm^{-1} 'amorphous' band and of carbonyl absorption at 1715 cm^{-1}.[432] Miller and co-workers[434] used polarized infrared spectra obtained by Fourier transform spectroscopy to study several absorptions of polyethylene crystallized by orientation and pressure in a capillary viscometer.

Tabb and co-workers[435] used Fourier transform infrared to study the effect of irradiation on polyethylene. Aldehydic carbonyl and vinyl groups decreased and the ketonic carbonyl and *trans*-vinylene double bonds increased on irradiation.

Infrared reflection was used in studies of oxidation of polyethylene at a copper surface in the presence and absence of an inhibitor, N, N-diphenyloxamide.[436]

Cooper and Prober[437] have used alcoholic sodium hydroxide to convert the acid groups to sodium carboxylate ($1563 \, cm^{-1}$, $6.40 \, \mu$) to analyse polyethylene oxidized with corona discharge in the presence of oxygen and oxone. This procedure requires 5 days, and has been found to extract the low molecular weight acids from the film.

Heacock[438] has described a method for the determination of carboxyl groups in oxidized polyolefins without interference by carbonyl groups. This procedure is based upon the relative reactivities of the various carbonyl groups present, in oxidized polyethylene film, to sulphur tetrafluoride gas. The quantity of the carboxyl groups in the film is then measured as a function of the absorption at $1835 \, cm^{-1}$ ($5.45 \, \mu$)

$$ RC\!\!\underset{CH}{\overset{O}{\diagdown}} + SF_4 \rightarrow RC\!\!\underset{F}{\overset{O}{\diagdown}} + HF + SOF_2 $$

The appearance in the spectra of irradiated (γ rays from ^{60}Co) polypropylene specimens, after storage in air, of strong bands in the region of $1710 \, cm^{-1}$ ($5.85 \, \mu$), corresponding to carbonyl groups, must be explained by reaction of oxygen with the long-lived allyl radicals, with formation of peroxide radicals which form carbonyl groups by decomposition.

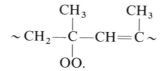

The intensity of the $1710 \, cm^{-1}$ ($5.85 \, \mu$) band (and consequently the degree of oxidation) increases sharply with time of storage of specimens in air. Irradiated amorphous specimens oxidize to a considerably smaller extent than isotactic polypropylene specimens. The degree of oxidation of specimens irradiated at $-196°C$ increases more rapidly than when the specimens are irradiated at $25°C$. All these facts indicate that the lifetime of the allyl radicals is longer in crystalline polypropylene, and that the concentration of these radicals is higher in specimens irradiated at a low temperature. The free radicals are destroyed only after heat treatment of the specimens in an inert atmosphere at $150°C$. After this heat treatment the intensity of the $1710 \, cm^{-1}$ band ceases to increase on storage, i.e. no further oxidation occurs.

Adams[439] has compared on a qualitative basis the non-volatile oxidation products obtained by photo- and thermal oxidation of polypropylene. He used infrared spectroscopy and chemical reactions. The major functional group obtained by photodecomposition is ester, followed by vinyl alkene, then acid. In comparison, thermally oxidized polypropylene contains relatively more aldehyde, ketone, and γ-lectone, and much less ester and vinyl alkene. Photodegraded polyethylene contains mostly vinyl alkene followed by carboxylic acid. Gel permeation chromatography determined the decrease in polypropylene molecular weights with exposure time. Adams determined that there is one functional group formed per chain scission; in thermal oxidation there are two groups formed per scission.

Adams[439] makes the following comments regarding the infrared spectrum of oxidized polypropylene.

Hydroxyl region. The hydroxyl absorptions in the infrared for polypropylene has a broad band centred at $3450 \, cm^{-1}$ ($2.90 \, \mu$) (associated alcohols) with a definite shoulder at $3610 \, cm^{-1}$ ($2.77 \, \mu$) (unassociated alcohols). At a similar extent of degradation, thermally oxidized polyolefins show hydroxyl bands of roughly half the absorbance values of the photo-oxidized polyolefins. Thus thermal oxidation produces about half as many hydroxyl groups as photo-oxidation in polyolefins.

A portion of the polypropylene hydroxyl absorption could be due to hydroperoxides. If so, then an exposed sheet, with the volatiles removed, heated in a nitrogen atmosphere for 2 days at 140°C, should show a decrease in the hydroxyl infrared band and an increase in the carbonyl band due to the decomposition of hydroperoxides under such treatment. The infrared spectrum of the photodegraded polypropylene sheet subjected to this thermal treatment showed a 20% decrease in the hydroxyl band. However, the broad carbonyl band at $1740 \, cm^{-1}$ ($5.75 \, \mu$) did not increase but showed a 5% decrease. The small γ-lactone ($1780 \, cm^{-1}$, $5.61 \, \mu$) and vinyl alkene ($1645 \, cm^{-1}$, $6.08 \, \mu$) bands did show a slight increase, however. Thus, these results are due not to hydroperoxide decomposition but to some carboxylic acids converting to γ-lactones and some terminal alcohols dehydrating to vinyl alkenes at the high temperature. While hydroperoxides are undoubtedly an intermediate in the photo-oxidation process, they decompose too rapidly under ultraviolet light to build up any significant concentration.

Carbonyl region. The polypropylene carbonyl band after 335 h exposure is broad, with few discernible features except for the vinyl alkene band at $1645 \, cm^{-1}$ ($6.08 \, \mu$). The broadness of the carbonyl band indicates a large variety of functional groups, and makes accurate quantitative analysis difficult. The large vinyl alkene at $1645 \, cm^{-1}$ ($6.08 \, \mu$) stands out clearly, and distinct carboxylic acid ($1715 \, cm^{-1}$ $5.83 \, \mu$) and γ-lactone ($1790 \, cm^{-1}$, $5.58 \, \mu$) spikes can be readily identified.

After the volatile products are removed by the vacuum oven, the carbonyl band for polypropylene decreases. Isopropanol extraction removes about 40% of the polypropylene carbonyl. The carbonyl band is then narrow and appears to centre at the ester absorption at $1740 \, cm^{-1}$ ($5.75 \, \mu$).

Treatment with base converts lactones, esters and acids to carboxylates (1580 cm^{-1}, 6.33 μ), leaving only a small band at 1720 cm^{-1} (5.81 μ), which is due to aldehyde and ketone.

Upon reacidification of the polypropylene, some of the original esters at 1740 cm^{-1} (5.75 μ) do not re-form but become carboxylic acids and γ-lactones. Curiously, the vinyl alkene band becomes less intense with each step and broader, shifting down to 1640–1600 cm^{-1} (6.10–6.25 μ), the vinyl groups may be isomerized into internal alkenes or become conjugated during the various treatments, although no such change occurs with either the polyethylene vinyl alkene or with the process-degraded polypropylene vinyl alkene. Wood and Statton[440] developed a new technique to study molecular mechanics of orientated polypropylene during creep and stress relaxation based on use of the stress-sensitive 975 cm^{-1} (10.25 μ) band and orientation-sensitive 899 cm^{-1} (11.12 μ) band. The far infrared spectrum of isotactic polypropylene was obtained from 400 to 10 cm^{-1} and several band assignments were made.[441] Isotacticity of polypropylene has been measured from infrared spectra and pyrolysis–gas chromatography following calibration from standard mixtures of isotactic and atactic polypropylene. The infrared spectrum of oxidized polypropylene indicated small amounts of OOH groups plus larger concentrations of stable cyclic peroxides or expoxides in the polypropylene chain.[441]

Grassie and Weir[442] described an apparatus for the measurement of the uptake of small amounts of oxygen by polystyrene with a high degree of precision. Grassie and Weir[443] investigated the application of ultraviolet and infrared spectroscopy to the assessment of polystyrene films after vacuum photolysis in the presence of 253.7 nm radiation using the apparatus mentioned above. During irradiation there is a general increase in absorption in the region 230–350 nm. Rates of increase are relatively much greater in the 240 and 290–300 nm regions, however, as shown in Fig. 74. Absorption in the 240 nm region is characteristic of compounds having a carbon–carbon double bond in conjunction with a benzene ring. Styrene, for example, has an absorption band at 244 nm.

Schole et al.[444] have applied an oxidative degradation technique to the study of polystyrene. In this technique the polystyrene sample is mixed with a support in a precolumn which is mounted at the inlet to a gas chromatographic column. Figure 75 shows a typical gas chromatogram obtained under these conditions.

Shaw and Marshall[445] have carried out an infrared spectroscopic examination of emulsifier free polystyrene which had been oxidized during polymerization. Evidence was found for the presence of surface carboxyl groups bound to the polymer chains, presumably formed by oxidation during polymerization. The band at 1705 cm^{-1} (5.86 μ) was assigned in part to the carbonyl stretching mode of dimeric carboxylic acid, formed by oxidation, in the polystyrene chains. Absorption at 1770 cm^{-1} (5.65 μ),

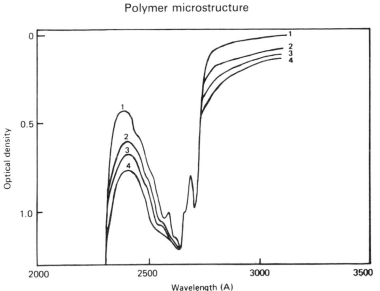

FIG. 74. Effect of irradiation on the ultraviolet spectrum of polystyrene: (1) unirradiated; (2) 60 min; (3) 120 min; (4) 180 min.

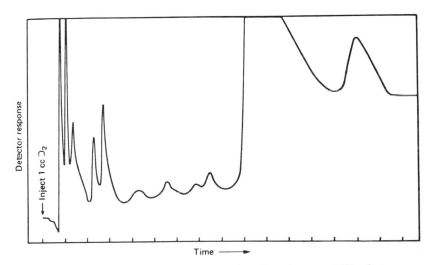

FIG. 75. Gas chromatogram of oxidation products for polystyrene. 15% polystyrene on 35- to 80-mesh Chromosorb P. Column length, 1 ft. Precolumn temperature, 240°C.

which was very weak, was tentatively attributed to the carbonyl stretching mode of the monomeric form of this acid. The structure of the acid end-group was not established, but the results obtained suggest that it was possibly a phenylacetic acid residue or a residue of standard (unoxidized) and of oxidized emulsion polymerized polystyrene in the region 800–

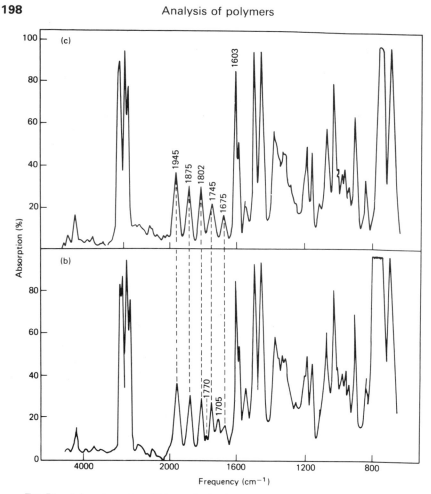

FIG. 76.   Infrared spectra of (a) standard polystyrene sample and (b) latex B1 in the range 800–4000 cm^{-1}.

4000 cm^{-1} (12.5–25.0 μ). The spectrum of latex B1, shown in Fig. 76(b) contains weak bands at 1770 cm^{-1} (5.65 μ) and 1705 cm^{-1} (5.86 μ).

5.7   End-group analysis

Below are quoted two examples of end-groups as they occur in polymers.

Polyethylene glycols.

$$\sim CH_2-CH_2OH + nCH_2CH_2O = \sim CH_2 - CH_2O(CH_2CH_2O)_{n-1}$$

ethylene glycol   ethylene oxide   polyethylene glycol

$$'CH_2CH_2OH$$

hydroxy endgroup

PVC

$$n\text{-}CH_2{=}CHCl = \sim (CH_2{-}CHCl)_{n-1}{-}CH_2{=}CH$$
$$|$$
$$Cl$$

vinyl chloride   PVC   double bond end-group

The end-groups hydroxy and a double bond, respectively, are a structural feature of the polymer and, as such, it may be important to identify and determine these groups. Various techniques have been used to determine end-groups. Obviously these methods must be highly sensitive as, particularly with high molecular weight polymers, the percentage of end-groups is very low. Thus if 1 mole (62 g) of diethylene glycol is reacted with 20 moles (880 g) of ethylene oxide according to the following equation, then the molecular weight of the product is $62 + 880 = 942$.

$$HO{-}CH_2CH_2OH + 20CH_2CH_2O{=}$$
$$HOCH_2CH_2(CH_2CH_2O)_9OCH_2CH_2O(CH_2CH_2O)_9CH_2{-}CH_2OH$$

Thus two hydroxy end-groups (34 g) occur per 997 g of polymer, i.e. 3.4% hydroxy group. Similarly if 100 moles of ethylene oxide reacted with 1 mol of diethylene glycol then the final product would have a hydroxy end-group content of 0.76%. One of the uses to which end-group analysis can be put is the determination of molecular weight. Thus if a polyethylene glycol–ethylene oxide condensate was found to contain 0.3% hydroxy end-groups.

$$HOCH_2CH_2OH + nCH_2CH_2O{=}HO(OCH_2CH_2CH_2)_{n/2-1}{-}O{-}$$
$$CH_2CH_2O(CH_2{-}CH_2O)_{n/2-1}OH$$

Percentage hydroxy $\quad = \dfrac{2 \times 17}{62 + n \times 44} = 0.3$

$$n = \dfrac{3400 - 18.6}{13.2} = 255$$

Molecular weight of polymer is

$$HO(OCH_2CH_2)_{n/2-1}OCH_2CH_2O(CH_2CH_2O)_{n/2-1}OH$$
$$= HO(OCH_2CH_2)_{126.5}OCH_2CH_2O(CH_2CH_2O)_{126.5}OH$$
$$= 11,226$$

End-groups in other polymers

Polyethylene glycol

Fritz et al.[446] have described a method for the determination of hydroxy groups in poly(ethylene glycols). The method has a sensitivity of 10^{-4} mol hydroxyl per kg polymer, which is achieved by using silylation with arylsilylamines and subsequent photometric measurement of the silylated polymer.

PVC

Unsaturated end-groups in PVC have been estimated from NMR spectra, using reference model compounds such as 1-chloropentene-2, 1,1-dichloropentene-2, and 1,2-dichloropentene-2.[447] Fourier transform proton NMR studies of PVC indicated that the unsaturated end-groups contained allylic chlorine atoms. The presence of the unsaturated end-groups would partially explain the PVC degradation.[448]

Polycarbonates

Terminal hydroxy groups were measured in polycarbonate following complexation with ceric ammonium nitrate; absorbance has been measured at 500 nm and at 530–540 nm.[449]

Polyesters

Smirnova and co-workers[450] employed trichloroacetyl isocyanate and trifluoroacetic anhydride acetylations to determine hydroxyl end-groups in polyester polyols. The isocyanate reagent measured proton resonance and was best suited for samples having molecular weights of less than 4500; the anhydride method was more sensitive and applicable to higher molecular weights than a ^{19}F NMR method.

Carboxy and hydroxy polybutadienes

Law[451] has also described infrared methods for the determination of hydroxy equivalent weight and carboxyl equivalent weight of carboxy and hydroxy polybutadienes.

Styrene–acrylonitrile oligomers

Infrared spectroscopy has been used to study styrene oligomers containing acrylonitrile active ends.[452] The addition of acrylonitrile to oligostyrene alkylmetal salt solutions resulted in attachment of acrylonitrile across the double bond of the oligostyrene and by formation of the carbanion C—HCN.

Polyisobutylenes

Proton NMR signals for olefinic end-groups, $-CH_2C(CH_3){=}CH_2$, $CH{=}C(Me)_2$, and $-CH_2C-(CH_2)CH_2-$ were observed in high molecular weight polyisobutylenes by Manaff et al.[453]

Butadiene–isoprene copolymers

Valuev *et al.*[454] separated oligomeric butadiene–isoprene block copolymers into fractions containing two, one, and no terminal hydroxyl groups, respectively, by thin-layer chromatography. Infrared spectroscopy was used for the determination of hydroxyl content.

Polyethylene terephthalate

Nissen and co-workers have described a method for carboxyl end-groups in poly(ethylene terephthalate). Hydrazinolysis led to formation of terephthalomono-hydrazide from carboxylated terephthalyl residues to provide a selective analysis for carboxyl groups via ultraviolet absorbance at 240 nm.[455]

Dye partition methods for end-group analysis of polymethylmethacrylate

This technique is best illustrated by an example concerning the determination of end-groups in persulphate-initiated polymethylmethacrylate.[456,457] Sulphate and other anionic sulphoxy end-groups were determined by shaking a chloroform solution of the polymer with aqueous methylene blue reagent. The greater the anionic content, the more the methylene blue phase is decolorized. The blue colour is evaluated spectrophotometrically at 660 nm. The quantity of anionic sulphoxy end-group present in the polymer is obtained by comparing the experimental optical density values with a calibration curve of pure sodium lauryl sulphate obtained by following a similar procedure.[458]

Results obtained by Ghosh *et al.*[456,459] in end-group analyses of polymethylmethacrylate indicate that all the polymer samples exhibit a positive response to methylene blue reagent in the dye partition test, indicating the presence of at least some sulphate (OSO_3^-) end-groups.

Maiti and Saha[460] have described a dye partition technique[461–463] utilizing disulphine blue for the qualitative detection and, in some cases, the determination of amino end-groups in the free radical polymerization of polymethylmethacrylate. They found only 0.01–0.62 amino end-groups per chain in polymethylmethacrylate made by the amino-azo-bisbutyronitrile system, whereas in polymer made by the titanous chloride and acidic hydroxylamine systems they found 1.10–1.90 amino end-groups per chain.

Ghosh *et al.*[464,465] have also described a dye partition method for the determination of hydroxyl end-groups in poly(methylmethacrylate) samples prepared in aqueous media with the use of hydrogen peroxide as the photo-initiator. In this method the dried polymers were treated with chlorosulphonic acid under suitable conditions whereby the hydroxyl end-groups present in them were transformed to sulphate end-groups. Spectrophotometric analyses of sulphate end-groups in the treated polymers was carried out by the application of the dye-partition technique, and thus a measure of hydroxyl end-groups in the original polymers was obtained (average 1 hydroxy end-group per polymer chain).

Saha et al.[466] and Palit[467,468] have developed a dye-partition method for the determination of halogen atoms in copolymers of styrene, methylmethacrylate, methylacrylate, or vinyl acetate with a chlorine-bearing monomer such as allyl chloride and tetrachloroethylene. The quaternized copolymers were quaternized with pyridine, then precipitated with petroleum ether or alcohol and further purified by repeated precipitation from their benzene solutions with a mixture of alcohol and petroleum ether as the non-solvent. The finally precipitated polymers were then washed with petroleum ether and dried in air. The test for quaternary halide groups in polymers was carried out with a reagent consisting of disulphine blue dissolved in 0.01 M hydrochloric acid and the colour evaluated spectrophotometrically at 630 nm. Saha et al.[466] and Palit[467,468] found that there may be some uncertainty in the quantitative aspects of this method.

Polystyrene

Stetnagel and Palit[469] applied dye partition end-group analysis procedures to the examination of sulphate end-groups in persulphate initiated polystyrene.

Ghosh et al.[470] have carried out end-group analysis of persulphate initiated polystyrene using a dye partition and a dye interaction technique. Sulphate and hydroxyl end-groups are generally found to be incorporated in the polymer to an average total of 1.5 to 2.5 end-groups per polymer chain.

Banthia and co-workers[471] determined sulphate, sulphonate and iso-thioronium salt end-groups in polystyrene by the dye-partition technique. Polymer polarity did not affect the results of the end-group determination. Nitrile groups incorporated in polystyrene by initiation or copolymerization have been detected and estimated by dye-partition techniques after reduction to amino groups with lithium aluminium hydride in tetrahydrofuran.[472]

Acrylic acid–acrylamide copolymers

Mukhopadhyay et al.[473] have reported a reverse dye-partition technique for the estimation of acid groups in water-soluble polymers such as copolymers of acrylamide and other carboxyl-bearing monomers (e.g.

acrylic acid), and the monomer reactivity ratio, r, has been determined by measuring the carboxyl group content in those copolymers.[474,475] The carboxyl contents of the purified copolymers were determined by the reverse dye-partition method.

5.8 Structural elucidation via free radical formation

Identification of the free radicals produced when polymers are irradiated, normally by gamma-radiation from ^{60}Co or by fast electrons from an electron accelerator tube, can sometimes give vital information regarding structural features of the original polymer.

Slovakhotova *et al.*[476] have applied infrared spectroscopy to a study of structural changes in polypropylene *in vacuo* with fast electrons from an electron accelerator tube (200 kV accelerating field) and with gamma-radiation from [60]Co. They found the infrared spectrum of irradiated polypropylene contains absorption bands in the 1645 cm^{-1} (6.08 μ), 890 cm^{-1} (1.23 μ) and 735–740 cm^{-1} (13.60–13.51 μ) regions. The first two

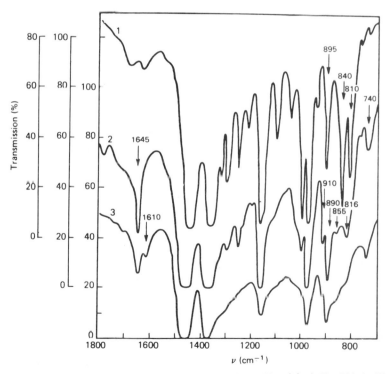

FIG. 77. Infrared spectra of isotactic polypropylene: (1) original ($d = 200\,\mu$); (2) irradiated by fast electrons at 25° C, dosage 500 Mrad ($d = 300\,\mu$); (3) dosage 400 Mrad ($d = 200\,\mu$).

bands correspond to $RR'C{=}CH_2$ vinylidene groups and the band in the 735–740 cm^{-1} (13.60–13.51 μ) region to propyl branches, $R{-}CH_2CH_2CH_3$ (Fig. 77). When polypropylene is degraded thermally these groups are formed by disproportionation between free radicals formed by rupture of the polymer backbone. Under the action of ionizing radiations the polymer backbone ruptures, with formation of two molecules, with vinylidene and propyl end-groups, at a temperature as low as that of liquid nitrogen, because the corresponding bands are found in the infrared spectrum of polypropylene irradiated at $-196°C$ and measured at $-130°C$. When polypropylene is irradiated with dosages greater than 350 Mrad a band appears at 910 cm^{-1}, corresponding to vinyl groups, $R{-}CH{=}CH_2$, i.e. degradation of polypropylene can also involve simultaneous rupture of two C—C bonds in the main and side chains. The strength of the vinyl-group band in the spectrum of irradiated polypropylene is lower than that of the vinylidene group, although the extinction coefficients of these bands are approximately the same.[477]

In the spectrum of amorphous polypropylene (Fig. 78) irradiated with a dosage of 4000 Mrad at $-196°C$ and measured at $-130°C$, in addition to the band at 1645 cm^{-1} (6.08 μ), a weaker band appears, with a maximum

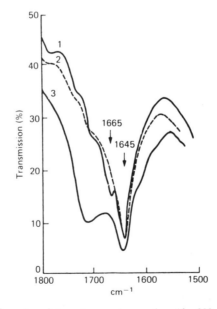

Fig. 78. Infrared spectra of amorphous polypropylene ($d = 300\,\mu$); (1) irradiated by fast electrons with a dosage of 4000 Mrad at $-196°C$ (spectrum recorded at $-130°C$); (2) the same specimen after heating (spectrum recorded at $25°$); (3) the same specimen 2 weeks after irradiation (spectrum recorded at $25°$).

near $1665\,cm^{-1}$ ($6.00\,\mu$), possibly due to internal double bonds.

$$\sim CH-CH=C-CH-C\sim$$
$$\underset{CH_3}{|}\underset{CH_3}{|}\underset{CH_3}{|}$$

When the spectrum of this specimen is recorded after it is heated to $+25°C$ this maximum disappears, leaving only a shoulder on the strong band at $1645\,cm^{-1}$ ($6.08\,\mu$). The extinction coefficient ε,[477] of the band at $1665\,cm^{-1}$ ($6.00\,\mu$) is less than that of the $1645\,cm^{-1}$ ($6.08\,\mu$) band by a factor of 6.7. Supplementary evidence of the formation of internal double bonds in irradiated polypropylene is provided by the presence in the spectrum of bands in the $815-855\,cm^{-1}$ ($12.27-11.69\,\mu$) region. In this region lie bands due to deformation vibration of CH at double bonds in

$$\underset{CH_3}{\overset{R}{>}}C=CHR$$

groups, existing in various conformations.[478] The appearance of bands at $815\,cm^{-1}$ ($12.27\,\mu$) and $855\,cm^{-1}$ (11.69μ) in the spectrum of irradiated polypropylene can be regarded as an indication of the formation of internal double bonds in the polymer.

A study of the ESR spectra of irradiated polypropylene[479] has shown that the alkyl radicals formed during irradiation at $-196°C$

$$\sim CH_2-\overset{\cdot}{C}-CH_2-CH\sim$$
$$\underset{CH_3}{|}\underset{CH_3}{|}$$

undergo transition to alkyl radicals when the specimen is heated; i.e. on heating the radical centres migrate to internal double bonds with the formation of stable allyl radicals. Irradiation at room temperature leads immediately to the formation of allyl radicals. It is very probable that the decrease in intensity of the internal double bond valency vibration band at $1665\,cm^{-1}$ ($6.00\,\mu$) and the broadening of its maximum after a specimen irradiated at a low temperature is heated to room temperature, is associated with the formation of allyl radicals because interaction of the π-electrons of the double bond with the unpaired electron of the allyl radical must have a marked effect on the vibration of the double bond. Conjugation of two double bonds gives rise to a band of lower frequency. It is possible that conjugation with an unpaired electron also lowers the frequency of the double-bond vibration. Comparison of the intensities of the terminal double-bond bands at $890\,cm^{-1}$ ($11.23\,\mu$) and $910\,cm^{-1}$ ($10.99\,\mu$) with the band at $1645\,cm^{-1}$ ($6.08\,\mu$) in the spectrum of irradiated isotactic polypropylene shows that the intensity of absorption in the $1645\,cm^{-1}$ ($6.08\,\mu$) region does not correlate with the intensity of absorption in the 890 cm^{-1} (11.23 μ) and 910 cm^{-1} (10.99 μ) regions. Thus, according to the known extinction coefficients for these bands,[477] the ratio of their optical densities should be

$$D_{890}/D_{1645} = 3.7, D_{910}/D_{1645} = 3.2 \text{ and } (D_{890} + D_{910})/D_{1645} = 3.5$$

In the latter case D_{1645} is the sum of the optical densities of the vinylidene and vinyl absorption bands in this region. An optical density ratio for these bands of approximately this value was found (1.75 to 3.3) for the products of thermal degradation of polypropylene. It is seen that only in the case of amorphous polypropylene irradiated with γ-radiation from ^{60}C is the ratio $(D_{890} + D_{910})/D_{1645}$ close to the value calculated from the extinction coefficients of these bands. In the spectra of irradiated isotactic polypropylene, however, the intensity of the 1645 cm^{-1} (6.08 μ) band is greater than would be expected if only vibration of terminal double bonds contributes to absorption in this region. This increase in absorption in the 1645 cm^{-1} (6.08 μ) region can be related to absorption by the internal double bond in the allyl radical, the vibrational frequency of which is lowered by conjugation of the π-electrons of the double bond with the unpaired electron of the radical. In amorphous polypropylene irradiated at room temperature the alkyl radicals can combine rapidly; therefore there is obviously little formation of allyl radicals. This explains the fact that the ratio of the optical densities of the terminal double bond bands in the 900 cm^{-1} (11.10 μ) and 1645 cm^{-1} (6.08 μ) regions is close to the calculated value.

The occurrence of conjugated double bonds in irradiated polypropylene is indicated by the following facts: (1) in the spectra of isotactic polypropylene irradiated with dosages of 2000–4000 Mrad at room temperature there is a band at 1610 cm^{-1} (6.21 μ), which is the region in which polyene bands occur, whereas this band is absent from the spectrum of isotactic polypropylene irradiated with the same dosages at $-196°C$; (2) in the electronic spectra of these polypropylene specimens the boundary of continuous absorption is shifted to a region of longer wavelength in comparison with the spectra of polypropylene irradiated with the same dosages at $-196°C$.

It has been shown from ESR spectra[517] that when specimens of isotactic polypropylene are heated above 80°C they contain polyenic free radicals

$$\sim \dot{C}H-(C{=}CH)_n-CH \sim$$
$$\qquad\quad | \qquad\qquad |$$
$$\qquad CH_3 \qquad\quad CH_3$$

and this also indicates the possibility of migration of double bonds along the polymer chain.

Free radical studies have been applied to a wide variety of polymers including polyethylene,[480–491] polypropylene,[492–498] polybutenes,[492]

TABLE 50

The determination	The technique	Reference
Propylene in 1-butene–propylene	^{13}C NMR	518
Butene–ethylene–propylene	NMR	519
α-Olefins in ethylene–butene	NMR	520
Methylinethylene–α-Olefin	IR	521
4-Methylpentene-1–1-pentene	NMR	522
Propylene–butene	NMR	523–525
Ethylene–vincyclohexane	IR	526
Vinylchloride–propylene	IR	527
	^{1}H NMR	528, 529
Vinylchloride in chlorinated PVC	^{13}C NMR	530
Vinyl groups in vinylchloride–vinylidene chloride	NMR	531, 532
Glycidylmethacrylate in methylmethacrylate–glycidyl methacrylate	IR	533
Methacrylic acid–methylmethacrylate	^{13}C Fourier transform NMR	534
Ethylene oxide units in polyethylene oxides	NMR	535
Acrylonitrile in styrene–acrylonitrile	IR	536
Vinyl acetate–polyethylene	IR	537
Chlorogroups in polystyrene	^{13}C NMR	538
4-methyl-1-pentene in 4-methyl-1-pentene–1-pentene copolymer	NMR	539
Butene in butene–propylene copolymer	NMR	523
Methylmethacrylate in styrene–methylmethacrylate copolymers	NMR	540–544
Butadiene in acrylonitrile-butadiene–styrene terpolymers	Osmium tetroxide oxidation	545

polystyrene,[499–501] PVC,[502,503] polymethylmethacrylate,[504–511] polyethylene glycol,[512] polycarbonate[513] and polyacrylic acid.[514–516]

Determination of copolymer composition

Apart from the question of determining the microstructure of a polymer it is often necessary to determine the ratios of the various monomers present in copolymers and terpolymers. Depending on the particular problem a wide variety of techniques have been employed to this end, some examples of which are presented in Table 50.

6

Thermal methods for the examination of polymers

THERMAL methods of analysis of polymers are important in that these techniques can provide information about the thermal stability of polymers, their lifetime or shelf-life under particular conditions, phases and phase changes occurring in polymers, and information on the effect of incorporating additives in polymers.

Thermal methods for the examination of polymers can conveniently be discussed under the following seven main headings. The technique of controlled pyrolysis followed by gas chromatography and/or spectrometry is, in effect, a thermal method of analysis and is discussed elsewhere in this book (Chapter 5).

Thermogravimetric analysis
Differential thermal analysis
Differential scanning colorimetry
Thermal volatilization analysis
Thermomechanical analysis
Thermal degradation studies
Oxidative stability studies

6.1 Thermogravimetric analysis

This is one of the fastest-growing analytical techniques for the evaluation of the thermal decomposition kinetics of polymeric materials. It involves continuous weighing of the polymer as it is subject to a temperature programme. This technique can provide quantitative information about the thermal decomposition of polymeric materials from which the thermal stability of the polymer can be evaluated. The lifetime or shelf-life of a polymer can be estimated from these kinetic data. Ozawa[546] and Flynn and Wall[547] observed that the activation energy of a thermal event could be determined from a series of thermogravimetric runs performed at different heating rates.

As the heating rate increased, the thermogravimetric change occurred at higher temperatures. A linear correlation was obtained by plotting the logarithm of the heating rate or scan speed against the reciprocal of the absolute temperature at the same conversion or weight loss percentage. The slope was directly proportional to the activation energy and known constants. To minimize errors in calculation, approximations were used to calculate the exponential integral.[547-549] It was assumed that the initial thermogravimetric decomposition curve (2–20% conversion) obeyed first-order kinetics. Rate constants and pre-exponential factors could then be calculated and used to examine relationships between temperature, time and conversion levels. The thermogravimetric decomposition kinetics could be used to calculate:

1. the lifetime of the sample at selected temperatures,
2. the temperature which will give a selected lifetime,
3. the lifetimes at all temperatures at known percentage conversion.

Figure 79 shows thermograms (percentage weight versus temperature) for a 10 mg specimen of PTFE obtained at four different heating rates – 2.5, 5, 10 and 20°C per minute in a dynamic air atmosphere or 20 cm³/min. From these data can be calculated the rate of decomposition of PTFE, the activation energy and the relationship between the rate constant or half-life and temperature.

The results shown in Fig. 79 agree with the theoretical prediction that, as the rate of heating increases, the thermograms are displaced to higher temperature. Activation energies at selected percentage conversion levels were calculated using the results. The activation energy was calculated

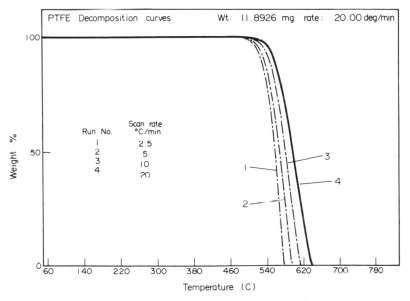

Fig. 79. Thermogravimetric curves of PTFE.

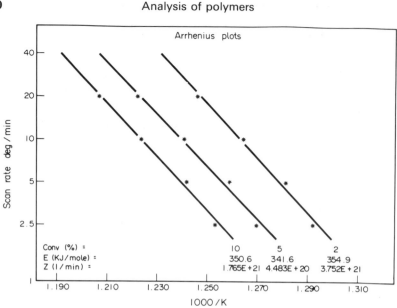

FIG. 80. Arrhenius plots for thermogravimetric analysis of PTFE.

from the slope of the graph of scan time against the inverse of the absolute temperature (Fig. 80). After the activation energy had been determined, the rate constants, half-lives and percentage conversions could be calculated for certain temperatures. In Fig. 81 the rate constant is plotted against temperature to provide information on the stability of the sample from ambient temperature to 800°C. The half-life can be calculated from these

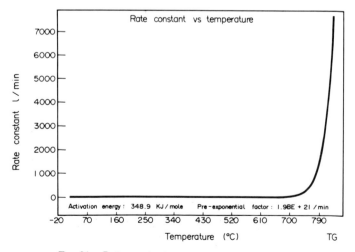

FIG. 81. Rate constant versus temperature curves.

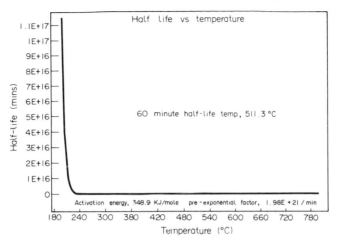

FIG. 82.   Half-life versus temperature curves.

kinetic data and a graph of half-life versus temperature plotted (Fig. 82).
The temperature equivalent to a half-life of 60 min is also determined.
These data can be printed out in tabular form (Table 51) to facilitate
quantitative comparisons. The rate of percentage conversion with respect
to time can be computed for any selected temperature. Figure 83 shows
percentage conversion versus time at 500°C, together with the half-life of
the sample, 130.88 min.

The results calculated using the thermogravimetric decomposition
kinetics can be used for comparative studies. However, if the lifetime at
specific temperatures and percentage conversion are known for the

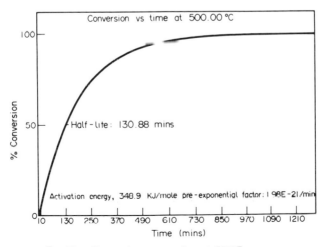

FIG. 83.   Conversion versus time at 500°C curves.

TABLE 51   *Thermogravimetric Analysed PTFE half-life table*

Temp. (°C)	Temp. (K)	1000/T (1/K)	K(T) (1/min)	Half-life			
				Years	Days	Hours	Minutes
400	673.16	1.485	1.667E-06	0	288	14	26
410	683.16	1.463	4.154E-06	0	115	20	57
420	693.16	1.442	1.007E-05	0	47	18	21
430	703.16	1.422	2.383E-05	0	20	4	36
440	713.16	1.402	5.504E-05	0	8	17	51
450	723.16	1.382	1.242E-04	9	3	21	0
460	733.16	1.363	2.740E-04	0	1	18	9
470	743.16	1.345	5.920E-04	0	0	19	30
480	753.16	1.327	1.253E-03	0	0	9	13
490	763.16	1.310	2.600E-03	0	0	4	26
500	773.16	1.293	5.296E-03	0	0	2	10
510	783.16	1.276	1.059E-02	0	0	1	5
520	793.16	1.260	2.081E-02	0	0	0	33
530	803.16	1.245	4.022E-02	0	0	0	17
540	813.16	1.229	7.647E-02	0	0	0	9

Activation energy, 348.9′ kJ/mole pre-exponential factor 1.98E + 21/min, Convs 2 5 10

sample, adjusted lifetimes can be calculated. This type of data can be obtained by elevated temperature measurements on 'bulk' samples.

Thermogravimetric analysis has been used to study degradation kinetics and various factors affecting thermal stability of polymers, such as crystallinity, molecular weight, orientation, tacticity, substitution of hydrogen atoms, grafting, copolymerization, addition of stabilizers, etc. More important systems investigated include cellulose,[550–554] polystyrene,[555,556] ethylene–styrene copolymers,[557] styrene–divinylbenzene-based ion exchanges,[558] vinylchloride–acrylonitrile copolymers,[559] poly(ethylene terephthalate),[560] polyesters such as poly(isopropylidene carboxylates) and polyglycollide,[561–563] Nylon 6 grafted with acrylonitrile, acrylamide, methylmethacrylate, and methylacrylate,[564] polypyromellitimides,[565] poly-N-naphthylmaleimides[566] and poly[benzobis (aminoiminopyrolenes)],[567] acrylics[568,569] PVC[570,571] acrylamide-acrylate copolymers and polyacrylic anhydride.[572]

6.2   Differential thermal analysis

In this technique the polymer sample is temperature-programmed at a controlled rate and, instead of determining weight changes as in thermogravimetric analysis, the temperature of the sample is continually monitored. Just as a phase change from water to ice or vice versa is accompanied by a latent heat effect, so when a polymer undergoes a

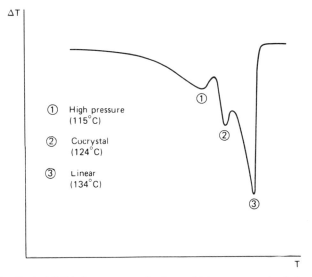

FIG. 84.   Typical DTA thermogram of a linear–high-pressure polyethylene blend.

phase change from for example, a crystalline to an amorphous form, heat is either produced or absorbed.

Figure 84 shows a typical DTA thermogram of a linear high-pressure polyethylene blend.[573] This polymer, upon heating, undergoes three phase changes from its high-pressure form (115°C) to cocrystalline form (124°C) to a linear form (134°C).

The 115°C peak was associated with the high-pressure polyethylene, whereas the 134°C peak was shown to be proportional to the linear content of the system. Clampett[574] also applied differential thermal analysis to a study of the 124°C peak, which he describes as the co-crystal peak. His results appear to indicate that there are two classes of co-crystals in linear–high-pressure polyethylene blends with the linear component being responsible for the division of the blends into two groups. The property of the linear component which is responsible for this division is related to the crystallite size of the pure linear crystal.

Differential thermal analysis has many other applications including: (i) measurement of crystallinity of random and block ethylene methylacrylate copolymers,[575] and of polybutene,[576] polyethylene terephthalate, 1,4-cyclo- hexylene dimethyl terephthalate and polypropylene;[577,578] (ii) examination of morphologically different structures of polyethylene ionomers;[579] (iii) glass transition measurements in polystyrene;[580] (iv) investigation of multiple melting peaks in isotactic polypropylene[581–583] and polyethylene terephthalate;[584] (v) first-order transitions in ethylene methylacrylate copolymers,[585] (vi) annealing of isotactic polybutene-1 and of the three polymorphic forms of polybutene-1;[586,587] (vii) examination of the effects of thermal history on polyethylene;[588] (viii) polymer stabilization studies,[589,590] (ix) polymer pyrolysis kinetics studies;[591,592] (x) effect of side-chain length in polymers on melting point;[593] and (xi) the effect of heating rate of polymers on their melting point.[594,595]

Differential thermal analysis has also been used to identify polymers including polyethylene,[596] polypropylene,[596] polymethyl methacrylate,[596] polyesters,[596] PVC[596] and polycarbonates.[596]

6.3 Differential scanning calorimetry

This is a technique whereby the temperature changes which occur whilst a chemical reaction takes place in a polymer, e.g. crystallization, are continuously monitored until the reaction is complete. This technique is of great value in carrying out kinetic studies on polymers.

This technique has been used to moniter the crosslinking rate of polyethylene–carbon black systems,[597] to investigate multiple melting peaks in polystyrene,[598-600] and in studies on polycarbonate.[601] Differential scanning calorimetery has been used to study[602] the kinetics of chemical reactions of polymers and as a check on experimentally determined activation energies and Arrhenius frequency factors. A combination of programmed and isothermal techniques has been used for characterizing unresolved multistep reactions in polymers.[602]

Differential scanning calorimetry has been used to study crystallization kinetics of many polymeric systems including amorphous cellulose,[603] polyethylene and chlorinated polyethylenes,[604-606] aliphatic polyesters,[607] Nylon 8,[608] Nylon 66 and 610.[609] Using this technique, kinetic studies have been made of polymerization of styrene,[610,611] methylmethacrylate,[611,612] vinyl acetate,[613] bis-maleinimides,[614] and phenolformaldehyde,[615] and also curing of epoxide resins[616-618] and polyesters.[619] A combination of differential scanning calorimetry and differential thermal analysis has been applied to the investigation of the kinetics of oxidation of isotactic polypropylene[620] and the decomposition of polyoxypropylene glycols,[621] and in studies on curing kinetics in diallylphthalate moulding compounds[622-624] and specific heat measurements on PVC composition.[625]

Differential scanning calorimetry and low-angle X-ray scattering have been used to draw phase diagrams of poly(ethylene oxide)–polystyrene block copolymers in the presence of diethyl phthalate as a preferential solvent of polystyrene.[626] Johnston[627,628] studied the effects of sequence distribution on the glass transition temperatures of alkyl methacrylate–vinyl chloride and α-methylstyrene–acrylonitrile copolymers by differential scanning calorimetry, differential thermal dialysis and thermomechanical analysis. Crystallinity values have been determined[629] for poly(p-biphenyl acrylate) and poly(p-cyclohexylphenyl acrylate) from both heat of fusion and heat capacity measurements by differential scanning calorimetry. A method based on differential scanning calorimetry measurement of heat of crystallization has been reported for the determination of the molecular weight of PTFE.[630] Differential scanning calorimetry measurements of the energy during melting of aqueous polymer solutions and gels yield heats of mixing and sorption.[631] The true melting points have been determined of crystalline polymers by plotting the differential scanning calorimetry melting peak temperatures as a function of the square root of heating rate and linear extrapolation to zero heating rate.

Differential scanning calorimetry has been used to study the crystal structure and thermal stability of poly(vinylalcohol) modified at low levels by various reagents and by grafting with other vinyl monomers.[632] It has also been used to study the degree of crystallinity of Nylon 6[633,634] and crosslinked poly(vinylalcohol) hydrogels submitted to a dehydration and annealing process.[635] The development of crystal modifications of poly-1-butene has been followed using differential scanning calorimetry.[636]

A kinetic study of isothermal cure of epoxy resin has been carried out.[637,638] Kinetic parameters associated with the crosslinking process of formaldehyde–phenol, and formaldehyde–melamine copolymers have been obtained from exotherms of a single differential scanning calorimetric temperature scan.[639] The heat of volatilization of polymers has been determined. Values obtained for poly-(methylmethacrylate) agreed well with calculated values.[640] Chemically crosslinked and oriented low-density polyethylenes have been investigated using thermal optical analysis and differential scanning calorimetry.[641,642]

6.4 Thermal volatilization analysis

In this technique, in a continuously evacuated system, the volatile products are passed from a heated sample to the cold surface of a trap some distance away. A small pressure develops which varies with the rate of volatilization of the sample. If this pressure is recorded as the sample temperature is increased in a linear manner, a thermal volatilization analysis (TVA) thermogram showing one or more peaks is obtained.

Thermal volatilization analysis (TVA) thermograms for various poly-methylmethacrylates are illustrated in Fig. 85.

As in the case of thermogravimetric analysis, the trace obtained is somewhat dependent on the heating rate. With polymethyl methacrylate the two stages in the degradation are clearly distinguished (Fig. 85). The first peak above 200°C represents reaction initiated at unsaturated ends formed in the termination step of the polymerization. The second, larger, peak corresponds to reaction at higher temperatures, initiated by

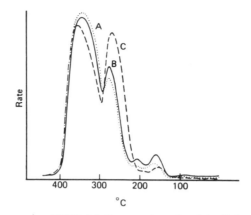

FIG. 85. Thermograms (10°C/min) for samples of poly(methylmethacrylate) of
various molecular weights; (a) 820,000; (b) 250,000; (c) approx. 20,000.

random scission of the main chain. It is apparent that as the proportion
of chain ends in the sample increases, the size of the first peak increases
also. These TVA thermograms illustrate very clearly the conclusions
drawn by Macallum[643] in a general consideration of the mechanism
of degradation of this polymer. The peaks occurring below 200°C can
be attributed to trapped solvent, precipitant, etc. These show up very
clearly, indicating the usefulness of TVA as a method of testing polymers
for freedom from this type of impurity.

The technique has been applied to a range of polymers including
polystyrene,[644,645] styrene–butadiene copolymers,[644] PVC,[644] polyisobutene, butyl rubber and chlorobutyl rubber and poly-α-methyl-
styrene.[644]

6.5 Thermomechanical analysis

In this technique[646,647] the sample is placed between a vitreous silica
platform and a movable silica rod, and subjected to a variable load
(Fig. 56). After equilibration the polymer is wetted with water and the
swelling behaviour followed by measurement of the movement of the
silica rod magnified optically and automatically recorded. In a typical
case there is a sudden contraction of the sample due to inhibition with
water followed by a parabolic swelling curve (Fig. 86). When measurements are performed against a standard reference material it is possible
by this method to obtain comparative swelling data for a range of
polymers.

When the specimen is held at various controlled temperatures in this
technique it is possible to obtain data on thermal relaxation phenomena
in polymer films, fibres and powders.[648] The technique has been used to

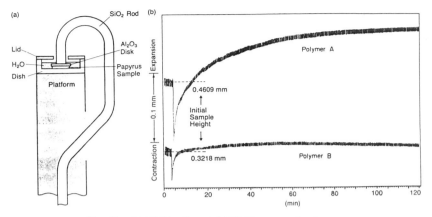

Fɪɢ. 86. (a) TMA cell and (b) TMA curves of polymers.

estimate the tensile compliance for polymers ranging from hard plastics to rubbers.[649] The results correlate well with tensile modulus measurements obtained by conventional techniques. The evaluation of the thermomechanical behaviour of polyurethane showed changes of rigidity due to structural changes occurring during polymer degradation.[650]

6.6 Thermochemical analysis

This is a generic name for a wide variety of processes in which the polymer is heated at controlled conditions of temperature and a chosen atmosphere ranging from inert (e.g. nitrogen on helium) to reactive (e.g. oxygen) and the breakdown products produced are examined by any one of a wide variety of analytical techniques, either qualitatively or quantitatively, or both. A classical example of thermochemical analysis is the pyrolysis–gas chromatography technique described elsewhere in this book. Some other examples of the application of the technique are given below.

In a classical example of the application of thermochemical analysis O'mara[651] studied the thermolysis of a PVC resin containing 57.4% chlorine by two techniques. The first method involved heating the resin in the heated (325°C) inlet of a mass spectrometer in order to obtain a mass spectrum of the total pyrolysate. The second, more detailed, method consisted of degrading the resin in a pyrolysis–gas chromatograph interlaced with a mass spectrometer through a molecule enricher.

Samples of PVC resin and plastisols (10–20 mg) were pyrolysed at 600°C in a helium carrier gas flow. Since a stoichiometric amount of hydrogen chloride is released (58.3%) from PVC when heated at 600°C, over half of the degradation products, by weight, is hydrogen chloride.

A typical pyrogram of a PVC resin obtained by this method using an SE32 column is shown in Fig. 87. The major components resulting from

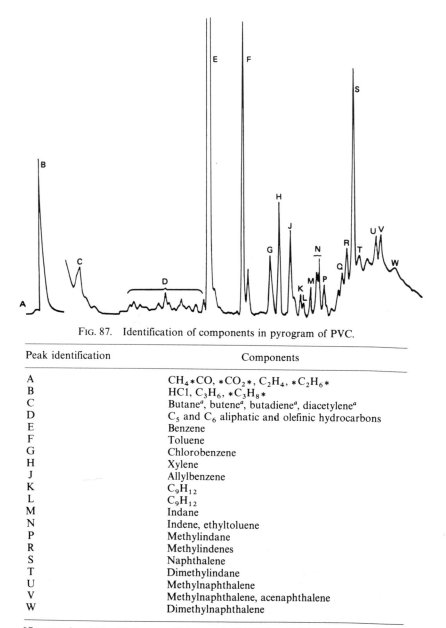

FIG. 87.   Identification of components in pyrogram of PVC.

Peak identification	Components
A	$CH_4 * CO$, $* CO_2 *$, C_2H_4, $* C_2H_6 *$
B	HCl, C_3H_6, $* C_3H_8 *$
C	Butane[a], butene[a], butadiene[a], diacetylene[a]
D	C_5 and C_6 aliphatic and olefinic hydrocarbons
E	Benzene
F	Toluene
G	Chlorobenzene
H	Xylene
J	Allylbenzene
K	C_9H_{12}
L	C_9H_{12}
M	Indane
N	Indene, ethyltoluene
P	Methylindane
R	Methylindenes
S	Naphthalene
T	Dimethylindane
U	Methylnaphthalene
V	Methylnaphthalene, acenaphthalene
W	Dimethylnaphthalene

*Separated and identified on an 8ft. Poropak QS. HCl ~ 58.3%, ash 3–4%.

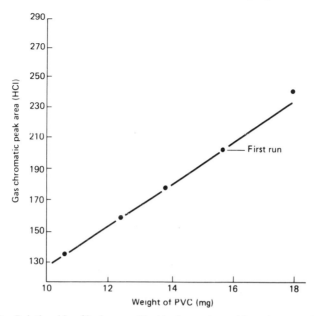

Fig. 88. Relationship of hydrogen chloride chromatographic peak area and amount
of PVC pyrolyzed at 600°C.

the pyrolysis of PVC are hydrogen chloride, benzene, toluene and
naphthalene. In addition to these major products, an homologous series
of aliphatic and olefinic hydrocarbons ranging from C_1 to C_4 are formed.

O'Mara[651,652] obtained a linear correlation between the weight of
PVC pyrolysed and the weight of hydrogen chloride obtained by gas
chromatography. (Fig. 88). Good agreement is obtained between the

TABLE 52   Pyrolytic analysis of HCl from PVC compounds, polyethylene and
chlorinated poly(vinylchloride)

Sample type	Percentage HCl	
	Found	Theoretical
PVC plastisol	30.1	29.1
PVC plastisol	37.0	37.3
PVC plastisol	44.7	44.5
PVC–vinyl acetate copolymer	53.5	52.9
PVC–vinyl acetate copolymer	46.3	46.3
PVC compound	38.5	38.6
PVC compound	47.5	47.7
Chlorinated polyethylene	34.9	34.3
Chlorinated polyethylene	40.7	41.1
Chlorinated poly(vinylchloride)*	63.9	69.2
Chlorinated poly(vinylchloride)	68.5	71.5

*Impure material, theoretical recovery not expected.

Analysis of polymers

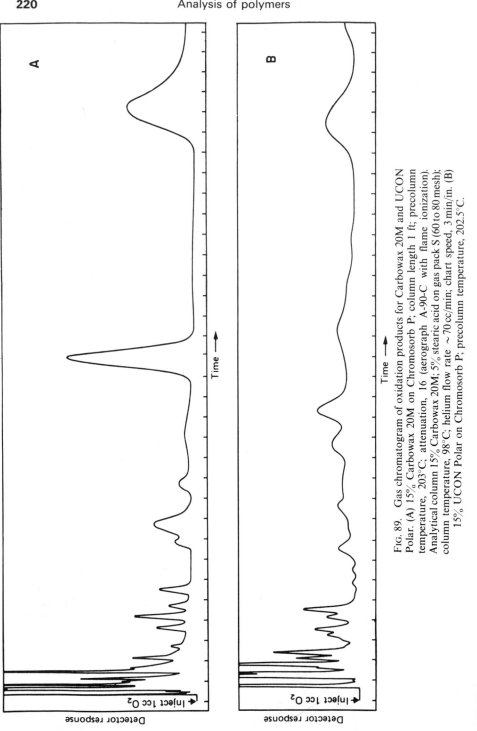

Fig. 89. Gas chromatogram of oxidation products for Carbowax 20M and UCON Polar. (A) 15% Carbowax 20M on Chromosorb P; column length 1 ft; precolumn temperature, 203°C; attenuation, 16 (aerograph A-90-C with flame ionization). Analytical column 15% Carbowax 20M; 5% stearic acid on gas pack S (60 to 80 mesh); column temperature, 98°C; helium flow rate ~ 70 cc/min; chart speed, 3 min/in. (B) 15% UCON Polar on Chromosorb P; precolumn temperature, 202.5°C.

expected and found hydrogen chloride contents by this procedure (Table 52).

The technique was then applied to various PVCs into which different inorganic filters had been incorporated ($CaCO_3$, CaO, $Al(OH)_3$, Na_2CO_3, Al_2O_3, $LiOH$,[653,654] TiO_2, SnO_2 and ZnO_2.[653] This work provided valuable information regarding reaction mechanisms that occur upon heating filled and unfilled PVC to high temperatures.

Mass spectrometry has been used quite extensively as a means of obtaining accurate information regarding breakdown products produced upon pyrolysis of polymers. This includes applications to polystyrene,[655-658] polyethers,[657,659,660] PVC,[657] polycarboxypiperazine and polyurethanes.[661]

The polymers that have been most extensively studied by thermochemical analysis are: PVC,[662-664] polystyrene,[665-668] styrene–acrylonitrile copolymers,[669-671] polyethylene and polypropylene,[672-677] polyacrylates and copolymers,[678-681,687-689] polyethylene terephthalate and polyphenylenes, and polyphenylene oxides and sulphides.[681-686] Whereas, in most instances, thermochemical analysis would be performed under an inert atmosphere such as nitrogen in order to avoid the production of secondary oxidation products, it is the case in some instances that the interest is in obtaining information on the nature of the products produced under oxidizing conditions.

6.7 Oxidative stability studies

Schole *et al.*[690] have reported on the characterization of polymers using an oxidative degradation technique. In this method the oxidation products of the polymers are produced in a short pre-column maintained at 100–600°C just ahead of the separation column in a gas chromatograph. The oxidation products are swept on to the separation column and detected in the normal manner.

Figure 89 shows the chromatograms of Carbowax 20M and UCON Polar polyglycols. The components are the oxidation products produced in the pre-column by the oxygen.

An alternative technique involves infrared spectroscopic examination of films of the polymer that have been subject to oxidative degradation at elevated temperatures.[691-697] Changes in concentrations of carbonyl and other oxygenated functional groups can be obtained by this method.

7

Fractionation and molecular weight

POLYMERS normally do not consist of a particular molecule with a unique molecular weight, but rather are a mixture of molecules with a molecular weight range which follows a distribution. With some types of polymers the picture is further complicated by the appearance of what are known as crosslinks. These are chemical bonds which link one polymer chain to another. Crosslinking will, therefore increase the molecular weight of a polymer and, incidentally, decrease its solubility in organic solvents. These are some of the features which make it possible to produce for a given polymer, say polypropylene, a range of grades of the polymer, each with different physical properties and end-uses and each characterized by a different molecular weight distribution curve and degree of crosslinking. The factors which control these parameters in a polymer are complex, and are linked with the details of the manufacturing process used. They will not be discussed further here. The measurement of the molecular weight is a task undertaken in its own right by polymer chemists, and is concerned with the development of new polymers and process control in the case of existing polymers. Additionally, however, it is necessary to separate a polymer not into unique molecules each with a particular molecular weight, but into a series of narrower molecular weight distribution fractions. This is required in order to obtain a more detailed picture of the polymer structure and these separated fractions may be required for further analysis by a wide range of techniques.

In the simplest case, discussed in section 7.1 below, it is required simply to separate, for example, the total gel fraction (i.e. crosslinked material) of a polymer from the total soluble fraction (non-crosslinked material). This is typified in the example discussed below on the separation of polystyrene into its gel and soluble fractions. In a more complicated case it may be required to carry out a separation of the original polymer into a series of fractions each with a narrower molecular weight distribution than the parent polymer. These methods are usually based on fractionation techniques, in turn based on solvent gradient fractionation or thermal gradient fractionation as discussed in section 7.2. Finally, there is the case where it is required to determine the molecular weight distribution of the polymer. Depending on the type of polymer being examined many methods exist for carrying out these measurements (section 7.3) based on gel permeation chromatography or size exclusion chromatography, to mention new techniques. Techniques that have been in existence for longer include turbidimetric methods, osmometry, viscometry, light-scattering methods, and sedimentation velocity.

7.1   Measurement of crosslinked gel content

Method for soluble–insoluble separations

A good example of this is the separation of the two fractions from a styrene–butadiene–acrylonitrile polymer and the determination of the monomer units in the separated fractions.

To remove non-polymer additives the polymer is dissolved (or dispersed) in chloroform. This solution is slowly poured into an excess of stirred methanol to reprecipitate the polymer and leave the soluble non-polymer additives in a clear chloroform–methanol phase which can be separated from the polymer by filtration. The polymer is then washed with methanol and vacuum-dried.

Contact with methyl ethyl ketone followed by high-speed centrifuging is an excellent method of separating the two fractions. Table 53 reports

TABLE 53   *Determination of monomer units in ABS terpolymers*

Determined	Cycolac T 1000 natural nibs	Cycolac H 1000 natural nibs		Kralastic MH nibs
Analysis of additive free polymer				
Butadiene (%)*	20.5	28.6		19.1
Acrylonitrile (%)[†]	23.8	20.9		20.9
Styrene (%)[‡]	54.0	48.5		
	(total = 98.3%)	(total = 98.0)		
Analysis of insoluble fraction				
Insolubles in polymer (%)	56.0*	51.5**	63.5	11.0
Butadiene (%)	30.8	31.2	38.2	49.9
Acrylonitrile (%)	22.1	20.6	19.2	10.7
Styrene (%)	48.0	48.0	—	—
	(total = 100.9%)	(total = 99.8%)		
Analysis of soluble fraction				
Solubles in polymer (%)	44.0	48.5	—	—
Butadiene (%)	10.3	7.0	—	13.9
Acrylonitrile (%)	27.6	28.7	—	21.5
Styrene	62.0	64.0	—	61.0
	(total = 99.9%)	(total = 99.7%)		(total = 96.4%)

*Iodine monochloride method.
[†]Kjeldahl method.
[‡]Infrared method.

complete gel and soluble fraction analysis carried out on various ABS polymers. In both the whole ABS polymer and in its soluble and insoluble fractions the sum of the determined constituents usually adds up to $100 \pm 2\%$.

Methods have been described for the determination of gel in PVC,[698-700] vinyl chloride–propylene copolymer,[700] polybutadiene–polyisoprene copolymer styrene–butadiene copolymer, acrylonitrile–butadiene, copolymer polyacrylonitrile[699] and styrene–acrylonitrile copolymer.[701]

7.2 Types of polymer fractionation procedures

7.2(a) Fractionation of polymers based on molecular weight

The time-consuming and laborious nature of the earlier fractionation procedures is illustrated well by the work of Nakajima[702] on the fractionation of polyethylene and its thermally degraded products, involving extraction with boiling hydrocarbons with increasing boiling points between 45°C and 95°C, and on the fractionation of polypropylene.[703] It was necessary to extract the polymer using a Soxhlet apparatus with 17 different hydrocarbon fractions based on normal paraffins with different boiling temperatures in the range from 35°C to 135°C. The extreme laboriousness of such procedures is self-evident.

Fractional extractions of polymers by the column technique is no less laborious. Two types of column extraction procedure are known. Gradient elution fractionation is achieved at a given temperature by making use of solvents with gradually increasing solvent power, or the increasing temperature fractionation is performed with a given solvent at increasing temperatures. According to the findings on column techniques by Wijga et al.,[704] the gradient of the polymer separates fractions according to molecular weight, whereas the increasing temperature method fractionates the polymer mainly according to tacticity.

7.2(b) Gradient elution with solvents of increasing solvent power at constant temperature

The polymer is packed in a column and a solvent mixture passed down the column. Initially the solvent is a poor one for the polymer. Then increasing proportions of a better polymer solvent are incorporated in the solvent mixtures utilizing a gradient elution technique. A series of fractions is thus obtained, containing different molecular weight fractions of the original polymer. Methods have been described for the large-scale elution fractionation of ethylene–propylene copolymers,[705] methyl methacrylate–styrene copolymers,[706] polyethylene,[707] and polypropylene.[708]

7.2(c)   Increasing (or decreasing) temperature fractionation

When a solid polymer packed in a column is contacted with a continuing flow of solvent in which it has a limited solubility at room temperature then, as the temperature of the column is raised in a controlled programme, fractions of the polymer with different tacticities will progressively dissolve. Collection of portions of the column eluate at various temperatures will provide a series of fractions.

Akutin et al.[708] have shown that by a thermal precipitation method the molecular weight distribution of low-density polyethylene could be deduced from the precipitation curves using simple calculations. Ogawa and Hoshino[709] compared fractionations of isotactic polypropylene by using temperature and solvent gradient methods. Comparison of results agreed fairly well on fractionation of polyethylene–polypropylene blends between hypothetical calculations and experimental data by solvent gradient method.[710]

7.2(d)   Fractionation by polymer freezing

Polystyrene dissolved in benzene has been fractionated by slowly freezing the solutions with dry ice and alcohol. A more detailed treatment of this method is given by Loconti and Cahill.[711,712] The polymer in the first frozen-out portions was of higher molecular weight than later. Ruskin and Parravano[713] were able to fractionate polymer dissolved in cyclohexane by both the zone-melting and freezing techniques.

Bryson et al.[714] investigated the fractionation of polystyrene by slow freezing of dilute benzene solutions of the polymer using ice-water mixtures instead of dry ice–alcohol. Bryson et al.[714] concluded that no fractionation of the polymer occurs according to molecular weight from benzene solutions.

7.2(e)   Precipitation chromatography

In this technique the polymer is dissolved in a solvent and then gradual additions are made of a non-solvent to the polymer. The polymers precipitated after each non-solvent addition are separately collected. An example is the case of vinylchloride–vinyl acetate copolymers using solvent–non-solvent systems such as acetone–methanol, acetone–heptane and tetrahydrofuran–water.[715]

Precipitation with methanol from benzene solution has been used to fractionate methacrylic acid–styrene copolymer,[716] polychloroprene,[717,718] polybutadiene,[717,718] and polyiosprene.[717,718] Other solvent–non-solvent systems that have been used include nitrobenzene–tetrachloroethane (polyethylene tetraphthalate[719]) and ethylene carbonate–ethylene cyanohydrin or methyl ethyl ketone–cyclohexane (styrene–acrylonitrile copolymers).[720]

7.2(f) Column absorption chromatography

Chromatography on silica gel columns

This is another technique for achieving fractionation of polymers. It has been applied to the fractionation of polyoxyethylene,[721] polyoxypropylene,[721] polytetrahydrofuran,[721] polyepichlorohydrin,[721] oligobutadienes,[721] and carboxy- and hydroxy-terminated polybutadienes.[722,723] Isotactic polymethylmethacrylate has been separated from polymethylmethacrylate using competitive absorption on silica gel from chloroform solution.

7.2(g) Thin-layer chromatography

Thin-layer chromatography is a useful technique for separating polymers into molecular weight fractions on a fairly small scale. It has been used to fractionate polyethylene terephthalate,[724–726] styrene–butadiene copolymers,[727] styrene–acrylonitrile copolymers,[728,729] polyoxypropylene glycols,[730] Nylon–styrene graft copolymers,[731] polymethylmethyacrylate[732–734] styrene–methacrylate polymers,[735] poly-α-methylstyrene,[736] polyvinyl acetate–styrene copolymers and polyvinyl alcohol–styrene copolymers.[737] Various solvents have been used for migration on thin-layer plates including benzene–acetone,[729] toluene–acetone,[728,729] chloroform–diethyl ether,[726] chloroform–ethyl acetate and chloroform.

7.3 Methods for measuring molecular weight distribution and molecular weight

7.3(a) Column fractionation methods

The most popular and, incidentally, newest column method of measuring molecular weight distribution is known as gel permeation chromatography or size-exclusion chromatography. Other techniques are turbidimetry and chromatographic methods.

    Methods for the determination of molecular weights of polymers, or on the fractions of polymers obtained in a fractionation, include viscometry, osmometry, light-scattering methods, sedimentation velocity measurements, field desorption, masspectrometry, laser mass spectrometry and end-group analysis.

7.3(b) Gel-permeation chromatography

This is a technique where a solution of the polymer is passed down a column packed with a gel, which separates the polymer into fractions based on molecular weight by reason of the small pores present in the gel which, being of molecular dimensions, tend to pass through the smaller polymer molecules and slow down the larger molecules. The next problem is to detect the increasing higher molecular weight fractions of the polymer as they leave the column. Two principal types of detector are employed; one based on measurement of dielectric constant and the other on measurement of refractive index.[738-741] An important characteristic of detectors in this application is that they are universal and fairly non-selective as occurs, for example, in the case of ultraviolet or infrared detectors[742] which operate at a particular wavelength and rely on all individual constituents and the sample absorbing at a particular wavelength. The dielectric constant detector is a bulk property detector that complements, rather than competes with, the refractive index detector. The dielectric constant detector offers specific advantages for certain analyses, while the refractive index detector is preferable in other applications. Bode et al.[743] used this technique for the examination of styrene–butadiene copolymers.

The dielectric constant detector has several advantages over the refractive index detector for some size-exclusion chromatography analyses. For example, the refractive index detector loses sensitivity faster as a function of temperature than does the dielectric constant detector. Therefore at elevated temperatures the dielectric constant detector may be more sensitive than the refractive index detector, even when the opposite is true at room temperature. This is caused by the reduced output of the light source and the reduced sensitivity of the photodiodes commonly used in refractive index detectors. As shown in Table 54, monomer ratios also greatly affect the response of the refractive index detector in some copolymers. In the case of styrene–butadiene copolymers, the refractive index response variation due to changes in the monomer ratio is approximately an order of magnitude higher than the variation observed with a dielectric constant detector. This effect can be

TABLE 54 *Detector response*

	RI	ΔRI*	ε	$\Delta\varepsilon$*
Polybutadiene	1.52	0.11	2.3	3.3
Polystyrene	1.60	0.19	2.6	5.0
Percentage		42%		5.6%

* Versus tetrahydrofuran; RI = 1.41, $\varepsilon = 7.6$.

significant for size-exclusion chromatographic analyses of copolymers with unknown or variable monomer ratio.

Figure 90 shows chromatographs of a polystyrene calibration standard. The composite chromatogram of Fig. 91 shows the chromatographic separations of four different styrene–butadiene copolymers. Figure 92 is a composite chromatogram showing the results obtained with a refractive

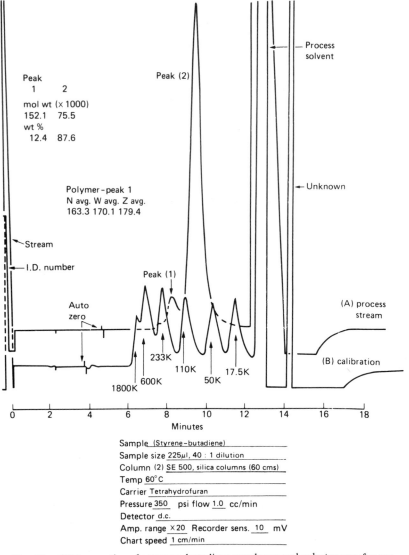

FIG. 90.   SEC separation of a styrene–butadiene copolymer and polystyrene reference standards.

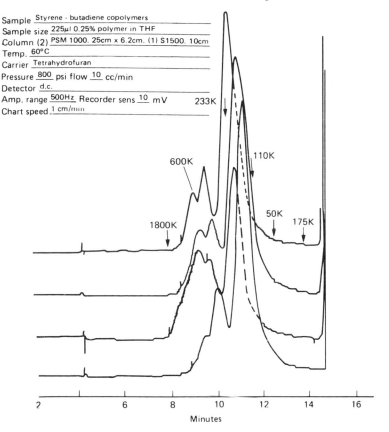

Sample _Styrene - butadiene copolymers_
Sample size _225μl 0.25% polymer in THF_
Column (2) _PSM 1000. 25cm x 6.2cm. (1) S1500. 10cm_
Temp. _60°C_
Carrier _Tetrahydrofuran_
Pressure _800_ psi flow _10_ cc/min
Detector _d.c._
Amp. range _500Hz_ Recorder sens _10_ mV   233K
Chart speed _1 cm/min_

600K

110K

1800K

50K   175K

2   6   8   10   12   14   16
Minutes

FIG. 91.   SEC separations of four different styrene butadiene copolymers.

index detector and dielectric constant detector in series. Some differences in the responses are evident. They both clearly show the bimodal molecular weight distribution of the copolymer, but the additive that elutes after the process solvent is barely seen by the refractive index detector. Note also the difference in response to the process solvent of the two detectors. The measurement of molecular weight averages and the distributions for polymers by gel-permeation chromatography requires the construction of a calibration curve using relatively monodisperse polymers such as a series of polystyrene in the molecular weight range of 10^3–10^6 with $\bar{M}_w/\bar{M}_n < 1.10$. However, to calculate the molecular weight averages and the distribution for any polymers other than polystyrene by this calibration curve, it is necessary to transform molecular weight units in the curve to those for the polymer specified. One of the techniques used to transform the molecular weight units is so-called 'universal calibration',[744] and this method has been shown to have wide applicability. The procedure, however, requires accurate values

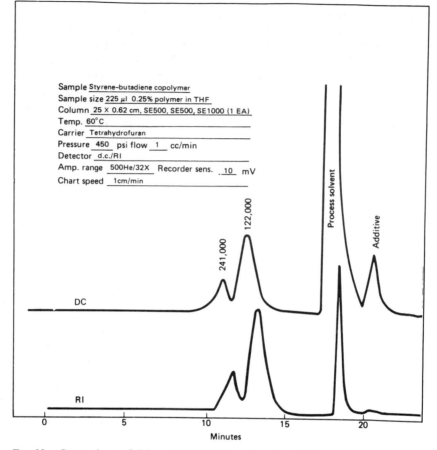

Fɪɢ. 92. Comparison of RI and DC detectors for the SEC separation of a styrene–butadiene copolymer.

of the Mark–Houwink parameters for both the sample and the standard used for calibration in the same solvent at the same temperature as the gel permeation chromatographic analysis, and therefore is not practical in most cases. The 'Q' factor method[745] also has many shortcomings, and should be used with caution. The 'Q' factor method can provide accurate results only when the Mark–Houwink exponents for both the sample and the standard are identical,[746] and when the Q factor is determined experimentally in advance in the same solvent at the same temperature.[747]

Many attempts have been undertaken to overcome the problem of the lack of suitable standards for each polymer type.[748–752] These attempts employ a known set of molecular weight averages of the broad molecular weight distribution polymers concerned. Some of the methods for calibration using polydisperse polymer samples were applied to linear calibration curves.[748-750] The calibration curve of a column given as a plot of log molecular weight versus the elution volume is generally a smooth curve with only slight curvature in the molecular weight range for which the column will be applied. Therefore, the calibration curve should be approximated by the third-order polynomial.[753] McCrackin presented a calibration method applied to the third-order polynomial,[751] but it requires considerable calculation. The method described by Mahabadi and O'Driscoll[752] demands a calibration curve for polystyrene standards, the Mark–Houwink parameters for polystyrene and intrinsic viscosity and an average molecular weight for polymer samples.

Weiss and Cohn-Ginsberg,[754] Belinskii and Nefedov,[755] and Kalinsky and Janca[756] reported the method to calculate size-exclusion chromatographic columns directly for the polymer under study.[754] The universal calibration curve, $\log [\eta]$ M versus elution volume, was first established from narrower molecular weight distribution polystyrenes, and its use was then extended to other systems to estimate the Mark–Houwink parameters for the sample from data on broad molecular weight distribution polymer (or polymers). Two polymers have different intrinsic viscosities, or one polymer for which the intrinsic viscosity and $\bar{M}_w$ (or $\bar{M}_n$) were known was required to calculate the parameters with gel permeation chromatography curves. Once the parameters were obtained, one could calculate molecular weight averages of the polymer under study. The validity of the method was estimated.[757] Recently, Hamielec and Omoridion[758] presented an improved version which includes peak broadening corrections, and they applied it to the non-aqueous gel-permeation chromatography of polyvinylchloride and the aqueous gel-permeation chromatography of polydextran.[758] A method for obtaining a calibration curve for polymer solvent systems where polystyrene standards are not soluble was proposed. The method requires the universal calibration curve based on polystyrene in A solvent, the integral distribution curves of elution volumes for the polymer under study in A and B solvents for the generation of a calibration curve of the polymer in B solvent.

Mori[759] has discussed a modification of the methods of Weiss and Cohn-Ginsberg[754] and Hamielec and Omoridion[758] for calibrating gel-permeation chromatographic columns.

The modified method utilizes polydisperse calibrating samples for which any molecular weight averages are known, and a calibration curve of log molecular weight versus the elution volume constructed with polystyrene standards. Assumption of any formula for the calibration curve is not required. This method does not require prior knowledge of the specific Mark–Houwink parameters for both polystyrene standards and the polymers under study, nor does it require the determination of instrinsic viscosities of both polymers.

Basic theory of determination of molecular weight distribution

Let us first assume that the molecular weight $(M_B)_i$ of a species of a polymer B is related to the molecular weight $(M_A)_i$ of a species of a polymer A eluting at the same elution volume i by the expression.

$$(M_B)_i = {}_s(M_A)_i^t \tag{1}$$

where s and t represent constants. Let polymer A be a polystyrene standard experimentally used to establish a primary calibration curve and polymer B be a polymer requiring analysis. Suppose that the weight average molecular weight $\bar{M}_w$ and the number average molecular weight $\bar{M}_n$ of polymer B with a broad molecular weight distribution are given. These values of $\bar{M}_w$ and $\bar{M}_n$ should be measured by absolute methods, such as light-scattering and osmometry. Let the chromatogram of a polymer sample be given by its height h_i as a function of elution volume i. The weight- and number-average molecular weights of the polymer B (secondary standard) may be computed from the chromatogram h_i of the polymer and a calibration curve for polystyrene standards by the formulas

$$(\bar{M}_w)_c = \sum (h_i s(M_A)_i^t)/\sum h_i \tag{2}$$

$$(\bar{M}_n)_c = \sum h_i/\sum (h_i/s(M_A)_i^t) \tag{3}$$

where the subscript c indicates calculated molecular weight averages. Guess value t first, then the molecular weight averages of the polymer B may be computed by eqns (2) and (3) from the chromatogram as follows:

$$(\bar{M}_w)_c = k_1 s \tag{4}$$

$$(\bar{M}_n)_c = k_2 s \tag{5}$$

where k_1 and k_2 are numerical values obtained from h_i and $(M_A)_i^t$. Let us assume

$$\bar{M}_w = (\bar{M}_w)_c; \quad \bar{M}_n = (\bar{M}_n)_c \tag{6}$$

then we obtain two values of s, which may be not equal in most cases. The goal of this calculation is to find a value t where a value of s from eqn (4) equals that of s from eqn (5). Trial value of t is assumed ar values of s are calculated by eqns (4)–(6). The process may be rej with other values of t until [s of eqn (4) − s of eqn (5)] turns (sufficiently small. The numerical calculations were performed desk-top microcomputer. A hand calculation is also applicable. O: obtain a final set of t and s in a few minutes. A calibration curve for polymer B can be constructed from eqn (1), a final set of t and s, and a polystyrene calibration curve. Besides one polymer sample with broad molecular weight distribution and its $\bar{M}_w$ and $\bar{M}_n$, we can use two polymer samples with broad molecular weight distribution given either two $\bar{M}_w$, two $\bar{M}_n$, or one $\bar{M}_w$ and one $\bar{M}_n$. The experimental substantiation of eqn (1) has been given by size-exclusion chromatography[760] of oligomers. For example, the following relation has been obtained between oligostyrene (A) and oligo (ethylene glycol) (B) in the range of molecular weight 200 and 3400

$$M_B = 0.967 M_A^{0.945}$$

Similar relations have also been obtained for n-hydrocarbon epoxy resin, and p-cresol novolac resin as oligomer B.

The theoretical basis of this calibration method can be derived from the principle of 'universal calibration'. If it can be assumed that all polymers at a given elution volume have the same hydrodynamic volume and the same value of $[\eta]M$, then we can write finally, for any particular elution volume i:

$$(M_M)_i = (K_A/K_B)^{1/(1+a_B)}(M_A)_i^{(1+a_A)/(1+a_B)} \tag{7}$$

where $[\eta]$ is the intrinsic viscosity in the solvent of the polymer having molecular weight M, the subscripts A and B refer to the polymer A and the polymer B, respectively, and K and a are the respective Mark–Houwink parameters (a coefficient and an exponent). As the values K and a are constants in the specified sample, solvent, and temperature, eqn (7) can be regarded as identical with eqn (1). Hence the values of the constants t and s are rearranged as

$$t = (1 + a_A)/(1 + a_B) \tag{8}$$

$$s = (K_A/K_B)^{1/(1+a_B)} \tag{9}$$

Thus, from eqn (1), having evaluated the constants s and t, it is possible, knowing the molecular weight M_A of a standard polymer A, to calculate the molecular weight (M_B) of a second polymer. It is then possible to calculate the weight average molecular weight $(\bar{M}_w)c$ of the polymer B and the number average molecular weight $(\bar{M}_n)c$ of the polymer B.

Figure 93 compares calibration curves, experimental and calculated, obtained for the seven polystyrene standards listed in Table 55. Though the curves, experimentally obtained and calculated by using the value of $\bar{M}_w = 257,800$, coincide partly at the middle of the curves, which corresponds to the centre of the chromatogram of NBS 706, their extreme parts clearly diverge, implying this discrepancy may arise from the effect of peak broadening.

If a sample includes only small (or larger) molecular weight species, as

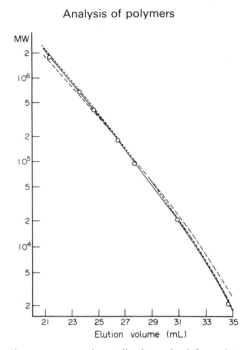

Fig. 93. Calibration curves experimentally determined for polystyrene standards (○) and calculated for polystyrene NBS 706 as $\bar{M}_w = 257\,800$ (dashed line) and $\bar{M}_w = 288,000$ (dotted line).

TABLE 55   *Molecular weight averages and intrinsic viscosity data of sample polymers**

Sample	$\bar{M}_w$	$\bar{M}_n$	$[\eta]$	Molecular weight range
PVC-1	132,000	54,000	1.097	$1800\text{–}1.3 \times 10^6$
PVC-2	118,000	41,000	0.937	$1000\text{–}1.1 \times 10^6$
PVC-3	83,500	37,400	0.749	$1500\text{–}7.3 \times 10^5$
PVC-4	68,600	25,500	0.640	$700\text{–}6.8 \times 10^5$
PMMA	60,600	33,200	0.303	$2000\text{–}3 \times 10^5$
PVAc	331,000	83,000		$2500\text{–}3.2 \times 10^6$
PS(NBS 706)	257,800 (LS)	137,000	0.815	$300\text{–}1.6 \times 10^6$
	288,000 (SD)			

*LS, measured by light-scattering; SD, measured by sedimentation equilibrium ultracentrifugation; $\bar{M}_w$ and $\bar{M}_n$, manufacturer's data; $[\eta]$, measured at 250°C in THF; molecular weight range is the lowest and the highest molecular weight of species included.

in the range of one extreme part, and has narrower molecular weight distribution than NBS 706, then the molecular weight average calculated will be higher (or smaller) than the true values. When the values $\bar{M}_w = 288,000$ was used as a standard value the results imply that the calibration curve might be shifted to some extent to correct the deviation of molecular weights for primary standards, rather than peak broadening.

To take the discussion further, consider a particular polymer, poly-propylene, in further detail. Crystalline polypropylene is mainly charac-terized by three factors: tacticity, molecular weight and molecular weight distribution. Molecular weight distribution is a very ambiguous factor, in spite of the amount of attention it has received. This is because the observed distribution curves depend on the determination method, and no definite method has been established.

The molecular weight distribution of crystalline polypropylene is generally determined by column fractionation or gel-permeation chro-matography. To obtain the molecular weight distribution curve by column fractionation takes many hours, and the polymer must be protected from thermal degradation. The determination of the molecular weight distribu-tion curve of polypropylene by gel permeation chromatography is, on the other hand, very easy.

Ogawa et al.[761] have compared column fractionation and gel-permea-tion chromatography methods for determining the molecular weight distribution of polypropylene. The calculated statistical parameters such as average molecular weight, standard deviation, skewness, and kurtosis for each distribution curve, and also the number-average and weight-average molecular weights were determined by osmometry and light scattering to compare with those from distribution curves. The results of their investigation on the comparison of the determination methods of molecular weight distribution for crystalline polypropylene are sum-marized as follows.

1. The molecular weight distribution curve obtained from column fractionation was narrower than that from gel-permeation chromato-graphy, and the D value from gel-permeation chromatography was closer to that from the absolute methods (osmometry and light-scattering).
2. It was confirmed that the distribution curve from column fractionation became similar to that from gel-permeation chromatography on correcting the overlapping of the distribution of fractionated polymers.
3. The broadening effect in gel-permeation chromatography and thermal degradation during fractionations were both found to be of little importance.
4. It was assumed by Crouzet et al.[762] that a broader distribution curve had been obtained from gel-permeation chromatography due to a broadening effect as compared with that from column fractionation, and that the distribution curve from column fractionation was more accurate than that from gel-permeation chromatography. However, Ogawa et al.[761] consider that the distribution curve obtained from gel-permeation is more accurate and reliable than that from column fractionation.

Gel-permeation chromatography has been applied to the determination of the molecular weight distribution of a wide range of polymers including low- and high-density polyethylene,[763-774] polypropylene,[776,777] *cis*-1-4-polybutadiene, ethylene–propylene copolymers,[775] polystyrene,[778-787] isotactic polystyrene,[788] styrene–butadiene copolymers,[789-793] butadiene and methyl styrene copolymers,[792,793] sulphated polystyrene,[794] PVC,[795-798] vinyl acetate–vinyl chloride copolymers,[800] vinyl chloride–vinylidene chloride copolymers,[801] vinylidene chloride–methylmethacrylate copolymers,[801] acrylate–acrylonitrile copolymers,[802] polyacrylonitrile,[803,804] acrylonitrile–butadiene–styrene terpolymers,[803] polyethylene terephthalate,[805-808] polymethyl methacrylate,[799,809-815] poly(2-methoxyethylmethacrylate).[815] polycarbonates,[816,817] polyethylene glycol oligomers,[799,818,819] polypropylene glycol,[819,820] polyvinylpyrrolidone,[804] polyvinyl acetate,[799] expoxy resins[799] and phenol novalac resins.[799]

7.3(c) Turbidimetry

Two principal turbidimetric techniques are used to determine the molecular weight distribution of polymers. The first of these is the temperature gradient method, which is very similar to the thermal gradient method, mentioned earlier, for the preparation of polymer fractions. This method has been applied to the determination of the molecular weight distribution of polyethylene.[821] In this method a solution of α-chloronaphthalene solvent and of a non-solvent containing a very low concentration of polyethylene is slowly cooled.

The high molecular weight species become insoluble and separate out, causing a small amount of turbidity. As the temperature continues to decrease, increasing amounts of polymer are precipitated out according to their molecular weight. Finally, a point is reached at which even the lowest molecular weight species become insoluble in the solution. At this point the turbidity is greatest, and ideally all of the polymer is precipitated but remains in suspension as very fine particles. If the increase in turbidity is plotted against the decreasing temperature, a cumulative plot is obtained which is similar to a cumulative wt% versus molecular weight. The increase in turbidity is related to the cumulative wt% and the molecular weight is related to the decrease in temperature. Figure 94 shows a comparison of typical results from a stirred cell and a divided cell. The ordinate in this is in terms of ΔE, the decrease in millivolts in phototube output from the starting value. Taylor and Tung[821] discuss the application to their results to procedures described by Morey and Tamblyn[822] and Claesson,[823] to convert the experimental turbidity into molecular weight distribution data and the difficulties they encountered in this work.

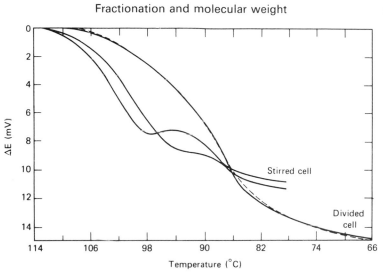

FIG. 94.   Comparison of turbidity, temperature curves from stirred and divided cells.

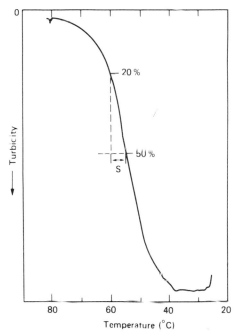

FIG 95   Chart record from photoelectric turbidimeter X–Y recorder, M_w/M_n of polymer = 8.0.

Gamble *et al.*[824] used a photoelectric turbidimeter for measuring molecular weight distribution of poly-α-olefins and ethylene–propylene copolymers. This instrument measures changes in turbidity as a function of temperature (Fig. 95).

A parameter designated as S was chosen by Gamble *et al.*[824] to be the difference in temperature between points representing 20% of the maximum turbidity and 50% of the maximum turbidity as suggested by Taylor and Tung.[249] This portion of the curve is essentially linear for measuring polydispersity. The parameter S is correlated with a parameter determined from the Wesslau equation.[825] Because of the fairly narrow working temperature range, 80–25°C, of the solvent–non-solvent mixture used (heptane-*n*-propanol) it was necessary to study the effect of polymer concentration to provide a suitable, turbid system. Curves in Fig. 96 demonstrate the effect of changing the concentration of ethylene–propylene rubber without changing the concentration of the non-solvent.

The second method of determining molecular weight distribution is turbidimetric titration. Beattie[826] developed an absolute turbidimetric titration method for determining solubility distribution (which is closely related to molecular weight distribution) of polystyrene. In this method polymer is precipitated from its solution in methylethyl ketone by addition of a non-solvent (isopropanol) of the same refractive index as that of the

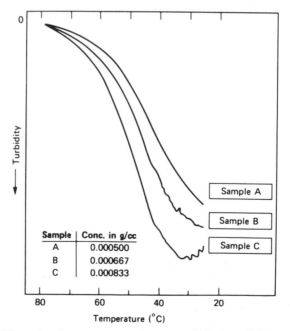

Sample	Conc. in g/cc
A	0.000500
B	0.000667
C	0.000833

FIG. 96.   Effect of polymer concentration on turbidimetry. Ethylene–propylene rubber.

solvent. He showed, by the use of light-scattering theory, that under these conditions the concentration of polymer which is precipitated can be calculated from the maximum turbidity on an absolute basis. He also discusses the effect of particle size and particle size distribution. Beattie concluded that, with polystyrene, under the specified conditions, the reproducibility of turbidimetric precipitation curves is very good, and that the method is accurate. Gooberman[827] also studied polystyrene but dissolved the polymer in butanone and titrated with isoproponal. He devised a correction procedure for the loss of precipitate during the titration and devised a method of location of the point of precipitation in relation to the weight-average molecular weight.

Tanaka *et al.*[828] have used the high-temperature turbidimetric titration procedure originally described by Morey and Tamblyn[822] for determining the molecular weight distribution of cellulose esters. This method has been applied to the measurement of the molecular weight distribution of polypropylene. They found that the type of molecular weight distribution of these polymers is a log-normal distribution function in a range of I(M), (cumulative wt%), between 5% and 90%. The effect of heterogeneity in the molecular weight distribution of polypropylene on the viscosity–molecular weight equation was examined experimentally; the results agreed with those calculated from theory. Strict temperature control ($\pm 0.15°C$) is necessary in these determinations.[829]

Taylor and Graham[830] have described a dual-beam turbidimetric photometer which they claim has distinct advantages over the earlier single-beam instruments,[831,832] and have applied it to the determination of the molecular weight distribution of polyethylene, polypropylene, ethylene–propylene rubbers and other polymers.[833]

Turbidimetric titration has been applied to a wide range of polymers including polyethylene,[826] polypropylene,[828] polystyrene,[829,827,824–836] polymethylstyrene,[837] PVC,[838] chlorinated PVC–styrene,[839] polyethylene oxide,[840] polyphenyleneoxide,[841] polyethylene terephthalate[842,843] and polyethylene glycol.[829]

7.3(d) Chromatographic methods

Molecular weight distributions of poly(ethylene glycols) have been determined by liquid–liquid extraction,[844] paper chromatography,[845] thin-layer chromatography[846] and gas chromatography.[847,848] Gas chromatography is, of course, limited to lower molecular weight polymers which can be volatilized.

Mikkelsen[847] reported a gas chromatographic analysis of polyethylene glycol 400; the original sample is injected directly into the gas chromatograph, a technique also employed by Puschmann.[848] Celedes Pacquot[849] converted the polyethylene glycols into methyl ethers in

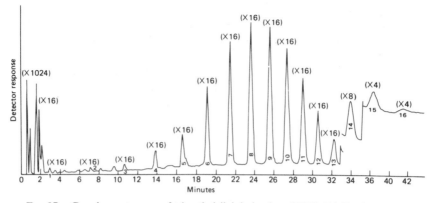

FIG. 97.   Gas chromatogram of trimethylsilyl derivative of PEG 400. Dual columns, 3 ft by 0.25 in. o.d.; 5% SE-30 on Chromosorb G. Temperature programmed from 100 to 370°C at 10°C/min, and then isothermal at 370°C.

a reaction with dimethylsulphate before injection into the gas chromatograph.

The rapid formation of trimethylsilyl ether derivatives of polyhydroxy compounds, followed by their separation and estimation by gas chromatography, has been described by Sweeley et al.[850] This most useful technique has been applied to the liquid polyethylene glycols 200, 300 and 400. Fletcher and Persinger[851] studied this application using poly(ethylene glycols) 200, 300 and 400 and reported the determination of response factors relative to ethylene glycol for conversion of peak areas to weight per cent when a thermal conductivity detector is used. A typical chromatogram of PEG 400 is shown in Fig. 97. *p*-Toluene sulphonic acid catalysed acetylation has also been used to produce volatile derivatives of propylene oxide-glycerol condensates which are suitable for gas chromatography.[852] Graphs relating initial retention distance on the gas chromatogram with molecular weight give straight line relationships.

7.3(e)   Viscosity measurements

The viscosity-average molecular weight ($\bar{M}_v$) of polymer sample may be computed from the chromatogram h_i of the polymer by the formula.

$$\bar{M}_v = \left(\sum h_i M_i^a / \sum h_i\right)^{1/a}$$

Substituting in the Mark–Houwink equation gives

$$[\eta] = K\left(\sum h_i M_i^a / \sum h_i\right)$$

where *a* and *K* are Mark–Houwink parameters and $[\eta]$ is the intrinsic viscosity of the polymer. It is thus possible to calculate average molecular weights from viscosity data on the weighed fractions obtained by chromatography, or on the original unfractionated polymer provided appropriate calibration data are available for the calculation of constants in the above equation.

Intrinsic viscosity measurements have been used extensively for the determination of the molecular weight distribution of poly-ethylene,[853-859] polypropylene,[860-863] PVC[864] and polyethylene tere-phthalate.[865,866]

7.3(f) Osmometry

Osmotic pressure measurement of solutions of a polymer in a solvent can be employed to determine the molecular weight of the polymer. The results in Fig. 98 show that after approximately 1 hour the osmotic pressure of solutions of poly-α-methylstyrene in toluene have become practically constant. From this osmotic pressure and the concentration of the polymer in solution it is possible to calculate the molecular weight

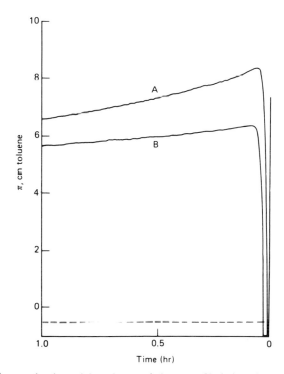

FIG. 98.   Low-molecular weight poly-α-methylstyrenes. Variation of osmotic pressure with time: (A) PAMS 1, 0.0913 g/100 ml; (B) PAMS 2, 0.104 g/100 ml. Pressures in cm toluene.

of the polymer from the following relationship:

$$M = \frac{aRT}{PV}$$

where M = molecular weight
        R = gas constant
        T = temperature (K)
        P = osmotic pressure
        V = volume of sample solution (ml)
        a = concentration of polymer in solution, $g\,l^{-1}$

Osmometry has been used to measure the molecular weight of a wide range of polymers including PVC,[867] polyethylene terephthalate,[868-870] Nylon,[869] polystyrene,[868] linear polyesters,[870] including polytetramethylene terephthalate, polypentamethylene terephthalate, polyhexamethylene terephthalate and polytetramethylene isophthalate.

Other methods of molecular weight measurement

These include light-scattering measurements on solvent solutions of polymers. Polymers studied by the light-scattering method include polystyrene,[871-874] styrene–methyl methacrylate copolymer,[874] water-soluble acrylic acid and acrylamide polymers[875] and poly-α-methyl-styrene.[876]

Sedimentation methods have also been applied to polystyrene[877] and polyglycols and polyesters.[878] End-group analysis is a useful method of measuring the molecular weight of certain types of polymers which are known to have one or two terminal end-groups per molecule. Thus Kalinina and Motorina[879] developed a method for the determination of the molecular weight of compounds having terminal hydroxyl groups, e.g. polyethylene glycols. This was estimated from treatment with ceric ammonium nitrate reagent in nitric acid followed by spectrophotometric measurement of the coloured complex.

Laser mass spectroscopy has been used[880] to determine the average molecular weight for polyethylene and polypropylene glycols. Field desorption mass spectroscopy has been applied[881] to the determination of molecular weight averages of polystyrene oligomers with molecular weights up to 5300 atomic mass units.

8

Additives in polymers

IN ORDER to appreciate fully the technique which have been developed for the analysis of additives in polymers, it is necessary to be familiar with the diffculties involved in such an undertaking, and also with the chemical and physical properties of the additives themselves.

Most of the analytical problems arise from three factors: the situation of the additive in a more or less insoluble polymer matrix, the high reactivity and low stability of many types of additives, especially antioxidants, and the low concentrations of additives present in many instances in the polymer matrix. The first factor severely limits the choice of analytical techniques that can be applied to the sample without prior separation of the additive from polymer, a procedure which is itself hindered by the nature of the polymer matrix. In addition, any extract of the polymer is liable to contamination by low molecular weight polymer 'wax' which may interfere with subsequent analysis and is difficult to remove.

The second and third factors mentioned above combine to make the handling of extracts an exacting job if quantitative information is required. Antioxidants, particularly, are labile unstable compounds, forming complex decomposition products; this considerably complicates interpretation of analytical data, and any loss of material by decomposition is liable to be significant since the quantities present are initially so low. The writer[882] and others, for example, have recommended that polymer extracts are kept in actinic glassware and used for subsequent analysis without delay. If any storage of solutions is necessary, this should be done under nitrogen, in the dark and in a refrigerator to minimize the effects of oxygen, light and heat on any labile compounds present. Lorenz et al.[883] have published data on sample changes during handling of antioxidant extracts, including losses during concentration by evaporation. Generally, however, this aspect of additive analysis does not seem to have received the consideration it deserves.

Apart from those factors which complicate the processing of the sample, there are others which complicate the interpretation of the data obtained, the principal ones being the wide range of additives used nowadays in polymer technology, which makes positive identification difficult by all but the most sophisticated analytical techniques, the presence of several types of additives in a single polymer formulation, e.g. plasticizers, ultraviolet stabilizers, slip agents and possibly two antioxidants, one for processing and one for service, may all be present in a single formulation, and finally, depending on processing history and age of the polymer, the possibility that additive decomposition products may also be present to complicate the analytical problem in hand. The latter type of additive decomposition should be distinguished from that occurring during analytical processing. For example, a particular type of polymer additive may undergo partial thermal degradation during extrusion operations involved in its manufacture, and then during analysis may degrade by another route under the influence of light.

To summarize, then, the determination of additives in polymers presents the analyst with some difficult problems. Only small concentrations are present, complex mixtures may be involved and, moreover, frequently the mixture is of compounds of completely unknown type. Most methods for the determination of additives in plastics come under one of three categories;

1. direct examination of polymer, i.e. non-destructive testing;
2. examination of solvent extracts of polymers;
3. examination of volatiles released upon heating the polymer

Direct examination of polymer

An example of direct examination is the examination of the polymer film by infrared or ultraviolet spectroscopy or of thicker sections of polymer by attenuated total reflectance infrared spectroscopy. Such techniques have severe limitations in that, because the additive is in effect heavily diluted with polymer, detection limits are usually well above the low concentration of additive present, and this method is only applicable if the additive has distinct sharp absorption bands in regions where the polymer itself shows little or no absorption. *In-situ* spectroscopic techniques are not likely to be of value, then, in the analysis of samples of unknown composition. If known amounts of additive can be incorporated into additive-free polymer, however, these techniques are likely to be extremely useful in the study of solvent extraction procedures, and the study of additive ageing processes (i.e. the effects of heat, light, sterilization, radiation, etc.), since the rate of disappearance or of decay can be measured directly by the decrease in absorbance of the sample at a suitable wave-length.

The analysis of polyethylene additives by means of direct ultraviolet spectroscopy is limited by excessive beam dispersion due to light-scattering from the polymer crystalline regions. Additives at low concentrations (0.1%) require sample thicknesses such that analysis must be performed in the presence of a high level of light-scattering which may change unpredictably with wavelength. At lower levels of concentration and corresponding greater sample thickness, unacceptable signal-to-noise ratios exist. Nevertheless, ultraviolet spectroscopy remains an attractive method for analysis for many additives. Principal advantages over infrared spectroscopy include greater sensitivity arising from higher extinction coefficients and a lack of interfering absorptions from the polymer matrix. These advantages can be realized, however, only if background scattering from the polymer can be reduced.

Solvent extraction procedures

Solvent extraction procedures are commonly used in additive analysis. In these procedures the polymer is refluxed with a solvent which either dissolves the polymer or softens it, so that additives dissolve into the solvent phase. A non-solvent from the polymer is then added to reprecipitate the polymer, leaving the additives in solution. A good example of this is the extraction of additives from polyethylene, in which the polymer is refluxed with toluene which partly dissolves the polymer. Upon the addition of absolute ethanol the polymer is reprecipitated, leaving the additives in solution in the toluene–ethanol solvent mixtures which is then filtered off and carefully evaporated to dryness to provide the additive extract which is essentially free from polymer. Such extracts, when prepared quantitatively, are extremely useful either for direct analysis for known additives or for subsequent separation of their components by one of a number of chromatographic techniques prior to examination of the individual separated compounds by spectroscopy for identification and for quantitation.

A list of soluble solvent extraction procedures for various types of polymers is given in Table 56.

Examination of volatiles produced upon heating

The third method, involving examination of volatiles released from the polymer upon heating, is obviously only applicable to additives or other non-polymer components such as monomers and residual solvents which are volatile. In this technique the polymer is heated under controlled conditions in a sealed container. A gas chromatographic carrier gas stream is then connected to the container to sweep the volatiles into a gas chromatograph for identification and for quantitation. A classical example of the application of this technique is the determination of styrene monomer and aromatic impurities in polystyrene. Up to 30 different volatiles have been identified in this way in polystyrene in amounts down to 5 ppm.

TABLE 56   *Solvent extraction methods of additive extraction from polymers*

Polymer type	Substances extracted	Extracting solvent	Comments	Reference
Polyethylene	cresolic and phenolic antioxidants	chloroform	heat at 50°C for 3 h in a closed container	884
Polyethylene	cresolic antioxidants	hexane	heat at 50°C for 24 h	885
Polyethylene	cresolic antioxidants	ether	in the dark at 20°C	886
Polyethylene	phenolic antioxidants	chloroform		887
Polyethylene	antioxidants	toluene	reflux to dissolve polymer in precipitate with methanol	888
Polyethylene	antioxidants	water	at 70°C under nitrogen	889
Polyethylene	phenolic antioxidants	carbon disulphide iso-octane		
Polyolefins	2, 6 di-*t*-butyl-*p*-cresol	cydohexane	reflux 30 min	890
PVC	diphenyl thiourea, 2-phenylindole dicyandiamide	methanol or ether		891
PVC	stabilizers, lubricants, plasticizers	ether		892
Rubbers	amine and phenolic antioxidants	ethanol/HCl	reflux, then steam-distil amines from extract	893
Rubbers	phenyl salicylate, resorcinol benzoate	ether		894
Rubbers	antioxidants	acetone		895 896
Rubbers	ketone–amine condensates, phenols, 2-mercaptobenz-imidazole	acetone		897
General	*p*-phenylenediamine derivatives	95% methanol or ethanol	reflux 16 h in an extraction cup	898

Methods for the determination of various classes of additives in polymers are reviewed below. These methods are only applicable when dealing with a known additive.

Firstly, an example is discussed of a direct method for the determination of additives in polymer film by a spectroscopic method. In the second section the application of solvent extraction procedures, followed by analysis of the extract by methods appropriate to various classes of compounds, is discussed. Methods based on the examination of volatiles released from the polymer upon heating are discussed in Chapter 9.

8.1 Direct spectroscopic analysis of polymer films

The difficulties involved in extraction of additives from polymers have led to a search for analytical techniques not involving a prior solvent separation of an additive extract. Of all the techniques tried, only those based on spectroscopy can claim any measure of success. Luongo[899] has tried ultraviolet examination of thin, hot-pressed polymer films. Using a double-beam spectrophotometer with air in the reference beam, he was able to estimate antioxidant levels ranging from 0.002 to 1.00% in polyethylene. Such a procedure is limited in that the polymer must exhibit a relatively flat absorption curve in the wavelength range used; also many antioxidants exhibit similar or identical spectra.

Miller and Willis[900] obtained infrared spectra of antioxidants from polymer films in a similar way, except that they compensated with additive-free polymer in the reference beam. Infrared spectroscopy is more specific than ultraviolet spectroscopy, but some workers[901] find that the antioxidant level in polymers is too low to give suitable spectra. Drushel and Sommers[902] combined specificity with simplicity by using spectro-fluorimetric and phosphorescence techniques. Again, they used a double-beam spectrophotometer; this time with a wedge of additive-free polymer in the reference beam. They admit that the method is only applicable if the antioxidant has distinct sharp bands, and if no other components exhibit intense absorption in the same region.

In-situ spectroscopic techniques are not likely to be of value, then, in the analysis of samples of unknown composition. If known amounts of additive can be incorporated into additive-free polymer, however, these techniques are likely to be extremely useful in the study of solvent extraction procedures, and the study of additive ageing processes (i.e. the effects of heat, light, sterilization, radiation, etc.), since the rate of disappearance or of decay can be measured directly by the decrease in absorbance of the sample at a suitable wavelength.

An example is given below of a method based on direct polymer film infrared spectroscopy for the determination of the ultraviolet absorber Cyasorb UV531 (2-hydroxy-4-*n*-octoxybenzophenone) at concentrations of 0.1 to 1% in unpigmented high-density polyethylene. Antioxidants such as Polygard and Santonox R do not interfere in this procedure.

Apparatus

Double-beam spectrometer covering the $15-17\,\mu$ region (e.g. Grubb Parsons GS2A). Hydraulic press with heated and water-cooled platens. Stainless-steel moulding plates (6 in. × 6 in. × 1/6 in). Shims 0.06 cm thick (circular 1 in. diameter or rectangular 1 in. long). Aluminium foil. Clear plastic rule calibrated in millimetres. Dial gauge calibrated in 0.01 mm divisions.

Procedure

Preparation of sample film. Cover two stainless-steel moulding plates with aluminium foil and place up to six 0.06-cm thick shims on one of these. Place approximately 0.2 g of polymer sample in the centre of each shim and carefully place the second moulding plate on top. Position the two plates in the press and apply contact pressure. Switch on the heating supply and set the thermostat to 120°C. When the temperature reaches 120°C increase the pressure to 3000 lb/in.², switch off the heating supply and water-cool to room temperature. Carefully strip off the aluminium foil from the polymer films and push out the films from the shims. Check the thickness of each film by means of the dial gauge. Six readings on each film should not vary by more than 0.03 mm. Reject any which has air bubbles, is uneven or is wedge-shaped. Shape the film to fit the spectroscopic sample holder and gently scrape one of the surfaces with a fine emery board to produce a series of fine parallel lines. This reduces the incidence of interference fringes.

Recording the infrared spectrum. Place the film in the sample holder and position in the infrared instrument so that the beam passes through the film at right angles to the scratch marks.

Record the infrared spectrum from 15.5 to 16.5 μ in accord with the spectrometer operating instructions using a scanning speed of $\frac{1}{2}\mu$ per minute.

Before removing the film from the instrument mark the position of the infrared beam. Remove the sample from the holder and measure the thickness to the nearest 0.01 mm by means of the dial gauge at six points within the marked area. Calculate the mean of these six measurements.

Measurement of absorbance. Remove the chart from the spectrometer and with a sharp pencil rule a base-line from approximately 15.8 μ to approximately 16.2 μ. With a ruler measure I and I_0 to the nearest 0.1 mm at the wavelength of the peak maximum (see Fig. 99). Calculate the absorbance at 15.94 μ and hence the absorbance per unit thickness by means of the following expression:

$$\text{Absorbance per unit thickness} = \frac{\log_{10} I_0}{\text{Film thickness (in cm)}}$$

Calibration. Prepare duplicate films from the standard sheets containing 0.1, 0.3, 0.5 and 1.0 wt%. UV 531 as above. Record the infrared spectrum as described above, and calculate the absorbance per unit thickness as described under measurement of absorbance. Construct a

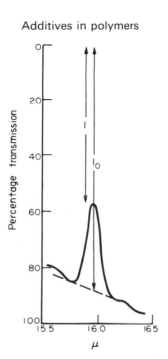

Fig. 99.   Determination of *UV* 531 in high-density polyethylene by direct film
infrared spectroscopy.

calibration curve by plotting absorbance per unit thickness against percentage weight UV 531 for each standard film. Use this calibration curve to obtain the UV 531 content of the polyethylene sample.

Albarino[903] has stated that analysis of polyethylene additives by means of ultraviolet spectroscopy is limited by excessive beam dispersion due to light-scattering from the polymer crystalline regions. Additives at low concentrations (0.1%) require sample thickness such that analysis must be performed in the presence of a high level of scattering, which may change unpredictably with wavelength. At lower levels of concentration, and correspondingly greater sample thicknesses, unacceptable signal-to-noise ratios exist. Nevertheless, ultraviolet spectroscopy remains an attractive method for analysis of many additives. Principal advantages over infrared analysis include greater sensitivity arising from higher extinction coefficients and a lack of interfering absorptions from the polyethylene matrix. These advantages can be realized, however, only if background scattering from the polymer can be reduced.

Albarino[903] demonstrated the feasibility of quantitative ultraviolet analysis of Irganox 1010 antioxidant in polyethylene at temperatures above the polymer melting point where the crystallites, which account for much of the scattering, are eliminated. Greater sample thickness and analytical sensitivity are possible compared to analysis of solid samples at room temperature. In this work, sample thickness was controlled by brass shims held between suprasil-grade silica windows (Amersil, Inc.) by a faceplate bolted to the cell body.

Polyethylene samples were prepared for analysis by calculating the weight required to fill the shim opening in the melt. Samples were inserted into the shim opening as pressed films cut to size; several layers were required for greater thicknesses. After gently tightening the faceplate, the cell was rapidly heated to 120–125°C by supplying about 65 W to the heater. By proper tightening of the faceplate the shim space was uniformly filled with polyethylene, after which the cell was transferred to the sample compartment of a spectrometer. Upon warm-up to the melt, an input power of 29 W maintained cell temperature within the limits given in Table 57 during scanning. Cell temperature was regulated only to the extent of maintaining the melt between 121 and 135°C. A small temperature increase, given by the intervals of Table 57, was generally allowed. Spectra were found to be insensitive to temperature in the intervals 128 ± 4°C to 145 ± 4°C; a thermometer in contact with woods metal was used to indicate initial cell temperature and temperature upon completion of spectra. Possible temperature gradients across the polyethylene melt were considered unimportant in view of the insensitivity of spectra to melt temperature.

Micrometer measurements of thickness were made on the solidified polyethylene samples. Errors due to polymer contraction on solidification were small, as the process of solidification generally results in a net volume change of the solid in the absence of constraints. As the polymer samples

TABLE 57   *Analysis of Irganox 1010 antioxidant in molten polyethylene*

	Composition Irganox 1010 in polyethylene (%)	Thickness (cm)	Temperature (°C)	Sample absorbance at 280 nm	Baseline absorbance at 280 nm	Antioxidant absorbance at 280 nm
1	0.101	0.030	121–129	0.212	0.032	0.180
2	0.101	0.058	122–126	0.347	0.044	0.303
3	0.101	0.081	122–125	0.511	0.053	0.458
4	0.101	0.112	123	0.678	0.065	0.613
5	0.051	0.218	124–125	0.692	0.107	0.585
6	0.051	0.056	122–127	0.217	0.043	0.174
7	0.051	0.109	124–127	0.378	0.064	0.314
8	0.051	0.165	124	0.548	0.086	0.462
9	0.010	0.274	123–128	0.278	0.129	0.149
10	0.010	0.508	122–124	0.486	0.222	0.264
11	0.010	0.612	125–128	0.585	0.264	0.321
12	0.010	0.780	127–135	0.753	0.330	0.423
13	0	0.058	123–124		0.058	
14	0	0.266	126–129		0.130	
15	0	0.508	124–132		0.218	
16	0	0.780	123–131		0.333	

were not constrained in any dimension, contraction occurred along the length and width of the specimen, as well as the thickness. That portion of the contraction resulting in a decrease in sample thickness was observed to be non-uniform across the face of the sample; micrometer measurements on this face were taken as true melt thickness. Shims designed to allow an outflow of excess molten polyethylene would facilitate thickness measurements as melt thickness would correspond to shim thickness.

Albarino[903] used standards consisting of polyethylene and Irganox 1010. These were made by milling at temperatures of about 127°C. Samples containing 0.051 and 0.010% Irganox 1010 were made from a master batch containing 0.101% Irganox 1010. These standards and an un-stabilized control were moulded into sheets 0.064–0.076 cm thick for use in the analysis.

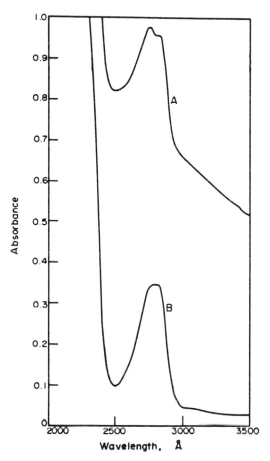

FIG. 100. Direct ultraviolet spectra of 0.101% Irganox 1010 in polyethylene: (A) 0.0045 cm, (B) 0.058 cm, 122–126°C.

A Cary 15 spectrometer was used for the study. Instrumental errors are stated as 0.002–0.005 absorbance in the 0–1.0 absorbance range and 0.008 in the 1–2 absorbance range.

The effect of sample melting on scattering is illustrated in Fig. 100. Figure 100A is the spectrum of a 0.045-cm polyethylene specimen with 0.101% Irganox 1010 at room temperature; Fig. 1013 was recorded at 122–126°C with a 0.058-cm specimen. A very substantial decrease in scattering has resulted with little change in the antioxidant absorption at 280 nm. The extent to which scattering may be reduced in the melt is indicated by Fig. 102D, where sample thickness was 0.780 cm and antioxidant concentration 0.010%. Spectral analysis on this sample in the solid state would not be possible because of its thickness.

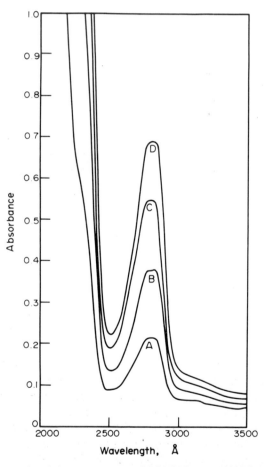

FIG. 101.   Direct ultraviolet spectroscopy of 0.051% Irganox 1010 in polyethylene: (A) 0.056 cm, 122–127°C; (B) 0.109 cm, 124–127°C; (C) 0.165 cm, 124°C; (D) 0.218 cm, 124–125°C.

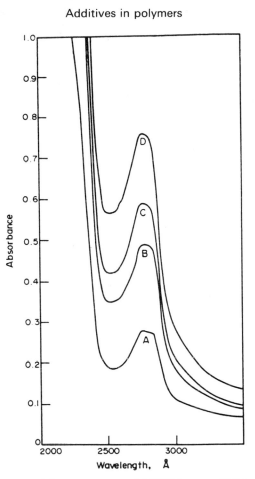

FIG. 102.   Direct ultraviolet spectroscopy of 0.010% Irganox 1010 in polyethylene:
(A) 0.274 cm, 123–128°C; (B) 0.508 cm, 122–124°C; (C) 0.612 cm, 125–128°C; (D)
0.780 cm, 127–135°C.

Application of the technique for the purpose of quantitative analysis
of additives requires proof of the validity of Beer's law ($\log I_0/I = A = abc$)
over the concentration range of interest. In the case of polyethylene
antioxidants it is particularly important to establish constant absorptivity
with concentration, as a fraction of the material is likely to exist in solution.

Spectra of molten polyethylene containing 0.051% (3.42×10^{-4} M)
nominal concentration of Irganox 1010 are given in Figs 101 and 102 as
a function of thickness. A similar set of curves was obtained for 0.101%
(6.78×10^{-4} M) antioxidant concentration. Four thicknesses were studied
at each concentration in order to establish linearity of absorbance with
sample thickness and molar absorptivity over the concentration range.

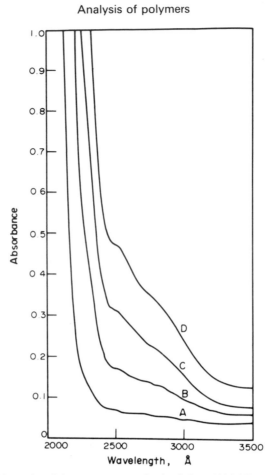

FIG. 103. Direct ultraviolet spectroscopy of polyethylene: (A) 0.058 cm, 123–124°C; (C) 0.508 cm, 124–132°C; (D) 0.780 cm, 123–131°C.

Control spectra of unstabilized polyethylene in the melt are given in Fig. 103 as a function of thickness.

Figure 104 is a plot of absorbance at 280 nm against thickness for the unstabilized polyethylene samples of Fig. 103. From Fig. 104 the contributions of polyethylene and the quartz cell windows to total sample absorbance at 280 nm were determined. The finite intercept of Fig. 104 represents scattering of the quartz windows at zero polyethylene thickness.

Graphs of total sample absorbance at 280 nm minus the baseline correction at 280 nm from Fig. 104 are given in Fig. 105 for the three stabilizer concentrations. Table 57 summarizes sample thickness, cell temperature, and absorbance data.

All graphs of Fig. 105 exhibit good linearity consistent with an intercept at the origin. In accord with Beer's law, the slopes of Fig. 105 divided by

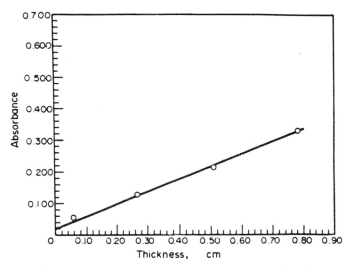

FIG. 104.   Direct ultraviolet spectroscopy of polyethylene. Absorbance versus thickness.

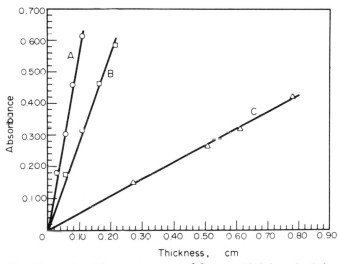

FIG. 105.   Direct ultraviolet spectroscopy of Irganox 1010 in polyethylene: (A) 0.101%; (B) 0.051%; (C) 0.010%.

respective molar concentrations give values of 8150 (0.101%), 8100 (0.051%), and 7950 (0.010%), with an average of 8070 (1000 cm^2/mole) for the molar absorptivity. These values are based on the nominal concentration of 0.101% and on the dilutions made from it, and are therefore subject to errors arising from stabilizer loss during initial compounding of the master batch and during subsequent dilutions to lower concentrations.

An estimate of ultimate sensitivity may be made by reference to Fig. 104, the baseline due to scattering, and Fig. 105, the absorbance at 0.01% antioxidant concentration. The slope for scattering in the melt predicts an absorbance of 1.0 for 2.54 cm of melted polyethylene. At this thickness, 0.01% Irganox 1010 would contribute to 1.4 absorbance for a total sample absorbance of 2.4. Reducing concentration to 0.001% would result in 0.14 sample absorbance, which with the same baseline at 1.0 gives a total of 1.14. Absorbance error in this range is stated as 0.008, or about 10% of the sample signal.

8.2   Examination of solvent extracts of polymers

8.2(a)   Chemical methods derivatization – visible spectrophotometric

Phenolic antioxidants. A very popular method of estimating antioxidants in polymer extracts is by coupling or oxidizing them to form coloured products and measuring the resulting absorbance in the visible region of the spectrum. This technique is not particularly specific for individual antioxidants, but is specific for phenolic antioxidants in general (also amine antioxidants, see below), and hence can often be applied without interference from other types of polymer additives.

In one procedure[905] the polymer sample in the form of a powder or thin film is extracted with ethanol and the extracted phenolic antioxidant coupled with diazotized *p*-nitroaniline in strongly acidic medium.

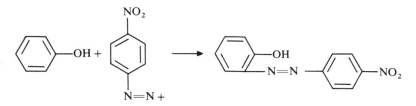

The solution is then made alkaline and the visible absorption spectrum determined. Many of the antioxidants studied have an absorption maximum at a characteristic wavelength. Hence, in some instances it is possible to both identify and determine the antioxidant, provided a pure specimen of the compound in question is available for calibration purposes.

Figure 107 shows absorption spectra in the 400–700 mu region of solvent extracts of five polystyrenes obtained by coupling with diazotized *p*-nitroaniline. Only polystyrene D shows clear evidence for the presence of a phenolic antioxidant – as is evidenced by the formation of a blue–violet coloration upon addition of the reagent.

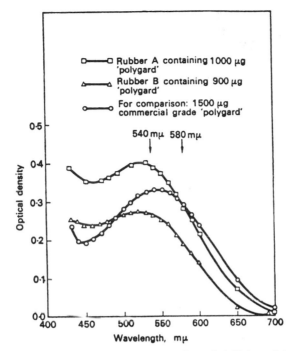

FIG. 106.   Comparison of absorption spectra of coupled 'Polygrad' impurity and extracts of 0.1 g of synthetic rubbers A and B.

Figures 106 and 107 show the absorption spectra obtained upon coupling extracts of various styrene-butadiene rubbers. The spectra of rubbers A and B (Fig. 106) are due to the presence in the rubber of nonyl phenol present as an impurity of decomposition products in the Polygard (*tris*(nonylated phenyl phosphite)) additive present in these rubbers. The phosphorus contents of these polymers indicated that they contained about 1% Polygard. Rubbers C and D (Fig. 107) did not contain a phenolic.

A British Standard method[906] for estimating the total phenolic antioxidant content of polyethylene involves a preliminary extraction of the polymer with hot toluene to extract the additive, followed by addition of ethanol to precipitate any dissolved polymer, then coupling the extract with diazotized sulphanilic acid to produce a colour which can be compared with standards by visible spectrophotometry. Dicresylol propane and Santonox R (4-4'-thiobis-(3-methyl-6-tert-butyl-phenol)), in particular, are mentioned as additives that can be determined by this technique.

Mayer[907] has described a quantitative method for estimating butylated hydroxy toluene (BHT) involving a colour-forming reaction with 2, 6-dichloro-*p*-benzoquinone-4-chlorimine. Glavind[908] and Blois[909] have

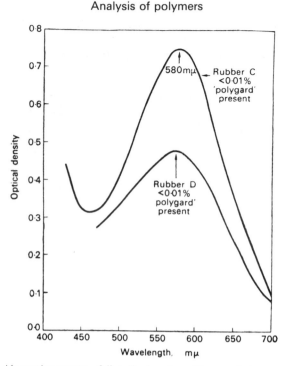

FIG. 107.   Absorption spectra of diazotized *p*-nitroaniline coupled extracts from 0.1 g
styrene–rubbers C and D.

devised methods for estimating total antioxidants, irrespective of type,
by coupling them with the free-radical α, α′-diphenyl-β-picrylhydrazyl.
The decrease in absorption of the reagent solution upon mixing with the
polymer extract is related to the amount of antioxidant present.

Redox   spectrophotometric   methods.   This   procedure[910]   involves
oxidation of the antioxidant (A) under controlled conditions with an
absolute ethanol solution of ferric chloride.

$$\text{A reduced} + Fe^{3+} = \text{A oxidized} + Fe^{2+}$$

followed by reaction of the ferrous iron produced with 2, 2′-dipyridyl to
form a coloured complex, the intensity of which is proportional to the
concentration of antioxidant present. This spectrophotometric procedure
has been applied to various phenolic and amine-type antioxidants, viz.
Succanox 18, butylated hydroxytoluene, Ionol (2, 6-di-teri-butyl-*p*-cresol),
and Nonox CI (*N*-*N*′-di-β-naphthyl-*p*-phenylenediamine). An advantage
of this redox procedure is that it determines only the antioxidant which
is present in the polymer in the reduced form, and does not include any
oxidized material produced, for example, by atmospheric oxidation during

polymer processing at elevated temperatures. Total reduced plus oxidized antioxidant can be determined by ultraviolet spectroscopic procedures and oxidized Santonox obtained by difference from the two methods.

Amine antioxidants. A wide range of amine antioxidants are used in plastic and rubber formulation. One method[916,964] for determining these includes coupling an ethanol extract of the polymer powder or thin film

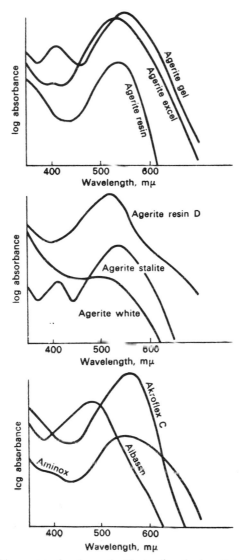

FIG. 108. Visible spectra of amine antioxidants, diazotized *p*-nitroaniline method.

with diazolized *p*-nitroaniline and evaluating the colour produced spectrophotometrically at the absorption maximum. Figure 108 shows the visible
spectra obtained by this procedure for a range of amine antioxidants.

In an alternative procedure recommended by the British Standards
Institution[965] for determining Nonox CI (*N, N'*-di-2 naphthyl-*p*-
phenylene-diamine) in low-density and high-density polyethylene, the
polymer is refluxed with toluene, reprecipitated with ethanol and filtered
to provide a clear extract. The antioxidant content of the filtrate is
determined colorimetrically by oxidation with hydrogen peroxide, in the
presence of sulphuric acid. This reagent produces a green colour with
Nonox CI, which gradually reaches a maximum intensity. The colour is
evaluated at 430 nm when the maximum depth of colour is reached.

The procedure[910] described under phenolic antioxidant, involving
oxidation of the antioxidant with ferric iron followed by spectrophotometric estimation of the ferrous iron produced using 2, 2'-dipyridyl,
is also applicable to the determination of amine antioxidants. Other
chromogenic reagents that have been used for the determination of amine
antioxidants include *p*-diazobenzene sulphonic acid,[947] α, α'-diphenyl-β-
picrylhydrazyl,[908] and benzothiazolin-2-one hydrazone.[966]

Tertiary phosphite antioxidants.   Nawakowski[973] has described a colorimetric method for determining Polygard (trisnonylphenyl phosphite)
based on hydrolysis to nonyl phenol, followed by coupling with *p*-
nitrobenzene-diazonium fluoroborate and colorimetric estimation at
550 nm.

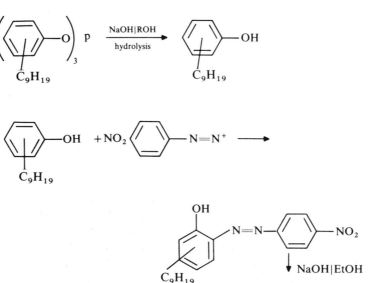

veiolet dye, absorption maximum 550 nm

Various other phenolic antioxidants produced dyes under these conditions, viz. Wingstay S, Agerite Superlite and Nevastain A.

The procedure was applied with good precision to the determination of Polygard in styrene–butadiene latexes. Good agreement was obtained between this procedure and direct determinations of phosphorus by elemental analysis.

8.2(b)   Titration methods

Organotin stabilizers

Organotin compounds in solvent extracts of polymers can be determined by potentiometric and manual titration procedures.[977,979] Potentiometric titration of non-sulphur-containing organotin compounds is achieved by titrating the carboxylate groups with sodium methoxide in pyridine by using antimony and calomel electrodes. The end-point roughly coincides with the change of colour when thymolphthalein is used as indicator. The titrant, i.e. sodium methoxide, is standardized against benzoic acid and the titration carried out under nitrogen as a precautionary measure. Dialkyl tin thio compounds can be titrated[977] with a solution of silver nitrate in $1+1$ isopropanol–water. The sample is dissolved in a solvent consisting of a $1+1$ mixture of benzene–methanol, containing sodium acetate trihydrate ($13.7\,\mathrm{g\,l^{-1}}$). The indicator electrode is a length of silver wire coated with sulphide. It is prepared by immersion of a silver wire into an alkaline solution of sodium sulphide followed by addition of a silver nitrate solution ($0.1\,\mathrm{N}$) and stirring. The electrode is wiped clean and polished with a clean cloth and is then ready for use. The reference electrode consists of a copper wire immersed in a pool of mercury covered with a solution of $0.1\,\mathrm{N}$ sodium acetate. Contact with the test solution is made through an agar-gel bridge. More than one end-point is sometimes produced. However, titration of reference samples under identical conditions helps in the interpretation of the results.

Phenolic antioxidants

Titrimetric methods for the determination of phenolic antioxidants include potentiometric titration with standard sodium isopropoxide in pyridine medium[904] or reaction of the antioxidant with excess standard potassium bromide–potassium bromate (i.e. free bromine) and estimation of the unused bromine by addition of potassium iodide and determination of the iodine produced by titration with sodium thiosulphate to the starch end-point.[887]

Diorganosulphide and tertiary phosphite antioxidants

Kellum[975] has described a method, based on selective oxidation, for the determination of diorganosulphide and tertiary phosphite types of secondary antioxidants in polyolefins. Some of these types of antioxidants are listed below:

> disteryl thiodipropionate
> dilauryl thiodipropionate
> 4, 4'-thiobis(6-tert-butyl-m-cresol)
> 1, 1'-thiobis(2-naphthol)
> triphenyl phosphite
> triethyl phosphite
> tri-p-tolyl phosphite
> tris(dinonyl phenyl)-phosphite
> 2, 2'-thiobis(6-tert-butyl-p-cresol)
> tri-isopropyl phosphite

These two classes of compounds were selectively determined in the presence of each other by oxidation using *m*-chloro-peroxy-benzoic acid to sulphones and phosphates. In this method, a heptane extract of the polyolefins containing the antioxidants is treated with a two-fold excess of the oxidant and allowed to react for 45 minutes before the unreacted oxidant is decomposed with sodium iodide to produce iodine which is estimated by sodium thiosulphate titration.

$$\text{(aryl)}-COOOH + 2\ NaI \longrightarrow \text{(aryl)}-COOH + H_2O + I_2.$$

This method has the advantage of being free from interference by hindered phenols, benzophenones, triazoles, fatty acid amides and stearate salts, all of which might be present in the polymer extract. Recoveries of the following antioxidants were usually in excess of 99%. Stabilizers and additives, other than secondary antioxidants, did not interfere in the procedure, these materials included phenolic antioxidants (2, 6-di-tert-butyl-p-cresol, 2, 2'-methylene-bis(6-tert-butyl-p-cresol), 4, 4'-butylidenebis(6-tert-butyl-m-cresol), benzophenone light stabilizers (2-hydroxy-4-octoxy benzophenone), substituted hydroxyphenyl benzotriazoles, erucamide (a slip agent), and calcium stearate (an anti-block agent).

Table 58 shows the results obtained by applying the oxidation method to blends of antioxidants in unstabilized polyethylene. The results are in excellent agreement with theoretical antioxidant content of these polymers.

TABLE 58   *Analysis of typical polypropylene samples with various stabilizers present*

Additive	Percentage added	Percentage found
Distearyl thiodipropionate	0.20	0.19
Distearyl thiodipropionate	0.40	0.36
Distearyl thiodipropionate	0.10	0.095
4, 4'-Thiobis (6-tert-butyl-*m*-cresol)	0.10	0.096
4, 5'-Thiobis (6-tert-butyl-*m*-cresol)	0.20	0.19
Distearyl thiodipropionate	0.20	0.39 total
4, 4'-Thiobis (6-tert-butyl-*m*-cresol)	0.20	0.39 total
Distearyl thiodipropionate	0.20	0.39 total

8.2(c)   Elemental analysis

A very useful preliminary to the identification of organotin compounds is the determination of the tin and sulphur content of an extract of the polymer which has been purified so that only the organotin compound is present. Tin can be determined[977] by wet oxidation of the sample followed by precipitation of tin with cupferron. When an accuracy of $\pm 0.2\%$ absolute is sufficient the sulphated ash procedure may be combined with spectrographic examination to check that other metals are absent.

Organotin compounds are widely used in the plastics industry as stabilizers for poly(vinyl chloride) compositions. The most important compounds used for this purpose are based on dialkyltin groups

$$\frac{R}{R} > Sn <,$$

especially where R = butyl or octyl.

It is convenient to divide tin stabilizers into those containing sulphur and those without sulphur. To the first group belong compounds, like dialkyltin mercaptides, mercapto-esters and mercapto carboxylates, to the second, dialkyltin carboxylates and their esters.

8.2(d)   Physical methods

Ultraviolet spectroscopy

Phenolic antioxidants. Straightforward ultraviolet spectroscopy is liable to be in error owing to interference by other highly absorbing impurities that may be present in the sample.[911-914] Interference by such impurities in direct ultraviolet spectroscopy has been overcome or minimized by selective solvent extraction or by chromatography.[912] However, within prescribed limits ultraviolet spectroscopy is of use and, as an

example,[915-918] procedures are discussed below for the determination of
Ionol (2, 6-di-tert-butyl-*p*-cresol) and of Santonox R (4, 4'-thio-bis-6-tert-
butyl-*m*-cresol in polyolefins.

Certain additives, e.g. calcium stearate and lauryl thiodipropionate, do
not interfere in the determination. Other phenolic antioxidants, e.g. Ionox
330, Topanol CA and Santonox R, do interfere.

Determination of Ionol antioxidant. Figure 109 shows the ultraviolet
spectrum of a chloroform extract of polyethylene indicating clearly the
absorption maximum at 278 nm.

Determination of Santonox R antioxidant. In an attempt to overcome the
difficulty of interference effects by other polymer additives in the ultraviolet spectroscopic determination of phenolic antioxidants Wexler[919]
makes use of the bathochromic shift exhibited by phenols on changing
from a neutral or acidic medium to an alkaline one. This shift is due to the
change of absorbing species because of solute–solvent interaction. Using a
double-beam recording spectrophotometer, he measured a difference
spectrum by placing an alkaline solution of the polymer extract in the
sample beam, and an identical concentration of sample in acid solution in
the reference beam. The resulting difference spectrum is a characteristic
and useful indication of the concentration and chemical identity of the

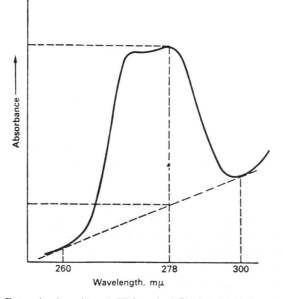

FIG. 109.   Determination of Ionol CP in polyolefins by ultraviolet spectroscopy.

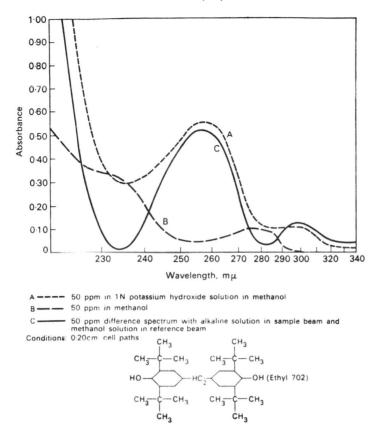

FIG. 110.   Ultraviolet spectra of 4,4'-methylenebis (2,6-di-tert butyl phenol) exhibiting bathochromic shift in alkaline medium.

phenolic substance. Possible interference due to non-ionizing, non-phenolic species is usually cancelled out in the difference spectrum, which should make the technique of interest to the polymer analyst. Typical spectra obtained for an antioxidant are shown in Fig. 110. The spectrum exhibits two maxima and two minima. Close adherence to Beer's law is usually obeyed by the difference peak spectra.

Organophosphorus antioxidants.   Brandt[974] has described an alternative method for the determination of Polygard (trisnonylphenyl phosphite) in styrene–butadiene lattices, which utilizes the bathochromic shift in the spectrum of phenols resulting from the formation of phenolate ions in alkaline solution.

The latex is flocculated by the addition of acid and the antioxidant extracted with iso-octane.

TABLE 59 *Determination of Polygard: comparison of ultraviolet with perchloric acid methods*

SBR latex type 6101	Perchloric acid (phosphorus) method		Ultraviolet method	
	Percentage			
Sample 1	1.31	1.30	1.18	1.30
Sample 2	1.26	1.28	1.24	1.36
Sample 3	1.23	1.24	1.36	1.32
Sheet rubber				
Type 1019	1.25	1.28	1.04	1.06
Type 1503	1.62	1.63	1.53	1.60
Type 1018	1.59	1.56	1.57	1.58
Type 1022	1.27	1.31	1.16	1.14

Polygard in iso-octane has an ultraviolet spectrum with a peak at 273 nm in neutral solution. By adding a strong base (tetrabutylammonium hydroxide) the Polygard is hydrolysed and the peak is shifted to 296 nm. The difference in absorbance at 299 nm between the neutral and alkaline solutions is directly proportional to the amount of Polygard present. By use of this bathochromic shift, interference of non-phenolic impurities is eliminated, and a background correction factor is not required. Results obtained by this procedure agree well with these based on direct determination of elemental phosphorus, (Table 59).

8.2(e) Fluorescence and phosphorescence methods

Ultraviolet absorbers and optical brighteners

Among the numerous additives commonly used in plastic materials, the ultraviolet absorbers are increasing in importance because they are often used in food-packaging materials to protect the plastics material, as well as the foodstuff packaged, from the actinic action of ultraviolet radiation. Actinic effects may cause discoloration of both the plastics material and the foodstuff, and on occasion also cause changes in taste and loss of vitamins in the food.

The ultraviolet absorbers can be divided in different groups (Table 60):

(a) benzephenone derivatives,
(b) salicyclic acid esters,
(c) resorcinol esters,
(d) benzotriazole compounds,
(e) coumarin derivatives.

Methods for the determination of ultraviolet absorbers and optical brighteners are based principally on fluorimetry.

TABLE 60   *Some ultraviolet absorbers for use in plastics materials*

Chemical formula	Trade name	Manufacturer
1. 2-hydroxy-4-methoxy- benzophenone	Uvinul M 40 Uvistat 24 Cyasorb UV9	General Aniline Co. Ward & Blenkinsop Cyanamid
2. 2,4-dihydroxy- benzophenone	Uvistat 12 Uninul 400	Ward & Blenkinsop General Aniline Co.
3. 2-hydroxy-4-methoxy-4- methyl-benzophenone	Uvistat 211	Ward & Blenkinsop
4. 2,4,5-trihydroxy butyrophenone	Inhibitor THBP	Eastman
5. 4-dodecyloxy-2-hydroxy benzophenone	Inhibitor DOBP	Eastman
6. 2-hydroxy-4-n-octoxy- benzophenone	Cyasorb UV531	Cyanamid
7. 2,2'dihydroxy-4-methoxy benzophenone	Cyasorb UV24	Cyanamid
8. 2,2'dihydroxy-4,4' dimethoxy benzophenone	Uvinul D49	General Aniline Co.
9. p-tert-butylphenylsalicilate		
10. resorcinol mono benzoate	Inhibitor RMB	Eastman
11. hydroxyphenyl- benzotriazole	Tinuvin P	Geigy
12. 7-diethylamino-4-methyl coumarin		Ward & Blenkinsop

In many instances visible fluorescence techniques are less subject to interference by other polymer additives present in a polymer extract than are ultraviolet methods of analysis. Therefore, in some instances visible fluorimetry offers a method of determining a polymer constituent without interference from other constituents, when this would not be possible by ultraviolet spectroscopy. Apart from specificity, fluorescence techniques are more sensitive than absorption spectroscopic techniques.

Uvitex OB has an intense ultraviolet absorption at a wavelength of 378 nm, which is high enough to be outside the region many potentially interfering substances present in the polymer extract would be excited to fluorescence. This is illustrated by a fluorimetric procedure for the determination of down to 10 ppm Uvitex OB in polystyrene. Antioxidants such as Ionol CP (2,6-di-tert-butyl-p-cresol), Ionox 330 (1,3,5-tri-methyl-2,4,6-tri(3,5-di-t-butyl-4-hydroxybenzyl)benzene), Polygard (tris(nonylated phenyl) phosphite), Wingstay T (described as a butylated cresol), and Wingstay W and many others, do not interfere in this procedure. In this procedure the polystyrene is dissolved in chloroform and the solution excited by ultraviolet radiation of wavelength 370 nm from a mercury vapour lamp and the fluorescence spectrum of the sample recorded over the range 400–440 nm. The reading from the fluorimeter is noted, and the Uvitex OB concentration in the polystyrene determined by reference to a prepared calibration graph.

The fluorescence given by optical brighteners under ultraviolet light on a thin-layer plate has been utilized as a means of estimating these compounds.[970] To estimate 7(-6 butoxy-5-methylbenzotriazol-2-yl) 3-phenyl-coumarin in polymer granules, the sample is extracted from the ground or chopped sample by heating under reflux with chloroform. The extract, together with a chloroform solution of an authentic sample of the brightener, is applied to two Kieselgel G plates, and chromatograms are developed with benzene–chloroform (2:3) and benzene, respectively; the spots are detected by their fluorescence in ultraviolet radiation. For spectrophotometric determination the sample is extracted by boiling under reflux for 1 h with 1, 2-dichloroethane. If the extract is cloudy it is treated with kieselguhr and filtered; if it contains insoluble pigment it is centrifuged. The extinction of the clear solution is then measured at 366 nm, and the optical brightener content is obtained by reference to the extinction of a solution of the authentic optical brightener in 1, 2-dichloroethane.

Phenolic and amine antioxidants

Aromatic amines and phenols are among the few classes of compounds in which a large proportion of their numbers exhibit sensible fluorescence. Parker and Barns[920] found that in solvent extracts of rubbers the strong absorption by pine-tar and other constituents masks the absorption spectra of phenyl naphthylamines, whereas the fluorescence spectra of these amines are sufficiently unaffected for them to be determined directly in the unmodified extract by the fluorescence method. In a later paper[921] Parker discussed the possibility of using phosphorescence techniques for determining phenyl-naphthylamines. Drushel and Sommers[902] have discussed the determination of Age Rite D (polymeric dihydroxy quinone) and phenyl 2-naphthylamine in polymer films by fluorescence methods and Santonox R (4, 4′-thio-bis-(6-tert-butyl-*m*-cresol)) and phenyl-2-naphthylamine by phosphorescence methods. They emphasize the freedom that such techniques have from interference by other polymer additives and polymerization catalyst residues. With practical samples of polyethylene film, difficulty was found in obtaining a reliable correlation between the concentration of stabilizer present in the film and its phosphorescence intensity at 77 K by this technique. This may be attributable to variations in the degree of crystallinity which affect the optical properties of the polyethylene film matrix in these samples.

Figure 111 shows the spectral characteristics of a 1000 ppm solution of Nonox CI antioxidant dissolved in diethylamine at 77K and at 293K. Calibration curves for both low-temperature luminescence intensity (phosphorescence plus fluorescence) and phosphorescence intensity versus

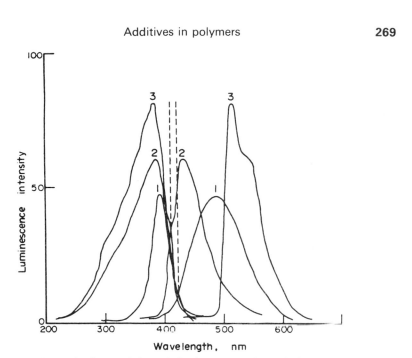

FIG. 111.   Spectral characteristics of luminescence observed for Nonox CI
(Sample 21): (1) excitation and emission spectra at room temperature, sensitivity scale
0.01; (2) excitation and emission spectra at 77K for total luminescence, sensitivity
scale 1.0; (3) excitation and emission spectra at 77K for phosphorescence, sensitivity
scale 0.01.

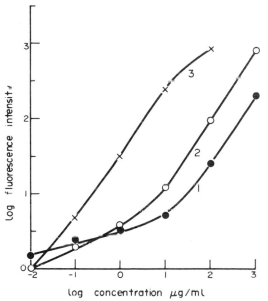

FIG. 112.   Working curves for room temperature fluorescence: (1) butylated hydroxy
anisole (●), (2) Agerite Superlite (○), (3) Nonox CI (×).

concentration of antioxidant in ppm of the compound in the solvent can be obtained.[992]

The wavelengths of maximum excitation and emission are set on the spectrophosphorimeter, and the relative intensity readings from the photo multiplier microphotometer recorded for a series of standard solutions containing 0.001–1000 ppm of each compound.

Figure 112 shows room temperature luminescence calibration graphs for three antioxidants including butylated hydroxy anisole and a phenolic antioxidant.

In conclusion, measurement of the phosphorescence characteristics of samples obtained after extraction of polymers with organic solvents yields useful information regarding the nature and concentration of the stabilizer compounds present. It should be possible to obtain good selectivity, with a sensitivity which compares favourably with that of ultraviolet absorption spectrophotometry, in the determination of two or more stabilizer compounds simultaneously by correct choice of excitation and emission wavelengths and phosphorescence speeds.

8.2(f)   Electrochemical methods

Phenolic and amine antioxidants

Electrochemical methods have been investigated but in general are only to be recommended where a simpler method is not available. Mocker[923-925] and Mocker and Old[926] have explored the use of polarography and find the technique to be more applicable to rubber accelerators than to antioxidants. They included both phenolic and amine types of antioxidants in their study. Difficulties arise because the dropping-mercury electrode cannot be used at potentials more positive than $+0.4$ V with respect to the saturated calomel electrode, and since many aromatic phenols (and amines) can only be oxidized at electrodes, positive voltages have to be applied in their analysis. Nevertheless, the polarography of some amines and phenols has been studied,[927-930] and whilst no electrode is as suitable for polarography as the dropping-mercury electrode, antioxidants have also been studied with other types of electrodes, notably the graphite[931-934] and the platinum[935,936] electrode. These methods include the determination by voltametric methods of N, N'-di-sec-butyl-p-phenylene-diamine and N-butyl-p-amino phenol[933] and of Ionol (2, 6-di-tert-butyl-p-cresol).[931] In addition, at least two commercially available antioxidants have been shown by differential cathode-ray polarography to exhibit reduction waves: Santonox R (4, 4'-thiobis-(3-methyl-6-t-butyl-phenol)) gives a poorly shaped wave at -0.6 V in an electrolyte consisting of ammonia and ammonium chloride in methanol–water[937] and 3, 5-di-t-butyl-4-hydroxytoluene gives a wave at -0.65 V in aqueous sodium or lithium hydroxide.[938] In both cases, 40 ppm of analyte gave a current which was adequate for quantitative analysis.

Budyina et al.[939] have described methods based on anodic voltametry for the determination of Ionol (2.6-di-t-butyl-p-cresol) and quinol in polyester acrylates. To determine Ionol and sample is dissolved in 25 ml of acetone and an aliquot (10 ml) is treated with 2.5 ml of acetone and 5 ml of methanol and diluted to 25 ml with a solution 0.1 M in lithium chloride and 0.02 M in sodium tetraborate. A polarogram is recorded with a graphite-rod indicator-electrode and a 0.5 cadmium sulphate–cadmium reference-electrode (Vasil'eva et al.).[940] To determine quinol, the sample (1–3 g) is dissolved in 80 ml of methanol or methanol: acetone (1:1) and the solution is diluted to 100 ml with the lithium chloride–sodium tetraborate solution. A polarogram is recorded under the same conditions. Concentrations are determined by the addition method. The $E_{1/2}$ values (versus the SCE) are 0.25 V for Ionol And 0.16 V for quinol.

Vasil'eva et al.[940] described a method for the determination of Ionol and 4, 4'-isopropylidenediphenol in PVC by anodic voltametry. The sample (1 g) is dissolved in 10 ml of dimethylformamide and the solution is treated with 5 ml of methanol to precipitate PVC and diluted to 25 ml with 1.5 M aqueous sodium acetate. A polarogram of an aliquot (10 ml) of the clear solution is recorded (graphite rod indicator-electrode and 0.5 M cadmium sulphate–cadmium reference electrode) to include the steps for 2, 6-di-t-butyl-p-cresol at 0.21 V versus the SCE, and for 4, 4'-isopropylidene-diphenol at 0.53 V versus the SCE. The application of voltametry to the determination of antioxidants has also been discussed by Barendrecht.[931]

Other procedures which have been described include the conversion of an antioxidant into a polarographically reducible form,[941] and a general method for antioxidants which involved measuring a decrease in the height of the wave, due to the reduction of dissolved oxygen by antioxidants.[936] Ward[942] has discussed in some detail the determination of phenolic and amine types of antioxidants and antiozonants in polymers by the chronopotentiometric technique, using a paraffin wax-impregnated graphite indicating electrode[943,944] and solutions of lithium chloride and lithium perchlorate and acetonitrile in 95% ethanol as supporting electrolytes. Precision obtainable for repeated chronopotentiometric runs in acetonitrile was found to be better than ± 1.0% in cases in which electrode fouling did not occur, and ± 1.7% when the electrode was fouled by electrolysis products.

Phenolic (and amine) antioxidants have been titrated electrometrically with lithium aluminium hydride, with platinum or silver electrodes.[945] Small amounts of water in the sample or analysis solvent have an influence on the results obtained by these procedures.

Electrophoresis is a technique worthy of further consideration for the analysis of antioxidants.[946] Sawada et al.[947] report successful separations by coupling the antioxidants with p-diazobenzene sulphonic acid before electrophoresis. Amine antioxidants are coupled in acetic acid and phenolic antioxidants in sodium hydroxide–ethanol. Electrophoresis was carried out in 1% w/v methanolic sodium borate.

Organic peroxides

Organic peroxides can occur in small amounts in some types of polymers such as polystyrene as a result of the fact that a peroxide has been used as a polymerization catalyst in polymer manufacture. Also, stable organic peroxides such as dicumyl peroxide have been used as synergists, in conjunction with bromine and or phosphorus-containing additives, to impart fire-resistance to cellular expanded polystyrene and other types of plastics.

In a solution of 0.6 M lithium chloride in 1:1 toluene: methanol benzoyl peroxide, p-tert butyl perbenzoate and lauryl peroxide have half-wave potentials respectively of 0.00, -0.95 and -0.15 V in the presence of ethyl cellulose as a maximum suppressor. Based on this, a procedure has been described[996] for determining down to 20 ppm of these substances in polystyrene in which a suitable weight of polymer is dissolved in toluene, and then an equal volume of 0.6 M lithium chloride in methanol is added. Precipitated polymer is removed by centrifuging and peroxides determined in the clear extract by cathode-ray polarography. Polymerization additives, styrene monomer or antioxidants in the polymer do not interfere in the polarographic procedure.

Organotin heat stabilizers

Tin can be determined in PVC by polarography and atomic absorption spectroscopy. In the polarographic method[978] the sample is decomposed with sulphuric acid–nitric acid (1:1), then hydrochloric acid is added and ammonia added until the solution is made 4 M to ammonium chloride. Tin was determined polarographically at $E_{1/2} = -0.52$ V. In the atomic absorption method, the sample is decomposed, and tin determined as the acetate using an air–acetylene flame at 224.6 nm. Both methods can determine tin in PVC in amounts down to 300 ppm.

Metallic stearate heat stabilizers

Other types of organometallic stabilizers have been used as the formulation of PVC.

Polarographic procedures

Mal'kova et al.[982] described an alternating current polarographic method for the determination of cadmium, zinc and barium stearates showed that Ionol (2, 6-di-t-butyl-p-cresol) and Santonox R (4, 4'-thio-bis-

ashed in a muffle furnace at 500°C, a solution of the ash in hydrochloric acid being made molar in lithium chloride and adjusted to
pH 4.0 ± 0.2. Alternatively, the sample solution can be prepared by boiling
with 2 M hydrochloric acid for 3 min, cooling and adjusting the pH of a
portion of the solution to 4 with 2 M lithium hydroxide. The solution
obtained in either instance is de-aerated by passage of argon and the
polarogram is recorded. Cadmium, zinc and barium give sharp peaks at
− 0.65, − 1.01 and − 1.90 V, respectively, versus the mercury-pool anode.
The first digestion procedure is recommended if the sample contains esters
of phosphorus acid. Polarography has also been used to determine
cadmium, lead and zinc salt in PVC[985].

8.2(g) Infrared spectroscopy

Due to interference effects by extraction solvents and other additives and
adventitious impurities in the polymer, direct infrared spectroscopy of
polymer extracts has found very limited application in the determination
of additives. When, however, infrared spectroscopy is proceded by a clean-
up procedure such as thin-layer chromatography (see later in this chapter),
then it is an extremely useful technique both for the identification and
determination of additives.

   Spell and Eddy[948] have described infrared spectroscopic procedures for
the determination of up to 500 ppm of various additives in polyethylene
pellets following solvent extraction of additives at room temperature. They
showed that Ionol (2, 6-di-*t*-butyl-*p*-cresol) and Santonox R (4, 4′-thio-bis-
(6-*t*-butyl-*m*-cresol) are extracted quantitatively from polyethylene pellets
by carbon disulphide in 2–3 h and by iso-octane in 50–75 h. The carbon
disulphide extract is suitable for scanning in the infrared region between
1075 and 1282 cm^{-1} (7.8 and 9.3 μ), whilst the iso-octane extract is suitable
for scanning in the ultraviolet between 250 and 350 mμ.

8.2(h) Thin-layer chromatography

Phonolic antioxidants

This technique, like gas chromatography, comes into its own when dealing
with mixtures of substances. However, both techniques have been
employed for the determination of phenolic antioxidants. In determining
Santonox R antioxidant in polyethylene, the polymer was refluxed with
toluene for 2 h, then absolute ethanol added to reprecipitate any dissolved
polymer. The filtrate was carefully evaporated to dryness and dissolved in
a known volume of chloroform. Portions of this solution, together with
standards, were applied to a 250 thick layer of Merck GF 254 silica gel and
the place eluted with 5:1 petroleum ether (40:60):ethyl acetate. After
development the plate was sprayed with a 2% ethanol solution of 2, 6-

dibromo-p-benzo-quinone-4-chlorime followed by 2% aqueous sodium tetraborate. The Santonox R content is obtained by comparing the intensity of the purple spot produced by the sample with those obtained with the standards. Van der Heide and Wouters[886] examined the thin-layer chromatography of some antioxidants in polyethylene. Slonaker and Sievers[885] and Waggon and Jehle[1026] did similar work, and were able to detect between 300 and 900 ppm of antioxidants in polyethylene. Schroeder was reviewed work on the application of thin-layer chromatography to phenolic antioxidants.[949]

Dilaurylthiodipropionate antioxidants

A procedure has been described[976] for the determination of dilauryl:β, β'-thiodipropionate antioxidant in polyolefins, ethylene–vinyl acetate copolymer, arylonitrile–styrene–butadiene terpolymer and polystyrene. In this procedure the total additives are first extracted from the polymer by extraction under nitrogen with a 1:1:4 mixture of chloroform, ethanol and n-hexane. To separate dilauryl-β, β'-thiodipropionate from its own oxidation products and from other additives, the extract was subject to thin-layer chromatography on a silica gel-coated plate. The spot containing dilauryl-β, β'-thiodipropionate is refluxed for 30 min at 80°C with methanolic potassium hydroxide to hydrolyse the ester to lauric acid, which is then

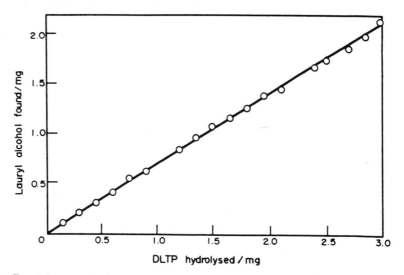

FIG. 113.   Relationship between the amount of lauryl alcohol found and the amount of dilauryl-thiodipropionate (DLTP) hydrolysed. Conditions of hydrolysis: temperature 80°C; reagent, 5 N; potassium hydroxide solution in methanol; time, 30 min.

extracted from the aqueous alcoholic phase with chloroform containing *n*-octadecene internal standard. This extract is gas chromatographed on an isothermal column packed with 1.5% fluorosilicone oil FS 1265 on 60 to 100 mesh Chromosorb W operated at 300°C and utilizing a flame ionization detector. Suitable reference solutions of dilauryl-β, β'-thiodiproprionate are run in parallel through the whole procedures for calibration purposes.

A series of hydrolyses at 80°C in 5 N methanolic potassium hydroxide solution for 30 min carried out within the concentration range of 0.15–3.00 mg of dilauryl-β, β'-thiodipropionate per 10 ml, gave the results shown in Fig. 113, in which the amount of lauryl alcohol formed by hydrolysis is plotted against the amount of dilauryl-β, β'-thiodipropionate hydrolysed.

The slope of the straight line, evaluated by the least-squares method, was 0.7028 (mean deviation of the experimental points from the calculated values was $\pm 2.5\%$ over the whole concentration range). The theoretical value of the slope, assuming that the complete hydrolysis of dilauryl-β, β'-thiodipropionate decomposed, is 0.7238. Determination carried out under these conditions therefore gives results that are consistently low by a factor of 0.97.

Ultraviolet light stabilizers

Thin-layer chromatography is useful for the determination of light stabilizers in polymers. In one procedure for determining Cyasorb UV 531 light stabilizer (2-hydroxy-4-*n*-octoxybenzophenone) in high-density polyethylene the stabilizer is extracted from the polymer by precipitation of the polymer with ethanol from a hot toluene solution as described above. An aliquot of the extract is applied to a GF245 silica-gel-coated thin-layer plate and chromatographed using methylene dichloride as an eluant. The zone corresponding to UV 531 is identified as a dark blue spot under ultraviolet light and then the silica get is removed from the plate and extracted with ethanol. The concentration of UV 531 is determined by measuring the ultraviolet absorption peak near 295 nm in ethanol solution and referring to a prepared calibration graph.

No interference is encountered from a number of additives, in the polymer viz. dilauryl thiodipropionate, Ionox 330, Ionol CP, Topanol CA, Santonox R, polygard.

Organic peroxide residues

Certain types of peroxides used in polymer formulations are extremely stable and unreactive. This applied to substances such as dicumyl peroxide used as an ingredient of some self-extinguishing grades of polymers.

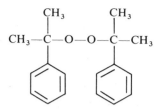

This substance cannot be determined by polarography, and will not react with many of the reagents normally used for determining organic peroxides. Brammer et al.[999] have described a method for determining dicumyl peroxide in polystyrene, which is not subject to interference by other organic peroxides or additives that may be present in the polymer. The dicumyl peroxide is extracted from the polymer with acetone and then separated from any other additives present by thin-layer chromatography on silica gel. The gel in the area of the plate containing dicumyl peroxide is then isolated and digested with potassium iodide in glacial acetic acid followed by titration of the liberated iodine by titration with very dilute sodium thiosulphate solution.

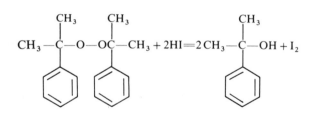

This procedure has a precision of $\pm 12\%$ of the determined value with polymers containing 0.25–0.5% dicumyl peroxide. It is a rather time-consuming procedure but has the advantage of avoiding all risk of interference from other types of peroxides present in the sample.

Organotin heat stabilizers

A quantitative thin-layer chromatography has been used[980] for the determination of organotin stabilizers.

Hexane–glacial acetic acid $(12 + 1)$ as eluting agent is satisfactory for thioacids and thiols. A small amount of the polymer extract dissolved in the elution solvent is applied to a thin-layer chromatographic plate coated with a 0.25-mm thick layer of Kieselgel G as the stationary phase, and eluted with the hexane–glacial acetic acid mixture. After drying, the stationary phase is sprayed with a 0.1% solution of catechol violet in 95% ethanol, blue spots appearing where tin compounds are present.

Separations have been achieved[981] of various organotin stabilizers on 0.25 mm layers of Merck Kieselgel GF 254 using butanol–glacial acetic acid (97 + 3) as the mobile phase and spraying the plate with catechol violet solution followed by irradiation of the plate for 10 min with ultraviolet radiation and a respray with catechol violet solution to detect the separated compounds. This procedure can be applied to the organotin compounds in PVC using a preliminary extraction of the additives from 5 g polymer with diethyl ether for 8 h. Ether is removed by evaporation and ethanol added to precipitate polymer, which is then filtered off. Down to a few parts per million of organotin can be determined in polymers.

8.2(i) Paper chromatography

This technique has been used to estimate cadmium, lead and zinc containing heat stabilizers in PVC.[984]

8.2(j) Column chromatography

Column chromatography has been used to determine barium, cadmium and zinc containing heat stabilizers in PVC.[983]

8.2(k) Gas chromatography

This technique has been used extensively to determine individual phenolic antioxidants in polymers, particularly the polyolefins and polystyrene (Table 61).

An example of the application of gas chromatography to the determination of phenolic antioxidants in polymers is a method for the determination of a very high molecular weight compound Ionox 330 (1,3, 5-tri-methyl-2,4,6 tri(3,5-di-t-butyl-4-hydroxy benzyl) benzene)) in poly propylene.[957] The solvent extract of the polymer is analysed under the following conditions:

Chromatograph	Aerograph Model 600-C with flame ionization detector.
Column	$\frac{1}{8}$ in. × 18 in. stainless steel with 5% SE-30, Silicone Gum Rubber (Methyl, General Electric Co), on 80/90 Anakrom ABS (Analabs, Inc.)
Temperatures:	
Injection port	330°C
Column	290°C
Detector	290°C
Carrier gas	65 ml/min helium
Retention time for Ionox 330	8.5 min

TABLE 61 *Separation of Phenolic Antioxidants—Gas Liquid Chromatographic Techniques*

Substances separated	Stationary phase	Column temperature (°C)	Other details	Reference
2,5-Di-t-butyl-p-cresol, 2-(2-Hydroxy-5-methylphenyl) benzotriazole	25% LAC-2R/446 (adipate ester) +2% H_3PO_4 on Chromosorb	135	H_2 carrier gas, F.I.D. error $\pm 1\%$	950
2,6-Di-t-butyl-p-cresol, (I) 2,6-Di-t-butyl-phenyl	10% Apiezon N on Celite 545	164	H_2 carrier, F.I.D. 10–3 M in presence of others can be detected	951
2,4,6-Tri-t-butyl phenol Diphenylamine 2,6-Di-t-butyl-p-cresol Phenyl-2-naphthylamine	Apiezon		F.I.D.	952
Halogenated bis-phenols	10% DC-710 Silicone oil on Chromoport 80–100 mesh	225–250	12 in glass column $\frac{1}{4}$ in. o.d. Carrier: 130 ml He/min	953
Low b.p. phenols	Capillary column coated with 10% xylenol phosphate	125	F.I.D.	954
Phenols and 5-t-butyl derivatives	Silicone oil 550-Carbowax 400 (3:2)	200	Mean deviation 0.4%	955
Phenols and cresols	5% W/W of various phosphate esters of phenols	110	120 cm × 4.5 mm column, Pye-Argon Chromatograph	956
Ionox 330	(a) 20% DC-710 Silicone oil on Chromosorb (b) 2% SE.30 Silicone gum on Chromosorb mesh	200–300 in. 10 min.	(a) 12 × 3/16 in. column (b) 12 × 1/16 in. stainless steel column	957
Low molecular weight phenol	Silicone-coated capillary column		Converted to trimethyl silyl esters before chromatography	958
2,6-Di-4-methylphenol	20% SE.30 on HMDS-tetrated 60 mesh Chromosorb W	200	E.C. detector	959

Calibration standards representing 0.1 g/l to 0.8 g/l Ionox 330 were analysed with an average repeatability of $\pm 2.0\%$. The calibration curve obtained as a plot of peak area versus concentration was linear and passed through the origin, indicating that no Ionox 330 is lost through decomposition in the hot injection port.

Other phenolix antioxidants that can be determined by gas chromatography of suitable extracts of the polymer include: Ionol (2, 6-di-t-butyl-p-cresol) in polystyrene,[960] 2-t-butyl-4-methyl phenol, 2, 6-di-t-butyl-4-methyl phenol and p-t-butyl phenol in polyethylene,[959,961] butylated hydroxy anisole and butylated hydroxy toluene[962] and Ionox 330.[963]

Amine antioxidants

Gaeta *et al.*[967] have described a gas chromatographic method for determining a number of antioxidants such as N-phenyl-2-naphthylamine and N, N'-sec-heptyl phenyl p-phenylene diamine in oil-extended synthetic polymers such as polybutadiene or styrene-butadiene rubber. This involves extracting the antioxidant from the polymer with ethanol, taking up the concentrated extract with the appropriate solvent and analysing the resulting solution by gas chromatography. They found – by the use of standard solutions of the antioxidants in carbon tetrachloride, acetone or carbon disulfide and the careful choice of chromatographic conditions – that the elution of these materials had little or no interference from the extending oil present. In addition, the oil was completely soluble in the solvent and all but the heaviest fractions eluted prior to that of the antioxidant.

The instrument used in this work was a dual-column flame ionization detector chromatograph equipped with 5 ft columns of $\frac{1}{2}$ in. stainless-steel tubing packed with 60–80 mesh Chromosorb Z support containing 5% Apiezon N grease. The detector temperature was 205°C while that of the injection port was 275°C with a carrier gas flow of 30 ml of helium per minute. The solvents used (carbon tetrachloride, acetone and carbon disulfide) were of reagent grade. A different solvent was needed for each antioxidant to ensure complete solubility.

Figure 114 shows the result of using this technique in the case of an acetone extract of polybutadiene containing N-phenyl-2-naphthylamine antioxidant.

The small peaks which appear near to that of the antioxidant are those due to the extending oil present. No elution of the oil occurred during that of N-phenyl-2-napthylamine as a chromatogram of the oil in acetone was obtained previous to that of the antioxidant. At column temperatures above 170°C the different component fractions of the oil are manifested as separated chromatographic peaks as they pass through the column. Since these polymer extenders are wide petroleum cuts there are some heavy

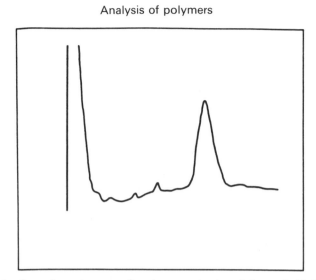

FIG. 114. Gas chromatogram of polymer extract solution of N-phenyl-2-naphthylamine in acetone at 250°C.

fractions which elute even after such high boiling materials as the substituted *p*-phenylenediamine antioxidants.

High-temperature gas chromatography has been used[968,969] for the analysis of mixtures of amine-type antidegradants in rubber. These workers used a separation column constructed of aluminium packed with 20% Apiezon L on 30–60 mesh Chromasorb W. Analysis was carried out on an acetone extract of the rubber sample, employing diphenylamine as an internal standard. Using column temperatures up to 310°C they were able to separate a range of antidegradants including 1, 2-dihydroxy-2, 2, 4-tri-methyl-6-ethoxyquinaline, N-isopropyl-N'-phenyl-*p*-phenylenediamine, N-phenyl-2-naphthylamine, N, N'-di-2-octyl-*p*-phenylene-diamine and N, N'-diphenyl-*p*-phenylenediamine. Near-quantitative determinations were obtained for all these substances. Apiezon L was found to be distinctly superior as a mobile phase to other substances tried. Thus Dow-Corning 710 Silicone fluid and butanediol succinate were too volatile at operating temperatures up to 310°C, whilst silicone rubber, although sufficiently non-volatile, did not give the high degree of resolution obtained with Apiezon L.

Ultraviolet light absorbers

Lappin and Zannucci[971] showed that a number of ultraviolet light absorbers could be quantitatively determined on an SE-30 column. SE-30 has a maximum operating temperature of about 350°C. This high temperature permits elution of some high molecular weight additives.

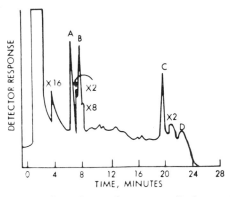

Fig. 115. Chromatogram of a polypropylene extract. Peaks are: (a) biphenyl; (b) butylated hydroxy toluene; (c) 4-(dodecyloxy)-2-hydroxybenzophenone and (d) unknown impurity.

They examined two methods of separating the additives from the polymer. Gas chromatography of a hexane extract of the polymer produced numerous extraneous peaks, probably owing to a decomposition of dissolved amorphous polymer in the injection port, rendering the chromatogram useless. When the polymer was dissolved in p-xylene and reprecipitated with an equal volume of p-dioxane, a relatively clean chromatogram was obtained from the filtrate. The position of decomposition peaks from unstable compounds, such as dilauryl 3, 3′-thiodipropionate and distearyl pentaerythritol diphosphite, are predictable and do not interfere with the determination of those additives studied by Lappin and Zannucci.[971] They found that the optimum polymer sample size was 3 g, although runs were possible on samples as small as 200 mg. Figure 115 is a chromatogram obtained for the extract of a polypropylene sample in which biphenyl is used as an internal standard. Benzophenone is another possible internal standard. Lappin and Zannucci[971] determined 4-(dodecyloxy)2-hydroxybenzophenone and 2, 6-di-tert-butyl-p-cresol in polypropylene by this technique (Table 62). The precision of the 2, 6 di-tert butyl-p-cresol determination is good but the quantity found (0.02%) was less than the amount added (0.05%), possibly due to some losses of volatile antioxidant during polymer compounding.

TABLE 62   Analyses of polypropylene for BHT and DOBP

Sample composition	Additive found (%)			
	Run 1	Run 2	Run 3	Run 4
0.05% 2, 6-di-tert-butyl-p-cresol	0.02	0.02	0.03	0.02
0.30% 4-(clodecyloxy) 2-hydroxybenezophenone	0.25	0.32	0.20	0.35

TABLE 63   *Separation of ultraviolet absorbers*

Ultraviolet absorber	Class of compound	Relative retention time (relative to Santonox R)	Column temperature (°C)
Tinuvin P	Benzotriazole	0.20	250
Tinuvin 326	Benzotriazole	0.64	250
Tinuvin 327	Benzotriazole	0.85	250
Cyasorb UV 531	Aromatic ketone	1.0	250
Tinuvin P	2-(2′-Hydroxy-5′-methylphenyl benzotriazole		
Tinuvin 326	2-(2′-Hydroxy-3′-t-butyl-5′-methyl ethyl phenyl-3-) chlorobenzo triazole		
Tinuvin 327	2-(2′-Hydroxy-3′, 5′-di-t-butylphenyl)-3-chlorobenzo triazole		
Cyasorb UV 531	2-Hydroxy-4-n-octoxybenzophenone		

4-(dodecyloxy)2-hydroxybenzophenone determinations ranged from 0.20 to 0.35% for a sample originally containing 0.30% of the additive.

Denning and Marshall[972] devised a method in which toluene extracts of polyethylene are examined for ultraviolet absorbers (and antioxidants) at two column temperatures in order to overcome the problem that, whereas some compounds have a relatively low retention time, others of higher molecular weight have very long retention time at 250°C. Santonox R was used as an internal standard in this method. Denning and Marshall[972] used a Pye Series 104, Model 64, dual-column chromatograph, equipped with a heated flame-ionization detector and an isothermal column oven. The oven was operated isothermally at 250 or 300°C. Relative retention data obtained for ultraviolet absorbers are given in Table 63.   Figure 116 is a diagrammatic representation of the separation of the group of antioxidants and ultraviolet absorbers identified at 250°C.

Plasticizers

Due to their volability the esters used as plasticizers in polymers such as flexible PVC can be determined by gas chromatographic analysis of a solvent extract of the polymer. Most published methods of analysis are based on this technique. The identification and determination of plasticizers is discussed further in Chapter 9.

Robertson and Rowley[986] have published an excellent detailed description of methods for the solvent extraction of plasticizers from polyvinylchloride and other polymers prior to their determination by weighing or gas chromatography of the extract. They state that the quantitative separation of plasticizers from the other ingredients is the first and most important step in the analysis of plasticized polyvinylchloride compositions. The most effective and convenient method of separation is by extraction with a suitable solvent, using the Soxhlet apparatus. The efficacy of this

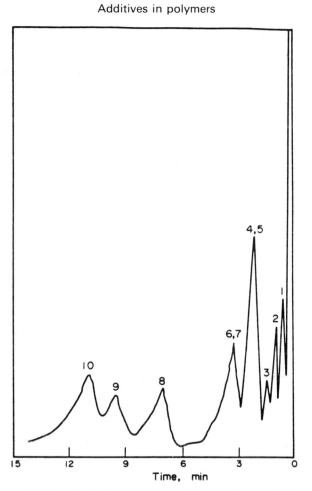

FIG. 116. Separation of antioxidants and ultraviolet absorbers at 250°C. Peaks are: (1) Topanol OC; (2) Polygard; (3) Irganox 1010; (4) Tinuvin P; (5) Nonox DCP; (6) Nonox DCP; (7) Polygard; (8) Tinuvin 326; (9) Tinuvin 327; and (10) Santonox R.

TABLE 64 *Multiple extractions*

Plasticizer	Concentration (%)	Extracts (%)	
		4 h ether, 5 h MeOH	4 h ether, 4 h CCl MeOH
Bisoflex 791	28.5	27.7	28.7
Tritolyl phosphate	28.5	25.8	28.9
Mesamoll	28.5	25.5	28.1
Bisoflex 795	28.5	28.1	29.0
Reoplex 220	28.5	20.4	29.3
Hexaplas PPA	23.5	11.0	17.4

procedure depends mainly on the choice of solvent. The ideal solvent would not dissolve any of the polyvinylchloride, but would remove all the plasticizer, and all, or none, of the other ingredients of the composition.

Robertson and Rowley[986] compared the efficacies of ether, carbon tetrachloride and methanol, in the Soxhlet extraction of a number of polyvinylchloride compositions. Binary azetropes of methanol with carbon tetrachloride, chloroform, 1, 2-dichlorethane, and acetone, and of diethyl ether with 1, 2-epoxypropane, were also compared.

Table 64 gives the results of following 4 h extractions with ether by extractions with other solvents. It can be seen that extraction for 4 h with carbon tetrachloride/methanol azeotrope gives better results than extraction for 15 h with methanol.

Haslam and Soppet[987] found that extraction with acetone, followed by precipitation of dissolved polymer with light petroleum, gave poor results: only 28.5% was recovered from a composition containing 31.8% tritolyl phosphate. Substitution of 1, 2-dichlorethane for acetone gave no improvement, but diethyl ether extracted 31.8% of the sample, and the extract contained a negligible amount of polyvinylchloride. For routine extraction it was recommended by Haslam and Soppet that the sample be stood overnight in cold ether, then extracted for 6–7 h in the Soxhlet apparatus. This procedure has also been described by Doebring,[988] who specified that anhydrous ether should be used. Thinius[989] compared the rates of extraction of dioctyl phthalate by ether, carbon tetrachloride, and petroleum, at room temperature. In 70 min the ether extract from a composition containing 40.0% plasticizer amounted to 39.7%; in the same time the carbon tetrachloride extract was 33.1% and the petroleum extract 29.0%. Extraction with carbon tetrachloride for 64 h gave only 35.6% extract. Thinius also found that mixtures of phthalate and phosphate esters could be completely removed by ether, but light petroleum gave only 65% of the expected yield. Toluene dissolved some polyvinylchloride at room temperature, and very much more in a Soxhlet extraction.

For compositions containing polypropylene adipate, which is only partly extracted by ether, Haslam and Squirrel[990] used a 6-h ether extraction, followed by an 18-h methanol extraction. The ether extract was 34.5% from a composition containing 36.1% of a mixture of equal parts of dioctyl phthalate and tritolyl phosphate, but only 33.1% from a composition containing 45.7% of a mixture of equal parts of dioctyl phthalate, tritolyl phosphate and polypropylene adipate. The combined ether and methanol extracts amounted to 34.7% and 44.3% respectively. Wake[991] quotes the results of some unpublished work carried out at the laboratories of the Rubber and Plastics Research Association: methanol extracted 40.5% and 36.8% from compositions containing 42.3% polypropylene sebacate and 36.4% polypropylene adipate respectively. The extracts contained polyvinylchloride equivalent to 0.6% and 0.9% respectively. Clarke and Bazill[992] extract with ether for 15 h, then with methanol

TABLE 65   *Gas chromatography, retention times for organic peroxides*

Compound	Column*		Temperature (°C)	Helium pressure[†] (psi)	Retention time (min)
tert-Butyl hydroperoxide	2	m-A	80	20	22.7
tert-Pentyl hydroperoxide	1	m-A	80	15	19.9
tert-Butyl peracetate	1	m-A	100	20	6.5
tert-Butyl peroxyisobutrate	1	m-A	100	20	15.3
Di-tert-Butyl peroxide	2	m-A	80	20	8.1
Di-tert-Pentyl peroxide	1	mA	80	15	15.4
2, 5-Dimethyl-2, 5-di(tertbutyl peroxy)-3-hexyne	1	m-O	138	15	4.9
2, 5-Dimethyl-2, 5-di(tert-butyl peroxy)hexane	1	m-O	138	15	6.9
n-Heptane	2	m-A	80	20	6.1
n-Dodecane	1	m-O	138	15	3.1
n-Pentane	1	m-A	100	20	0.4
n-Nonane	1	m-A	100	20	4.9

*Length (metres) and type of column used.
[†]Inlet pressure.
A, didecyl phthalate on diatomaceous earth;
O, silicon grease on diatomaceous earth.

for 8 h. They state that ether removes plasticizers whose molecular weight is less than 1000.

Organic peroxide residues

Gas chromatography has also been used to determine certain types of organic peroxides. Bukata *et al.*,[997] for example, describe procedures involving chromatography of heptane solutions of peroxide (Table 65) on phthalate/diatomaceous earth or silicone/diatomaceous earth columns, and using helium as the carrier gas. No doubt this type of procedure could be easily adapted to the examination of solvent extracts of polymers.

Hyden[998] describes a gas chromatographic procedure for the determination of di-tert-butyl peroxide. This is based on the thermal decomposition of the peroxide in benzene solution into acetone and ethane when the solution is injected into the gas chromatographic column at 310°C. The technique is calibrated against standard solutions of pure di-tert-butyl peroxide of known concentration.

8.2(I)   X-Ray fluorescence spectrometry

Organotin heat stabilizers

In the X-ray fluorescence method the polymer extract is diluted with 2-ethyl-hexanol to give a tin concentration of 0.9–1.7%, and about 40 ml of the solution is taken for the determination. The Sn K radiation (25.2 keV) excited by low-energy gamma-radiation from a ^{241}Am source is measured by means of a NaI(Tl) crystal, a single-beam γ-spectrometer being used for evaluation.

8.2(m)   Nuclear magnetic resonance spectroscopy

Plasticizers

Wide-line nuclear magnetic resonance spectroscopy has been used (Masfield)[995] for the determination of the di-iso-octyl phthalate content of polyvinylchloride. The principle of the method is that the narrowline liquid-type NMR signal of the plasticizer is easily separated from the very broad signal due to the polymer; integration of the narrow-line signal permits determination of the plasticizer. A Newport Quantity Analyser Mk I low-resolution instrument, equipped with a 40-ml sample assembly and digital read-out, has been used to determine 20–50% of plasticizer in polyvinylchloride. The sample may be in any physical state without significantly affecting the results; e.g. sheet samples are cut into strips 50 mm wide, which are rolled up and placed in the sample holder. A curvilinear relationship exists between the signal per g and the percentage by weight of the plasticizer. For highest precision it is necessary to know the type of plasticizer present; use of the appropriate calibration graph gives a precision of $\pm 0.5\%$. However, one general calibration graph can be used; the precision is then approximately $\pm 3\%$. As the NMR signal is temperature-dependent the temperature of calibration and of analysis should not differ by more than 4°C.

8.3   Analysis of additive mixtures in polymers

A necessary prerequisite to the methods discussed earlier in this chapter is that the analyst has a full knowledge of all the types of additive present in the polymer. This is necessary so that, in selecting a method for determining a particular constituent, due allowances can be made for other types of additive constituents present or of any decomposition products of additives present. Whilst this information might be to hand if an analyst is examining materials of known origin, this would not always be so. In such cases it is mandatory that the first step must be to completely identify the additives present, before any consideration can be given to the problem of selecting or devising a method quantitative analysis for any constituent present in the polymer.

The problem resolves itself into the preparation of a total solvent extract of the polymer in which all additives are completely recovered, followed by separation of the mixture into pure single components by a suitable form of chromatography and, finally, by identification of each separated pure component by suitable means, usually involving visible infrared, ultraviolet, mass or nuclear magnetic resonance spectroscopy and perhaps microanalysis for elements present. Only after this stage can the details of the quantitative determination of particular polymer constituents be considered.

It is advisable, when commencing the analysis of a polymer for unknown additives, to determine first its content of various non-metallic and metallic elements. Any element found to be present must be accounted for in the subsequent examination for, and identification of, additives. Hence elemental analysis reduces the possibility of overlooking any additive which contains elements other than carbon, hydrogen and oxygen. The analytical methods used to determine elements should be sufficiently sensitive to determine about 10 ppm of an element in the polymer, i.e. should be able to detect, in a polymer, a substance present at 0.01% and containing down to 10% of the element in question.

This requirement is met for almost all the important elements by use of optical emission spectroscopy and X-ray fluorescence spectrometry. Using these two techniques, all metals and non-metals down to an atomic number of 15 (phosphorus) can be determined in polymers at the required concentrations (Cook *et al.*,[1000] Hank and Silverman.[1001] Mitchell and O'Hear[1002] and Bergmann *et al.*,[1003]). Nitrogen is determinable at these levels by micro-Kjeldahl digestion techniques.

Apart from gas chromatography three forms of chromatography are worthy of serious consideration for the separation of additive mixtures, viz. column, paper and thin-layer chromatography. Of these, thin-layer chromatography is by far the most useful general technique.

Gas chromatography

Gas chromatography is particularly useful for characterization and determination of move volatile components of plastics such as fatty acids, alcohols, esters or hydrocarbons. Some examples of the application of this technique are discussed below.

Identification of plasticizer esters in PVC

Figure 117 shows a gas chromatogram of a mixture of plasticizers of the type used in PVC formulations. The polymer extract was obtained[1004] by dissolving the PVC in tetrahydrofuran, then adding 4 volumes of methanol to precipitate the polymer and leaving the plasticizers in solution. Many of the commonly used plasticizers are well separated and can be identified by the retention times.

Although the relative retentions of the plasticizers are helpful for the

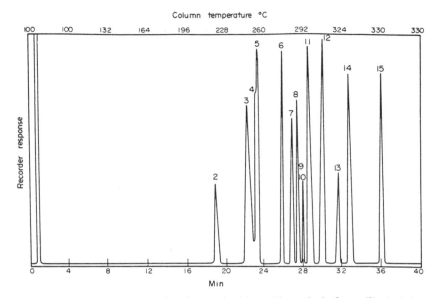

FIG. 117.   Gas chromatography of ester plasticizers: (1) tetrahydrofuran; (2) triethyl citrate; (3) methylphthalylethyl glycolate; (4) Ethylphthalylethyl glycolate; (5) Dibutyl phthalate; (9) butylbenzyl phthalate; (10) trioctyl phosphate; (11) di (2-ethylhexyl)adipate; (12) di (2-ethylhexyl)phthalate; (13) di (2-ethylhexyl); (14) di (2-ethylhexyl)sebacate; (15) di-n-decyl phthalate. Stainless-steel column 6 ft $\times \frac{1}{8}$ in. packed with 10% UCW-98 on 60–80 mesh Diatopont S. Initial column temperature 100°C, after 4 min of isothermal operation, temperature programmed at 8°C per min to 330°C flame ionization detector.

identification, the complexity of mixtures, normally encountered, necessitates hydrolysis and esterfication to obtain information on the components of plasticizers. A definite identification of the original plasticizer can be obtained only after the identification of the constituent alcohols and acids has been made. To hydrolyse the esters a freshly cut 0.1 g piece of lithium metal was added to the methanol solution of the plasticizers; and the mixture was refluxed for 2 h. At the end of this time the solution was allowed to cool, and then it was carefully acidified by dropwise addition of concentrated sulphuric acid. When the solution became acidic, as noted by pH indicating paper, it was refluxed again for 1 h in order to convert the acids to their methyl esters. After cooling, the solution was neutralized by addition of dry sodium carbonate, and the pH was checked. A suitable sample of this solution was injected into the gas chromatograph for detection of the alcohols and methyl esters of the acids. The gas chromatographic conditions used were identical to those for chromatographing the original plasticizers (Figure 118).

This scheme is helpful in identifying these products. Since the gas chromatographic conditions used for the alcohols and the methyl esters

Column temperature °C

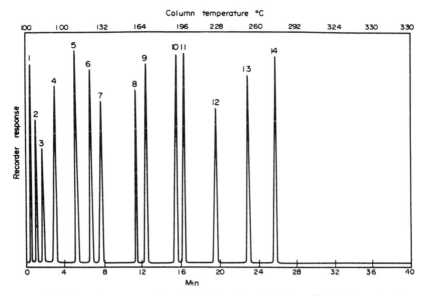

FIG. 118.   Gas chromatography of alcohols and methyl esters of acids: (1) methanol; (2) butanol; (3) pentanol; (4) hexanol; (5) heptanol; (6) 2-ethylhexanol; (7) octanol; (8) dimethyl adipate; (9) decanol; (10) dimethyl-*o*-phthalate; (11) dodecanol; (12) tetradecanol; (13) methyl palmitate; (14) methyl stearate.

of the acids are identical to those used for the original plasticizers, it is very easy to establish whether the plasticizers have been completely reacted. The presence of non-hydrolyzable components in a mixture can also be detected by examining the gas chromatograms of plasticizers before and after hydrolysis, as the original gas chromatographic peak will still be observed

Expanding agents in polystyrene

Gas chromatography has been used for the quantitative determination of expanding agents such as normal and isopentane in expandable grades of polystyrene.

These gas chromatographic analyses are carried out on solutions of the polymers in the presence of internal standards. To avoid interferences in the analysis it is essential for the solvent and the internal standard to have retention times different from those of the volatile compounds being determined in the polymer.

An available procedure for doing this analysis involves the gas chromatographic analysis of a solution of the polymer in propylene oxide in the presence of 2,2-dimethyl butane as an internal standard. This method is entirely satisfactory for analysing grades of expandable polystyrene in which it is known that isopentane and n-pentane are the only expanding agents present. However, if other types of expanding agent have been used in the polymer formulation then it is possible that, under the selected conditions, their retention times might coincide with those of propylene oxide or 2, 2-dimethyl butane, with the result that the analysis would be invalidated.

Column chromatography

One of the difficulties of column chromatography is the problem of identifying the fractions in which the separated compounds are concentrated. This can be achieved by the laborious process of examining all the fractions, for example by infrared or ultraviolet spectroscopy or by evaporating to dryness and weighing the residues: or by the less laborious process of monitoring the effluent as it leaves the chromatographic column so that solute containing fractions from the fraction collector can be picked out from the fractions which do not contain any substances. Several types of effluent monitors are available, based on the measurement of the ultraviolet absorption, conductivity, etc. These have the disadvantage of being too specific for dealing with mixtures of compounds of unknown type. For example, compounds which do not either absorb in the ultraviolet or do not ionize would be missed using these detectors. The most useful general-purpose monitors are those based on the measurement of refractive index and on thermal effects. The latter operates on the principle that as each separated compound moves down the column it is accompanied by heat of absorption and desorption due to interaction between solute molecules and the stationary phases. These heat pockets (i.e. separated compounds) are detected by a thermistor at the column outlet and recorded on a strip chart which can be operated in conjunction with a fraction collector. Thus, separated fractions can be readily located and bulked if necessary for further examination.

The most recent development in liquid chromatography, namely high-pressure liquid chromatography, combines the advantages of built-in detectors with improved resolution in the separation of mixtures due to improved column packings and operation at a derated pressure.

Recent advances have improved the speed and efficiency of high-pressure liquid chromatography to the point that they are now rapidly approaching the limits achievable by gas chromatography.

Analysis times in liquid chromatography can be shortened considerably without loss of peak resolution by optimizing the parameters of column length and diameter, flow rate, sample size, and support particle size. The theoretical groundwork for high-efficiency liquid chromatographic separations has been established by a number of investigations.[1005-1010] Because of their porous nature conventional liquid chromatographic absorbents of small particle diameter give rise to poor rates of mass transfer of solutes under rapid flow conditions giving poor column efficiency. Also porous absorbents tend to impact under the high pressures needed to achieve adequate flow. A number of high-efficiency liquid chromatographic supports have recently been introduced. These include Zipax (DuPont's CSP support[1011,1012]) Corasil I and II, and Durapak[1013] (Waters Associates). With the exception of Durapak, these materials, in the micron particle range, consist of particles with a solid core and thin porous coating. This unique combination gives very high coefficients of mass transfer. Durapaks consist of conventional liquid phases, such as β, β'-oxy-dipropionitrile, chemically bonded to a rigid porous bead. Textured glass beads for liquid–liquid chromatography similar to those reported for gas chromatography[1014] are being developed by Corning.[1015]

The Zipax material has been reported to have a relatively inert surface,[421] whereas the Corasil support is also recommended for liquid–solid column chromatography.[1016] Little has been reported on the use of these solid-core silicaceous backboned Corasil supports with a polar liquid coating in liquid–liquid chromatography. Kirkland[1011,1012,1017] has studied the support in some detail for several model systems, and has reported much-improved performance for this material in liquid–liquid chromatography when compared to coated glass beads or diatomaceous earth. A number of separations have been described by Halasz[1013] on Durapak-type supports. Majors[1018] describes some separations obtained on several of these commercially available support materials using a liquid chromatographic system capable of operation up to 5000 psi. Because of the nature of these solid-core supports, unusually large pressure drops across the chromatographic column do not occur unless one uses several columns in series, or extremely fast flow rates.

A schematic diagram of a typical high-pressure liquid chromatographic system is shown in Fig. 119.[1018] The system employs a microregulating high-pressure feed pump with an inlet pressure of 5000 psi. The solvent reservoir is placed a few feet above the pump since a slightly positive inlet pressure is required for operation. The degassed solvent is slowly stirred by means of a magnetic stirrer and is heated externally slightly above room temperature to keep the solvent degassed. A valve is placed on the high-pressure side of the pump to aid in priming the pump or to shut the solvent flow when required. Downstream from the pressure gauge, the entire system is connected with 0.04 in. i.d., 0.063 in.o.d. stainless-steel tubing to ensure a minimal dead-volume in the system, which is particularly desirable when changing solvents or solvent programming.

When necessary, the liquid flow can be split between the analytical column and the reference column. A valve placed before the reference column permits the flow rate through that column to be completely shut

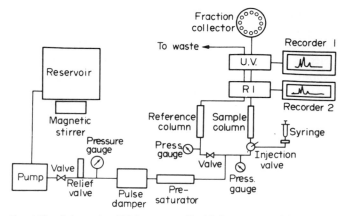

FIG. 119.   Schematic of high-pressure liquid chromatographic system.

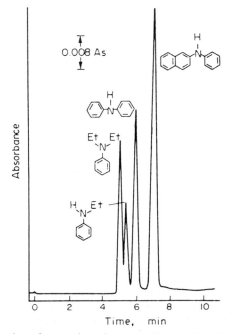

FIG. 120.   Separation of aromatic amine antioxidants using Zipax. Column 1000 mm × 2.1 mm i.d.; packing: 0.5% β, β'-oxydipropionitrile on 20–37 μ Zipax support; carrier-iso-octane; flow rate 0.31 ml/min; sample 10.6 μl of a mixture of 9.5 μg/ml each of N, N-diethylaniline and N-ethylaniline, 29 μg/ml of diphenylamine, and 52 μg/ml of N-phenyl-2-napthylamine in iso-octane.

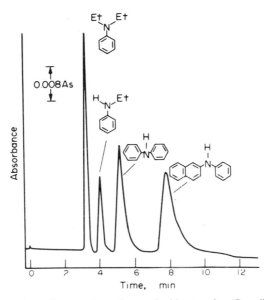

FIG. 121. Separation of aromatic amine antioxidants using Corasil 1. Column 1000 mm × 2.1 mm, i.d., packing: 0.5% β, β'-oxydipropionitrile on 37–50 μ Corasil 1 support; carrier-iso-octane; flow rate 0.50 ml/min.

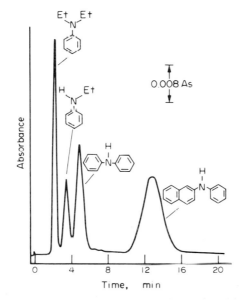

FIG. 122. Separation of aromatic amine antioxidants using OPN/Durapak. Column 1000 mm × 2.1 mm, i.d.; packing: 3.7% OC₂H₄CN on 36–75 μ Porasil C; carrier-iso-octane; flow rate 2.24 ml/min; sample, same as in Fig. 120.

off or varied, depending on the requirements of the system. This is particularly useful when using the refractive index monitor at higher flow rates. For column flow rates up to 2 ml/min good stability could be obtained by careful balance of flow. The reference column was either filled with the same liquid–liquid support employed in the analytical column or merely filled with uncoated glass beads. With the ultraviolet detector the reference column was normally not used.

Figures 120–122 show high-speed chromatograms obtained for different column materials with mixtures of N, N, diethylaniline, N-ethylaniline, diphenylamine and N-phenyl-2-naphthylamine amine antioxidants used in rubber manufacture.

A number of comparisons can be made for the supports. Although the relative elution order is the same on all three columns, the selectivity for each peak relative to N, N-diethylaniline appears to be affected. Selectivity for each solute (elution volume of solute divided by that of N, N-diethylaniline) was the greatest on Durapak and the least on β, β'-oxydipropionitrile/Zipax.

Figures 123 and 124 show the separation achieved of three plasticizers used in polyvinylchloride. These plasticizers cannot be directly determined by gas chromatography but must first be saponified. The separation of

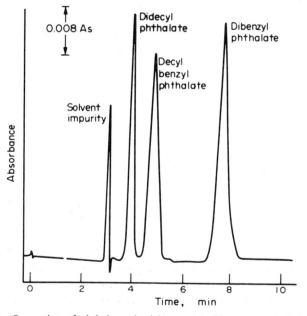

FIG. 123. Separation of phthalate plasticizers using Zipax support. Column and carrier, same as Fig. 120; flow rate 0.50 ml/min; sample, 10.6 μl of a mixture of 0.40 μl/ml each of didecyl phthalate and decyl benzyl phthalate and 0.35 mg/ml of dibenzyl phthalate in heptane.

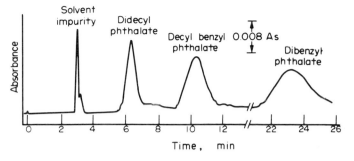

FIG. 124. Separation of phthalate plasticizers using Corasil 1 support: column, packing carrier, and flow rate same as in Fig. 121; sample same as Fig. 123.

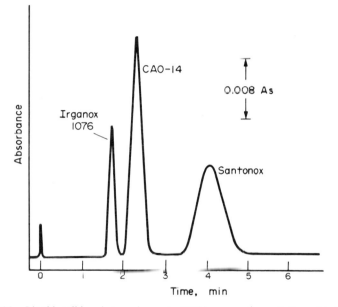

FIG. 125. Liquid–solid column chromatographic separation of three hindered phenolic antioxidants using Corasil II. Column 1000 mm × 2.1 mm, i.d.; packing: 37–50 μ Corasil II (activated at 110°C) carrier 1% (v/v) isopropanol in hexane; flow rate 0.95 ml/min; sample: 10.6 μl of a mixture of 0.54 mg/ml Irganox 1076, 0.81 mg/ml CAO-14, and 1.4 mg/ml Santonox R dissolved in carrier.

all three was obtained in less than 8 min on Zipax. The technique is also useful for the separation of hindered phenolic antioxidants (Fig. 125).

Figure 126 shows the linearity of both the peak height and the peak area measurements at 254 nm versus micrograms of antioxidant CAO-14 injected. The other antioxidants behaved similarly.

An additional advantage of conventional or high-pressure column chromatography is that it can separate a mixture of compounds into a

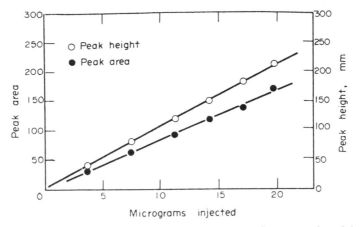

FIG. 126.   Calibration curve for CAO-14. Column, carrier flow rate, volume injected
same as in Fig. 125.

series of fractions which can then be separately examined by an appropriate technique for unknown components. In a procedure for identifying
the type of plasticizer present in PVC[1019] an ether extract of the polymer
is separated on a column of celite–silica gel and the fractions weighed
after removal of the ether. Those fractions which contained any weighable
amount of material were examined by infrared spectroscopy, which
enabled an identification to be made (Fig. 127).

Figures 128 and 129 contrast the type of chromatograms obtained by
the two types of column chromatography. In the conventional labour-
intensive manual method[1020] (Fig. 128), the individual fractions are each
made up to a standard volume and analysed. In the high-pressure
method[1021] the detector picks up each separated component to provide
a chromatogram of the type shown in Fig. 129.

The correct selection of elution solvents is very important when
attempting to separate mixtures on a column. This is illustrated in Table
66, which shows how a system using three solvents can be used to separate
nine plasticizers.

Thin-layer chromatography–infrared spectroscopy

Thin-layer chromatography using plates coated with $250\,\mu$ absorbent
is an excellent technique for efficiently separating quantities up to 20 mg
of total additive mixtures present in polymer extracts, individual components. This technique provides a few milligrams of each component,
sufficient to prepare a recognizable infrared or ultraviolet and mass spectra
which can be compared with the spectra of authentic known compounds.
However, the technique does not conveniently handle larger quantities,

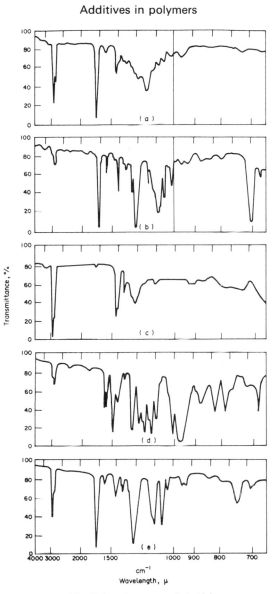

FIG. 127.   Infrared spectra of plasticizers.

although preparative thin-layer chromatography using thicker coatings will achieve larger scale separations with some loss of resolution.

Numerous works have been published on the experimental technique of thin-layer chromatography, which will not be discussed further except in so far as is relevant to its application to additive identification. Dohmann[1022] carried out an excellent short review of current techniques.

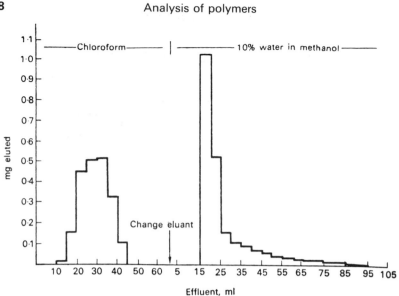

FIG. 128. Elution chromatogram of Eutylated hydroxy toluene and Santonox R polyethylene samples extracted with chloroform. Separation on alumina column. Antioxidant content of each separated fraction measured by ultraviolet spectroscopy.

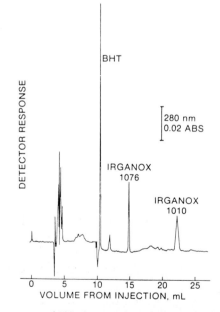

FIG. 129. Chromatogram of 100-μl extract from 2.07 g of polyethylene. Chromatographic conditions as in text. Separation on alumina column. Antioxidant content of each separated fraction measured by ultraviolet spectroscopy.

TABLE 66   *Column chromatography of PVC plasticizers elution sequence for plasticizer mixture–column 1:1 by weight Celite 545 (100–200 mesh): silica gel (100–200 mesh)*

(Ether extract of PVC) eluant	Fraction	Plasticizer found
Carbon tetrachloride	1	Cereclor (chlorinated hydrocarbon)
	2	—
	3	—
	4	Mesamoll (alkyl ester of a sulphonic acid)
Carbon tetrachloride/di-isopropyl ether (2%)	5	—
	6	Tritolyl phosphate
	7	Di-*n*-butyl phthalate
	8	
Carbon tetra chloride/di-isopropyl ether (5%)	9	Di-*n*-butyl sebacate
	10	—
	11	—
	12	—
Carbon tetrachloride/acetone (2%)	13–16	—
Carbon tetrachloride/acetone (5%)	17	Diethylene glycol
	18	Dibenzoate
	19	Abrac 'A' (epoxidized) vegetable oil
	20	—
Carbon tetrachloride/acetone (7.5%)	21	—
	22	Polypropylene secacate
	23	Polypropylene sebacate (trace)
	24	

He discusses thin-layer chromatography in the normal sense of the word, i.e. with plate layers up to 250 μ thick and 20 cm $\times$ 20 cm or 20 cm $\times$ 8 cm in area and also discusses preparative layer chromatography which, with some loss in resolution, can separate considerably larger quantities of compounds on plate layers up to 2 mm thick and 100 cm $\times$ 20 cm in area

Polymer extraction procedures using organic solvents do not extract all types of organic additives from polymers; many inorganic compounds and metalorganic compounds (e.g. calcium stearate) are also insoluble. The presence of metals will have been indicated in the preliminary examination of the polymer. Most types of organic polymer additives, however, can be readily extracted from polymers with organic solvents of various types. The first step is to solvent-extract the total additives from the polymer in high yield and with minimum contamination by low molecular weight polymer. Extracts should be used for analysis without delay, as they may contain light- or oxygen-sensitive compounds. When delay is unavoidable storage in actinic glassware under nitrogen in a refrigerator minimizes the risk of decomposition.

Total internal plus external additives can be extracted from low- and high-density polyethylene and polystyrene by procedures involving solution or dispersion of the polymer powder or granules (3 g) in cold redistilled sulphur-free toluene (50–100 ml), followed in the case of polyethylene by refluxing for several hours. Rubber-modified polystyrene does not completely dissolve in toluene if it contains gel. Methyl ethyl ketone or propylene oxide are alternative suitable solvents for polystyrene. Dissolved polymer is then reprecipitated by the addition of methyl alcohol or absolute ethanol (up to 300 ml), and polymer removed by filtration or centrifuging. The additive-containing extract can then be concentrated to dryness as described previously. Alternative procedures for the extraction of polyethylene and polypropylene involve refluxing with chloroform for 6 h or contacting with cold diethyl ether for 24 h or Soxhlet extraction with diethyl ether, methylene dichloride, chloroform or carbon tetrachloride for 6–24 h followed by concentration of the extract. Methylene dichloride is a particularly good solvent for polypropylene extractions because of its high volatility and also its small extraction of atactic material from the polymer, compared with other solvents. In addition to additives, most solvents also extract some low molecular weight polymer with subsequent contamination of the extract. To overcome this, Slonaker and Sievers[1023] have described a procedure for obtaining polymer-free additive extracts from polyethylene based on low-temperature extraction with n-hexane at $0°C$. This procedure is also applicable to polypropylene and polystyrene.

Thin-layer chromatographic grades of silica gel usually contain traces of organic impurities. If, during development of a chromatogram, these impurities migrate to regions of the plate which coincide with the R_f values of separated additives, then the impurities will interfere in the interpretation of the plate following spraying with aggressive detection reagents, such as concentrated sulphuric acid and antimony pentachloride, as the organic impurities in the adsorbent also react with these reagents and show up on the sprayed chromatograms. Also the impurites absorb strongly in the ultraviolet region, especially below $250 m\mu$. The adsorbent impurities do not have an appreciable adsorption in the infrared region. As these impurities are extracted from the silica gel with organic solvents such as diethyl ether, ethanol, acetone, benzone, chloroform and many others, they could occur as contaminants in some of the fractions of separated additives isolated from the plate by solvent extraction of the gel and consequently interfere in the interpretation of the spectra of the additives, particularly in the case of additives which absorb below $250 m\mu$ in the ultraviolet. For this reason an identical blank chromatogram (only sample absent) should always be run in parallel with the sample chromatogram in order to check whether such interference effects exist.

Ultraviolet absorbing impurities in adsorbants are influenced by the nature of the migration solvent. Depending on the polarity of the migration solvent used the impurities migrate to a greater or lesser extent up the plate towards the solvent front, with the result that the lower part of the chromatogram nearer the baseline is cleared of impurities, and the impurities become concentrated in the upper section of the plate nearer the solvent front. Subsequently, when a section of the absorbent is removed and eluted with a further solvent to recover a separated compound the amount of impurities contaminating the compound will depend on the location of the compound on the chromatogram (R_f value) and contamination may range from negligible to substantial.

In circumstances where slow-moving compounds are being separated the impurities may move away from the polymer additives towards the solvent front, and thus not interfere in the subsequent examination of the separated compounds. This behaviour leads to a convenient method for moving the impurities beyond the section of the chromatogram to be used for the separation of polymer additives by migrating the chromatogram with appropriate washing solvents before applying and migrating the sample mixture. If this premigration washing covers a longer distance on the plate than is to be used in the sample migration, then it is possible to move interfering impurities out of the way.

The polymer extract is then applied as a band along the base of the plate and the plate developed with suitable solvents.

Finding a chromatographic development solvent or a mixture of solvents for the separation of unknown mixture of additives is not always easy. In some cases a complete separation is not obtained with a single solvent or solvent combination, but necessitates the preparation of several chromatograms using different solvents.

An unknown mixture should first be chromatogaphed on 20 cm × 5 cm plates with solvent of different polarities to obtain an idea of the types of compounds present in the sample and to reduce the possibility of missing any of the sample components. Solvents of low polarity, such as n-hexane, tetrachlorethylene and carbon tetrachloride, cause polar sample constituents to migrate more readily. Solvents of intermediate polarity such as toluene, benzene, chloroform and methyl cellosolve have a greater effect on polar sample components, whereas highly polar solvent such as dioxan, methylene dichloride, ethyl acetate, nitromethane, acetone, lower alcohols and water elute polar sample constituents towards the solvent front, i.e. R_f values near unity. Mixtures of 40/60 petroleum spirit and up to 10% (v/v) ethyl acetate are very useful general solvents for the separation of unknown mixtures.

Detection techniques should be carried out immediately after the chromatogram has been developed, in order to reduce to an absolute minimum any opportunity for volatile sample constituents to be lost by evaporation from the plate. Detection of the separated compounds on the plate is achieved by examination under ultraviolet light which locates some, but not all, types of compounds; and by spraying with a range of general or specific spray reagents (Table 67 and 68).

A further general test for organic compounds on the plate involves

TABLE 67   *General spray reagents for location of compounds on 20 cm × 20 cm silica gel coated thin-layer chromatography plates*

A. Reagents applied to GF 254 plate without subsequent heating*

1. Potassium permanganate   (0.1 N) in aqueous sodium carbonate (5% w/v)
2. Potassium permanganate   (2% w/v in aqueous sulphuric acid (6% v/v)
3. Potassium permanganate   (0.1% w/v) in sulphuric acid (96%)
4 Antimony pentachloride   (2% w/v) in carbon tetrachloride
5. Phosphomolybdic acid   (3% w/v) in ethanol, then expose plate to
                                       ammonia vapour

B. Reagents applied to G254 plate with subsequent heating*

	Heat treatment of plate
1. Sulphuric acid aqueous (20% w/v)	5–15 min at 120°C, then 5 min at 150°C
2. As (2) under A	5–15 min at 120°C, then 5 min at 150°C
3. Phosphoric acid (10%) methanolic	5–15 min at 120°C, then 5 min at 150°C
4. Perchloric acid (2%) methanolic	5–15 min at 120°C, then 5 min at 150°C
5. As (4) under A	5–15 min at 120°C, then 5 min at 150°C
6. Phosphomolybdic acid (20% w/v) in methanol or methyl cellosolve, then expose plate to ammonia vapour.	5–15 min at 120°C

*Spray 20 cm × 2 cm wide sections of plate with each reagent using an aluminium or glass mask with suitable aperture.

TABLE 68   *Chemical analysis of additives in plastics; specific spray reagents for location of compounds by thin-layer chromatography*

Additive type	Spray reagent	Reference
Phenolic antioxidants	1. 2, 6-dichloro-benzoquinone chlorimine (1–2% in ethanol followed 15 min later by 2% borax in 40% aqueous ethanol).	1024 1025, 1026
	2. α, α'-diphenyl picryl hydrazyl (0.1% in 95% aqueous ethanol).	1026
	3. Palladium chloride (mix 150 ml palladium chloride with 100 ml of 2 N hydrochloric acid).	1026
	4. Diazotized p-nitroaniline. (Mix 5 ml 0.5% p-nitroaniline in 2 N hydrochloric acid with 0.5 ml 5% sodium nitrite until colourless, and 15 min later add 15 ml 20% sodium acetate).	1025
Amine antioxidants	Diazotized p-nitroanilne (mix 5 ml 0.5% p-nitroaniline in 2 N hydrochloric acid with 0.5 ml 5% sodium nitrite, until colourless and 15 min later add 15 ml 20% sodium acetate).	
Dialkyl thiodi-propionates	Potassium platinoiodide (mix 5 ml 5% platinum tetrachloride in 1 N hydrochloric acid with 45 ml 10% potassium iodide and 100 ml water).	1026
Phthalate ester plasticizer	Resorcinol. (Spray with 20% aqueous resorcinol in 2% aqueous zinc chloride and heat to 150°C. Then spray with 4 N sulphuric acid and heat for 20 min at 120°C. Spray with 40% potassium hydroxide to produce orange spots). Phthalic acid and phthalates also react.	1027, 1028

TABLE 68   *Contd.*

Additive type	Spray reagent	Reference
Acids and bases	Bromocresol green ⎱ Bromocresol purple ⎰ (0.5% in 50% Bromophenol blue ⎱ aqueous Methyl red ⎰ ethanol).	1029, 1030 1029, 1031 1029
Carboxylic acids Aliphatic (primary, secondary tertiary) amines, long chain quaternary salts and amine oxides	Sodium dichlorophenolindophenol (1% ethanolic). Cobalt thiocyanate 10 g Co(NO$_3$)$_2$6H$_2$O and 10 g ammonium thiocyanate made up to 100 ml. Produces blue colour.	1029, 1032 1033
Alkanolamines	Ninhydrin (Heat plate for 5 min at 110°, spray with 0.2% ninhydrin in acetone and heat 5 min at 110°C to produce colours. Further colours then produced upon spraying plate with 0.2% alizarin in acetone).	
Alkyl phenols	Phenols coupled as p-nitrophenol azo dyes applied to plate of silica gel impregnated with alkali. Separated azo dyes located as yellow/red colours upon exposure of plate to ammonia vapour.	
Carbonyl compounds	Carbonyl compounds in sample converted to 2,4-dinitrophenylhydrazones, applied to thin- layer plate, and plate developed. Separated 2,4-DNPH compounds located as yellow or brown colours upon spraying plate with 2% sodium hydroxide in 90% ethanol.	1034
Organic peroxides	Hydriodic acid. (Spray plate with a reagent comprising 40 ml glacial acetic and 0.2 g zinc dust added to 10 ml of 4% aq. potassium iodide, then spray with fresh 1% starch solution.) Peroxides (and certain other types of oxidizing agents) revealed by liberation of free iodine. Alternatively use 2,6-dibromo-benzoquinone chlorimine.	1035

*Spray 20 cm × 2 cm wide sections of the plate with the various reagents using an aluminium or glass mask with suitable aperture.

holding an electrically heated 25 cm long copper wire, set at red heat, at a few millimetres above the plate along the length in which the chromatogram has been developed. After a few seconds' exposure, many types of organic compounds reveal themselves by charring or otherwise discolouring. Having full information on the R_f values of the different sample components using the preferred development solvents, the next stage is to run a chromatogram on fresh 20 cm × 20 cm plates using a suitably sized sample (say 1 ml of a 1% solution) and mark off with a sharp stylus the bands corresponding to the known positions of the separated compounds. No detection reagents are applied to these plates although they may be examined under the ultraviolet light to precisely locate any components which show up under these conditions. The simplest method of removing the zones containing the separated compounds (after allowing solvent to evaporate from the plate) is to hold the plate vertically, its side resting on a sheet of paper and to scrape off the desired zone with a spatula. Each separated adsorbent band can be bottled off in 5-ml polythene stoppered tubes and retained for further examination.

The separated portions of adsorbent, each, hopefully, containing a pure constituent of the original polymer extract, are now extracted with suitable solvents to isolate the additive preparatory to identification by physical and chemical methods.

Each portion of adsorbent is transferred from the storage bottle to a separate small sintered glass extraction thimble (Fig. 130) and the organic compounds leached out with a suitable solvent such as anhydrous absolute ethanol, diethyl ether or methylene dichloride. This solvent must:

1. be a good solvent for the additive;
2. be sufficiently polar to desorb the additive from the absorbent (successive desorption with different solvents may be necessary at this stage);
3 have a low boiling point to facilitate subsequent removal of solvent and reduce to a minimum evaporation losses of any volatile sample constituents; and/or

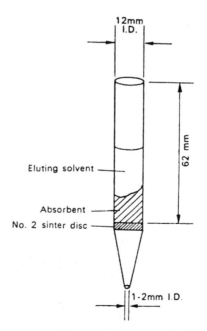

FIG. 130. Filtration apparatus for extracting separated additives from adsorbent isolated from thin-layer chromatography plates.

4. not interfere in the subsequent spectroscopic examination of extracts.

Provided the desorption solvent is sufficiently powerful and polar, it should recover between 50 and 100% of the additive present in the silica gel fraction and provide sufficient material for examination by ultraviolet or infrared spectroscopy, or mass spectroscopy.

McCoy and Fiebig[1036] described a technique using a cavity microcell for obtaining infrared spectra of 50–100 mg of organic compounds dissolved in a suitable spectroscopic solvent.

Alternatively, especially if the compound is insoluble in the usual spectroscopic solvents, the solid can be dispersed in well-ground solid dry potassium bromide using a dental mixing machine, and the mixture pressed into a 1 mm thick, 5 mm diameter disc. This disc can then be mounted in a cardboard or plastic holder and used to prepare a spectrum. To prepare ultraviolet spectra the portions of adsorbent from the thin-layer plate are separately dissolved in a suitable spectroscopic solvent and made up to volume in 1-ml flasks.

When quantitative determinations are desired using these techniques, it is necessary to know, or to determine, the absorptivities of the particular compounds. This is done by preparing and measuring solutions of known concentration of the pure compounds. It is also advisable to chromato-graph known amounts of the pure compounds to verify the applicability of the technique to the particular compounds. This is recommended because unexpected errors can occur if compounds have enough volatility to escape from the adsorbent, or are unstable and change during the chromatography and drying.

(a) Spray reagent; 20% aqueous sulphuric acid, plate heated for 15 minutes at 150°C

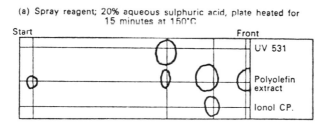

(b) Spray reagent; 2% ethanolic 2:6 dibromo benzoquinone-4-chlorimine; then 2% aqueous borax

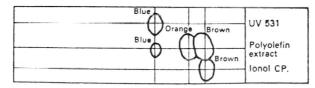

FIG. 131. Thin-layer chromatograms of solvent extracts of polyolefins. Plate: Merck silica gel GF 254, development solvent: 40:60 petroleum spirit:ethyl acetate (9:1 v/v).

An example of the application of this technique to the identification of additives in polypropylene is discussed below.

The polymer was extracted with diethyl ether to isolate total additives and a portion of a chloroform solution of the extract and of various known light-stabilizers and antioxidants run in parallel on a silica gel coated plate.

The chromatograms in Fig. 131 show that the polymer contained two additives appearing at R_f 0.6 and 0.85, which coincided in R_f value and in the colour obtained with 2, 6-dibromo-benzoquinone-4-chlorimine with known specimens of UV 531 (2-hydroxy-4-n-octoxy benzophenone) and Ionol CP (2, 6-di-tert-butyl-p-cresol). Spraying the plate with 2, 6-di-bromo-benzo-quinone-4-chlorimine also revealed an additional orange-coloured spot at R_f 0.8 which did not coincide with any of the known additives examined. Next, an attempt was made to identify unequivocally these three polymer components by comparing their infrared spectra with those of authentic specimens of the suspected compounds. Chloroform solutions (1 ml) containing 15–30 mg of the polymer extract, and of authentic UV 531 and Ionol CP, were applied along the edge of three 20 cm × 20 cm plates and the chromatograms developed using 40/60 petroleum spirit:ethyl acetate (9:1 v/v). The three ultraviolet adsorbing bands on each plate were then marked off and the silica gel corresponding to these zones removed from the plate and the additive extracted from each portion of gel with anhydrous diethyl ether. After removing ether the residues were intimately mixed with dry potassium bromide, and small discs prepared for infrared spectroscopy. As a control a further blank chromatogram was developed, omitting the addition of chloroform solution of sample. The gel from this plate corresponding in R_f value and area to the R_f 0.6 and 0.8/0.85 bands observed in the polymer extract, were isolated and ether extracted. Figures 132 and 133 show the infrared

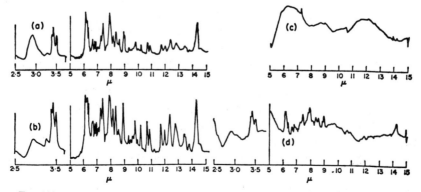

Fig. 132.   Infrared spectrum of UV 531 (2-hydroxy-4-n-octoxy benzophenone) isolated from polyolefin (potassium bromide discs).

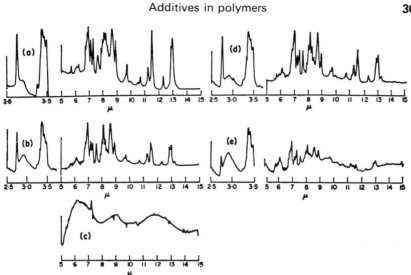

FIG. 133.   Infrared spectrum of Ionol and its degradation product isolated from polyolefin (potassium bromide discs).

spectra in the 2.5 to 15 μ region of the authentic additives (direct spectrum a and spectrum after separation on the plate b), the blank run c and the corresponding extract of the polymer d, e. The spectra a and b of authentic UV 531 are identical, as are spectra a and b in the case of Ionol CP, i.e. it is valid to compare the spectra of these additives after chromatography with their direct infrared spectra, indicating that contact with silica gel does not produce any structural alternation of these substances. Also, the blank spectra in Figs 132 and 133 show that only minor infrared absorptions due to plate impurities occur at 6.1 μ (water), and 8–10, 10.5, 13 μ (silica gel) and 7.2 μ (grease from glassware). These absorptions would not interfere in the interpretation of the additives spectra.

Comparison of Figs 132 b and d reveals that the compound at R_f 0.6 is identical or very similar to UV 531. The light-stabilizer in the polymer extract is certainly a substituted benzophenone, although it may differ from UV 531 in the length of the alkoxy substituent which is known to have little or no influence on the infrared spectrum of compounds of this class.

Comparison of Figs 133 b and d confirms that the R_f 0.85 compound in the polymer extract is Ionol CP, and comparison of d and c shows that the R_f 0.8 component of the polymer extract has a spectrum very similar to that of authentic Ionol CP, suggesting that it is a breakdown product produced, presumably, by partial degradation of Ionol CP during polymer processing.

Table 69 summarizes the results obtained in some experiments carried out to determine the recovery of compounds adsorbing at shorter and

TABLE 69 *Reproducibility of recovery of Ionox 330 and di-n-butyl phthalate from thin-layer plate (silica gel G254)*

Compound	Absorbance maximum	Number of plates prepared	Sample size (μl)	Sample concentration (% w/v)	Development solvent	Plate drying time (h)	Section of adsorbent removed from plate R	Solvent used to desorb compound from adsorbent	Adsorbtivity of standard solution of compound (litres g cm)	Recovery of compound 2‡	Standard deviation
Ionox 330	277 μ	10	1	1	4:1 cyclohexane:benzene	18†	0.03–0.14	Methanol (extract made up 1 ml)	8.0	100	1.2
Di-n-butyl-p-phthalate*	22 μ	5	1	5	9:1 isooctane: ethyl acetate	18	—	Methanol (extract made up to 1 ml)	29.0	99.6	1.2

Applied by Hamilton PR600 repeating sample dispenser.
* Plate premigrated once with methyl alcohol then reconditioned for 1 h at 120°C before application of sample.
† To allow complete evaporation of benzene which would interfere in subsequent ultraviolet spectroscopy.
‡ Based on ratio of theoretical absorptivity and absorptivity of extract from thin-layer plate, methanol used in reference cell in all experiments.

longer ultraviolet wavelengths, respectively, di-*n*-butyl phthalate (222 mμ) and Ionox 330 (277 mμ). These compounds were carried through the whole series of operations involving application of sample to a silica gel plate, solvent development, separation of adsorbent from plate and, finally, solvent extraction of the compound from the adsorbent. Ionox 330 has a low R_f value with 4:1 cyclohexane:benzene development solvent (Table 69) and hence, during solvent development, ultraviolet-absorbing adsorbent impurities are swept well away from this compound to the solvent front. Premigration of the plate with methyl alcohol was unnecessary and not used, therefore, prior to application of Ionox 330 to the plate. Premigration of the plate with methanol before sample application was, however, carried out in the case of di-*n*-butyl phthalate. This was because this compound has a fairly high R_f value with the 9:1 iso-octane:ethyl acetate development solvent used, with consequent possible contamination of the di-*n*-butyl phthalate band with ultraviolet-absorbing adsorbent impurities near the solvent front. It is seen in Table 69 that satisfactory recoveries of both compounds were obtained in this procedure.

9

Adventitious volatiles in polymers

POLYMERS, in addition to deliberately added non-polymeric additives such as described in Chapter 8, usually contain low concentrations of volatile constituents arising from their method of manufacture. The major types of substances in this category include unreacted monomers, non-polymerizable components of the original charge stock, residual polymerization solvents and water. The concentrations of these substances usually range from a few tens of parts per million to several hundred parts per million. Frequently, complex mixtures are present. Thus, the non-polymerizable fractions of styrene monomer (usually 1–2%) consist of several dozen aromatic hydrocarbons such as ethyl benzene. The polymerization solvent used in the manufacture of high-density polyethylene and polypropylene by the low-pressure catalysed route is usually a crude petroleum distillation cut with a complex composition.

It is important to be able to determine the concentration of these substances for many reasons, two examples of which are the effects they have on the mechanical properties of polymers and the risk of tainting in the case of foodstuff- or beverage-packaging grades of polymers.

Polymers often contain substances of medium volatility such as residual monomers, residual polymerization solvents and expanding agents. In addition, when polymers are heated they may release volatiles as a result of the thermal degradation of either the polymers themselves or their additives or catalyst residues. These volatiles can have an important bearing on such properties as processability, the tendency to form voids and, in the case of foodstuff-packaging grades, the possible tendency to impart taste or odour to the packed commodity.

One way of identifying non-polymeric constituents of polymers is to extract the polymer with a low-boiling-point solvent, remove the solvent from the extract by evaporation or distillation and analyse the residue. This procedure is, of course, inapplicable to the analysis of extracted polymer constituents which are volatile enough to be lost during the solvent-removal stage. Alternatively, an extract or a solution of the polymer may be examined directly for volatile constituents by gas chromatography, in which case losses of volatiles are less likely to occur. In such a procedure, however, the large excess of solvent used for extraction or solution might interfere with the interpretation of the chromatogram, obscuring some of the peaks of interest. Trace impurities in the solvent may also interfere with the chromatogram. Of course none of these procedures is suitable for studying the nature of volatile breakdown products which are produced only upon heating a polymer. For the identification and/or determination of residual solvents in polymers it is mandatory therefore to use solventless methods of analysis, i.e. there must be no risk of confusing solvents in which the sample is dissolved for analysis with residual solvents in the sample. Most methods for the determination of residual solvents are based on the technique of heating the solid polymer and examining the head-space over the polymer by gas chromatography.

9.1 Gas chromatography

Due to their volatility and complex composition it is not surprising that methods based on gas chromatography have emerged as being the most suitable way of analysing for these parameters. Basically, three different approaches have evolved in the application of gas chromatography:

1. solution of the polymer in a solvent and injection into the gas chromatography;
2. heating the dry polymer and sweeping the volatiles released into a gas chromatograph using the carrier gas;
3. head-space analysis, i.e. heating the polymer in a closed system, then withdrawing the head-space with a syringe for direct injection into the gas chromatograph.

9.1(a) Monomers

Determination of monomers in a solvent solution

Crompton *et al.*[1037,1038] have applied the gas chromatographic technique to the determination in polystyrene of styrene and a wide range of other aromatic volatiles in amounts down to the 10 ppm level. In this method a weighed portion of the sample is dissolved in propylene oxide containing a known concentration of pure *n*-undecane as an internal standard. After allowing any insolubles to settle an approximately measured volume of the solution is injected into the chromatographic column which contains 10% Chromasorb 15–20 M supported on 60–70 BS Celite. Helium is used as carrier gas and a hydrogen flame ionization detector is employed. Figure 134 shows a device[1038] which is connected to the injection port of the gas chromatograph in order to prevent the deposition of polymeric material in the injection port of the chromatograph with consequent blockages. When a solution of polystyrene is injected into the liner,

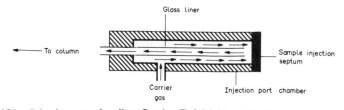

FIG. 134. Injection port glass liner fitted to F. & M. Model 1609 gas chromatograph.
The glass liner measures 60 mm × 40 mm o.d. × 2 mm i.d. and is very loosely packed
with glass fibre.

polymer is retained by the glass fibre, and volatile components are swept
on to the chromatographic column by the carrier gas.

Figure 135 illustrates, by means of a synthetic mixture, the various
aromatics that can be resolved, and Fig. 136 illustrates a chromatogram
obtained with a polystyrene sample, indicating the presence of benzene,
toluene, ethyl benzene, xylene, cumene, propyl benzenes, ethyl toluenes,
butyl benzenes, styrene and α-methyl styrene.

Table 70 illustrates the wide variety of solution–GLC methods that
have been used in the determination of monomers in polystyrene and
acrylate copolymers.

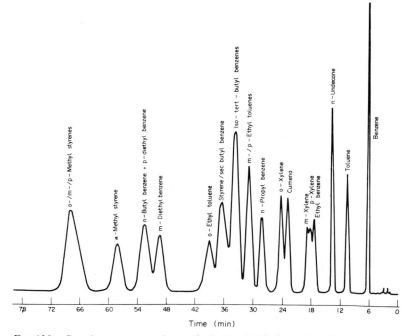

FIG. 135. Gas chromatogram of a synthetic blend of hydrocarbons likely to occur
in polystyrene on a Carbowax 15–20 M column at 80°C.

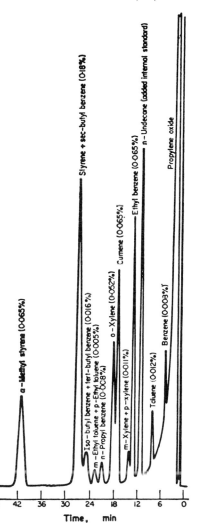

FIG. 136. Analysis of anhydrous propylene oxide solution of polystyrene for residual aromatic hydrocarbons, using a Carbowax 15–20 M column and a glass liner in injection port. Internal standard *n*-undecane. Concentrations are expressed as %w/w in the polymer.

Solution head-space gas chromatographic methods

In this procedure a solution of the polymer in a suitable solvent is placed in a closed container and allowed to equilibrate at a controlled temperature so that volatile monomers or other impurities dissolved in the polymer solution partition between the solution and the gas phase. Subsequent analysis of the gas phase enables the concentration of

TABLE 70   *Solution-GLC method for monomers in polystyrene and polyacrylates*

Monomer polystyrene	Method	Reference
Styrene ethyl benzene	solution in o-dichlorobenzene or methylene dichloride, GLC	1939
Styrene acrylonitrile	solution in DMF, GLC	1037, 1040
Styrene α-methyl styrene benzene toluene ethyl benzene xylene cumene propyl benzenes ethyl toluenes butyl benzenes	solution in propylene oxide, GLC	1037 1038
Styrene	solution in THF, GLC	1041 1042, 1041
Styrene	solution in DMF, GLC	1040
Styrene acrylonitrile butadiene	solution in DMF, GLC	1040
Styrene	solution in benzene GLC	1043
Polyacrylates Ethyl acrylate styrene	distilled in presence of toluene	1044
Methyl acrylate ethyl acrylate	solution in methylene dichloride, GLC	
2-ethylhexyl acrylate butyl acrylate ethyl acrylate vinyl propionate	solution, GLC	1041, 1046
2-ethylhexyl acrylate vinyl acetate	solution in propyl acetate, cyclohexanol, GLC	1047
butyl acrylate methyl acrylate methacrylic acid	solution in isopropanol, GLC	1048
methyl methacrylate ethyl acrylate,	solution, GLC	1049
styrene		1044
Ethyl acrylate styrene	distilled in presence of toluene, GLC	1044, 1045
vinyl acetate 2-ethyl hexyl acrylate	solution in propyl acetate, GLC	1049

monomers to be calculated. In a variant of this method the solid polymer is allowed to equilibrate with the head-space gas

Although the solid head-space method, discussed later, provides about 10-fold more sensitivity than the solution head-space method (assuming a 10% sample solution), the solid method may be applied only to sample systems where equilibration with the head-space is rapid and complete. For example, residual styrene monomer in polystyrene does not reach equilibrium with the head-space after 20 h at[1050] and thus may not be determined by the solid head-space method. Furthermore, even if equilibration between the solid and head-space is obtained, the partition coefficient must also be determined for the component of interest in each type of sample matrix.

The solution head-space approach is applicable to a much wider range of samples than the solid approach. When working with sample solutions, head-space equilibrium is more readily attained and the calibration procedure is simplified. The sensitivity of the solution method depends upon the vapour pressure of the constituent to be analysed and its solubility in the solvent phase. Vinylchloride, butadiene, and acrylonitrile, are readily promoted from polymer solutions into the head-space by heating to 90°C. The head-space/solution partioning for these constituents is not appreciably affected by changes in the solvent phase (viz. addition of water) since the more volatile materials favour the head-space at 90°C. Less volatile monomers such as styrene (b.p. = 145°C) and 2-ethylhexyl acrylate (b.p. = 214°C) may not be determined using head-space techniques with the same sensitivities realized for the more volatile monomers. By altering the composition of the solvent phase to decrease the monomer solubility, the equilibrium monomer concentration in the head-space can be increased. This resulted in a dramatic increase in the detection sensitivity for styrene and 2-ethylhexyl acrylate. Based on these principles, a procedure is described below, for the gas chromatographic analysis of residual vinylchloride, butadiene, acrylonitrile, styrene and 2-ethyl hexyl acrylate in polymers by head-space analysis.

The more volatile monomers vinylchloride, butadiene, and acrylonitrile can be determined by dissolution of the polymer in N, N'-dimethylacetamide in closed vials and analysis of the equilibrated head-space above the polymer solution. By this method it was possible to determine vinylchloride and butadiene at the 0.05 ppm level, and acrylonitrile down to 0.5 ppm. The injection of water into polymer solutions containing styrene and 2-ethylhexyl acrylate monomers prior to head-space analysis greatly enhanced the detection capability for these monomers, making it possible to determine styrene down to 1 ppm and 2-ethylhexyl acrylate at 5 ppm. Incorporation of polymer into the calibration standards compensates for the effect which the polymer matrix has upon the equilibrium partitioning of the monomer between the solution and head-space. The relative precision and error in the determination of these monomers near the quantitation limit was found to be less than 7%.

In this method weighed portions of the polymer were dissolved in septum-sealed vials containing measured aliquots of N, N'-dimethylacetamide. The vials were heated to 90°C to aid dissolution of

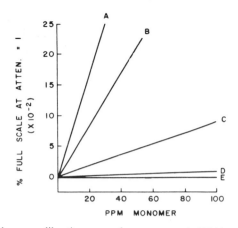

Fig. 137. Head-space calibration curves for monomers in DMA-polymer solutions: (A) butadiene, (B) vinylchloride, (C) acrylonitrile, (D) styrene, (E) 2-ethylhexyl acrylate.

the polymer. When solution was complete, the vials were cooled to room temperature. The solutions were swirled to mix and an aliquot of distilled water was forcibly injected into each polymer solution in order to decrease the solubility of monomers. The vials were shaken briefly to assure complete mixing of the water with the organic phase, and to prevent the precipitated polymer from forming a film on top of the solution. The vials were equilibrated at 90°C for 60 min prior to head-space sampling and analysis by flame ionization–gas chromatography. Standards comprising monomer-free polymer and known additions of standard monomer solutions were run in parallel. Figure 137 shows a typical head-space calibration curve.

Greater sensitivities and shorter analysis times were obtained using the

TABLE 71   *Comparison of quantitation limited* for residual monomers using con-ventional and head-space GC methods*

Monomer	Boiling point	Direct solution injection[†]	Solution head-space	Modified solution head-space
Vinylchloride	−13°C	1–2 ppm	0.05 ppm	—[‡]
Butadiene	−4°C	5 ppm	0.05 ppm	—[‡]
Acrylonitrile	76°C	10 ppm	0.5 ppm	—[‡]
Styrene	145°C	10 ppm	20 ppm	1 ppm
2-ethylhexyl acrylate	214°C	200 ppm	1,000 ppm	5 ppm

*The quantitation limit is defined as the monomer concentration necessary to produce a peak at least three times the baseline noise or 3% of full scale.
[†] Injection of a 10% polymer solution into a gas chromatograph.
[‡] A 2- to 3-fold increase in monomer peak height resulted from the injection of water into the polymer solution. A baseline disturbance due to elution of water negated any real improvement in detection limit for these monomers.

head-space analysis methods than were possible by the direct injection of polymer solutions into a gas chromatograph (Table 71).

Various methods have been described for the determination of styrene monomer in polystyrene by solution-head space analysis.[1050,1051] One technique has been automated. In the automated procedure the polymer (0.2 g) is dissolved and the solution is maintained at 70°C. Standard solution of the compounds to be analysed, dissolved in dimethyl-formamide, are used for calibration. The detection limit of styrene is 10 mg per kg of polystyrene. The column (2 m × 3 mm) used 15% of Reoplex 400 on Embacel (60 to 100 mesh) and was operated at 100°C, using nitrogen as the carrier gas and flame ionization detection.

Solid polymer–head-space analysis

This technique has been applied to the determination of styrene monomer in polystyrene and its copolymers. The polymer (2 g) is placed in a 250 ml

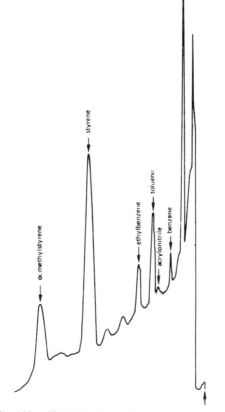

Fig 138.   Chromatogram of head-space vapour.

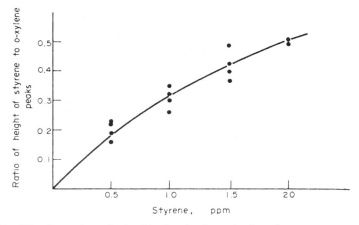

Fɪɢ. 139.   Internal standard calibration for determination of styrene monomer.

screw-top glass jar with a Teflon seal and left in an oven for 4–6 h at 110°C prior to withdrawal of a portion of the head-space for gas chromatographic analysis using a flame ionization detector. The chromatogram in Fig. 138 shows the volatiles from a sample of the terpolymer polystyrene-α-methylstyrene-acrylonitrile.

A sample of monomer-free polystyrene was spiked with styrene monomer at concentrations of 0.5, 1.0, 1.5 and 2.0 ppm and 2 ppm of an internal standard of o-xylene was added. The ratio of the peak height of styrene to o-xylene was used to form a calibration curve (Fig. 139). This curve has some scatter, but allows adequate accuracy for this concentration range.

In addition to gas chromatography, other techniques – particularly ultraviolet spectroscopy and polarography – have been applied to the determination of monomers in polymers.

9.1(b)   Volatiles other than monomers

Qualitative gas chromatography

Solid polymer head-space analysis.   A simple and inexpensive apparatus has been described[1037] for liberating both existing volatiles in polymers and those produced by thermal degradation from polymers by heating at temperatures up to 300°C, in the absence of solvents, prior to their examination by gas chromatography. The technique avoids the disadvantages resulting from the use of extraction or solution procedures.

The apparatus illustrated in Fig. 140 consists of a glass ignition tube, supported as shown in a Wade $\frac{1}{4}$ in. diameter brass coupling nut, covered with a silicone rubber spectrum and sealed with a Wade $\frac{1}{4}$ in. brass

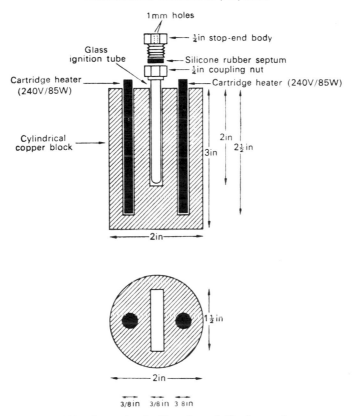

FIG. 140. Apparatus for liberating volatiles from polymers

stop-end body. The stop-end body has two 1-mm diameter holes drilled through the cap. The whole unit is placed in a slot in a cylindrical copper block (3 in. long × 2 in. dia.) which is heated by two (240 V, 85 W) cartridge heaters and controlled at temperatures up to 300°C from a variable transformer. The temperature is measured with a thermocouple capable of accurately measuring temperatures in the 100–300°C temperature range with a maximum error of ± 5%. The thermocouple is inserted in the slot adjacent to the ignition tube; it has been shown that under these conditions the thermocouple records the true temperature of the contents of the tube. The provision of a slot in the copper block enables more than one ignition tube to be heated simultaneously if required.

A sample of the polymer (0.25–0.50 g) is placed in an ignition tube and sealed with Wade fittings and a septum, as described. If necessary the tube is then purged with a suitable gas by inserting two hypodermic needles through the septum via the holes in the cap of the stop-end body and passing the gas into the tube through one hole and allowing it to vent through the other. After purging, the two needles are removed simultaneously and the tube is then heated in the copper block for 15 min. at required temperature. A sample (1–2 ml) of the head-space gas is withdrawn from the ignition tube into a Hamilton gas-tight hypodermic syringe via the septum, and injected into a gas chromatograph. It is advisable to fill the syringe with the gas used initially in the ignition tube and to inject this into the tube before withdrawing the sample. This facilitates sampling by preventing the creation of a partial vacuum in the ignition tube or the syringe or both. It also minimizes any undesirable entry of air into the ignition tube.

With the apparatus, a polymer may be heated under any desired gas and, while this may frequently be the carrier gas used with the gas chromatograph, it is also possible to carry out studies in oxidizing or reducing atmospheres. A polymer may also be heated to any temperature and samples of the head-space gas may be withdrawn at intermediate temperatures and times to determine under what conditions any particular volatile is liberated.

By using gas chromatography detectors of suitable sensitivities and selectivities, it is possible to examine polymers for the presence or formation of volatiles at both the percentage and the parts-per-million levels. For example, traces of organic halogen compounds lend themselves to analysis with an electron capture detector. Thermal conductivity cells of the hot-wire or thermistor type are suitable for the detection of inorganic

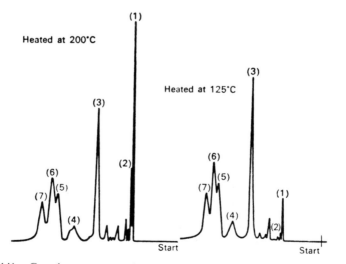

Fig. 141. Gas chromatogram of volatiles liberated from polyethylene heated at different temperatures for 15 min in air. Chromatographed on 200 ft × 1/16 in. i.d. dibutyl phthalate coated copper column at 30°C and 100 ml/min helium flow, with flame ionization detector.

volatiles, and a helium ionization detector could be used for analysing trace amounts of permanent gases.

Figure 141 shows some results obtained by applying this technique to a sample of solid polyethylene at 125°C and 200°C. Evidently low molecular weight paraffin solvents are present.

Food and drink containers extruded or moulded from polyethylene sometimes possess unpleasant odours which are likely to taint the packaged product and are unacceptable to the consumer. In one such case it was found that, by heating a sample of an odour-producing polyethylene for 15 min at 200°C under helium, the chromatogram of the liberated volatiles contained certain peaks which were absent from the corresponding chromatogram from a polyethylene which produced non-odorous food containers. The temperature of 200°C was chosen to stimulate extrusion temperature. The two chromatograms are shown in Fig. 142, from which it may be seen that components A, B, D and I are present in the odorous sample but are absent from the non-odorous

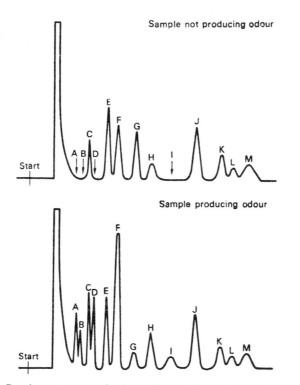

FIG. 142.   Gas chromatograms of volatiles liberated from odorous and non-odorous polyethylenes at 200°C for 15 min in helium. Chromatographed on 200 ft × 1/16 in. i.d. dibutylphthalate coated copper column at 30°C and 100 ml per min helium flow, with flame ionization detector.

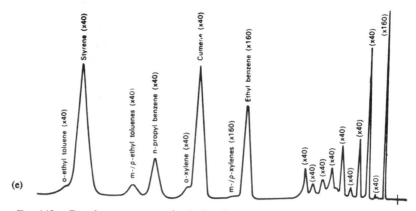

FIG. 143. Gas chromatograms of volatiles liberated from different polystyrenes at 200°C for 15 min in helium. Chromatographed on 15 ft × 3/16 in. 10% Carbowax 15–20 M on 60–72 Celite at 90°C and 100 ml/min helium flow with flame ionization detector.

sample. These substances were always associated with the odorous polyethylene.

Figure 143 shows a gas chromatogram obtained when this technique is applied to a sample of polystyrene heated to 200°C. A wide range of aromatic volatiles are present.

Quantitative gas chromatography

Determination of polymer volatiles in a solvent solution. Benzene, toluene, ethyl benzene, *n*-propyl benzene, cumene, isobutyl benzene, *o*-xylene, *m*- and *p*-xylenes, *m*- and *p*-ethyl toluenes, styrene/sec-butyl benzene and α-methylstyrene can all be quantitatively determined in amounts down to 10 ppm in polystyrene by a method based on solution of the polymer in propylene oxide containing *n*-undecane as internal standard and gas chromatography under the following conditions (Fig. 144)

Column	Copper tube (15 ft × 3/16 in i.d.) packed with 10% wt/wt Carbowax 15–20 M on 60–72 BS mesh acid-washed Celite.
Gas flows	Helium, 30 psig, rotameter = 10.0 (100 ml/min). Hydrogen, 12 psig, rotameter = 10.0 (75 ml/min). Air, 7 psig, rotameter = 10.0 (650 ml/min).
Temperatures	Injection, 155°C Column, 80°C Detector, 125°C Flame, 200°C

Solid polymer-head-space analysis. In an alternative procedure the polymer is heated and the head-space atmosphere analysed by gas

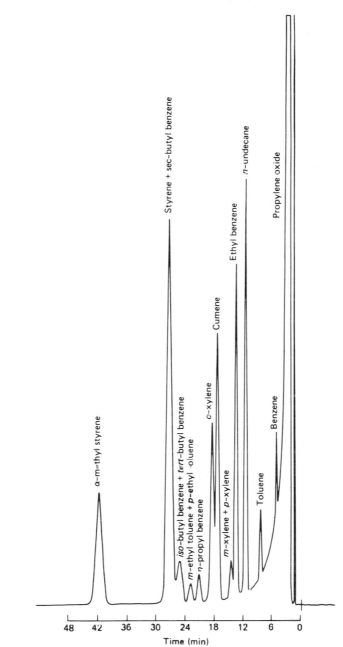

FIG. 144.   Gas chromatogram of blend of aromatic hydrocarbons in propylene oxide.

chromatography.[1068-1070] The film or sheet, together with internal standard, is placed in a 250 ml sealed container and heated at 100°C for 90 min.

The following results were obtained for the determination of toluene and ethyl acetate on adjacent pieces of a polythene adhesive-laminated to polypropylene double-coated with saran. The solvents originate from the adhesive.

									Mean	S.D.
Toluene/mg m^{-2}	127	135	119	127	140	122	130	119	127	7
Ethyl acetate/mg m^{-2}	136	145	124	141	136	133	138	128	135	6

Results as follows were obtained for the determination of toluene on adjacent pieces of printed film (polypropylene with a single saran coating). The toluene originated from the printing ink.

Toluene/mg^{-2} 11.9, 11.4, 12.3, 11.9, 12.5, 11.9, 12.3
Mean 12.0 mg m^{-2};   standard deviation 0.34 mg m.$^{-2}$

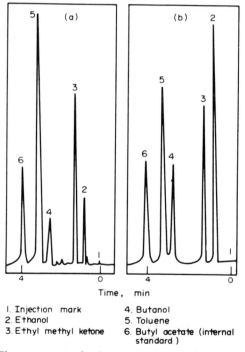

Time, min

1. Injection mark   4. Butanol
2. Ethanol   5. Toluene
3. Ethyl methyl ketone   6. Butyl acetate (internal standard)

FIG. 145. (a) Chromatogram of solvents from a sample of printed polythene–polypropylene laminate. Chromatograph, Pye 104 with flame-ionization detector; column, glass 5 ft × ¼ inch packed with 9% silicone oil and 3% UCON HB2000 on Chromosorb W; oven, 80°C; attenuation × 20,000; and chart speed, 1 cm min^{-1}. (b) Calibration jar prepared for sample shown in (a).

This method has been used to determine many different solvents in several different substrates. The solvents include ethanol, ethyl acetate, ethyl methyl ketone, 2-ethoxyethanol, propan-1-ol and toluene. The substrates include polythene, polypropylene and cellophane, which occur individually, coated with saran or combined in laminates.

Figure 145 (a and b) shows typical chromatograms obtained from a film and a calibration jar.

Gas chromatographic methods have been used for the determination of volatilies in styrene–butadiene,[1071,1072] PVC,[1073,1074], α-methylstyrene,[1072,1075] polycarbonates[1076]-polyethylene,[1037,1077,1078] polypropylene,[1079] and in rubber adhesives.[1080–1084]

GLC water (and volatiles). Jeffs[1085] has described a rather complicated piece of equipment for the determination of water and other volatiles in vinyl, acrylic and polyolefin powder polymers. The instrument is shown diagrammatically in Fig. 146, and consists essentially of a sample tube, forming an external loop, connected to a gas chromatograph. This loop

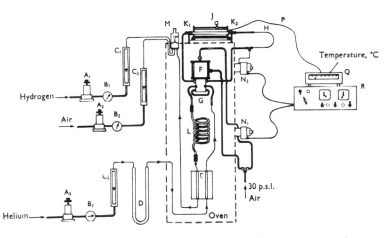

A_1, A_2 and A_3 = Edwards VPC I pressure controller

B_1, B_2 and B_3 = Pressure gauges 0 to 30 p.s.i.

C_1, C_2 and C_3 = Rotameter-type flow gauges

D = Clear plasticised PVC tubing (5 foot long × ¼ inch bore), packed with copper sulphate crystals, $CuSO_4.5H_2O$, >44 mesh

E = Katharometer

F = Pneumatic sample valve (Pye Cat. No. 12900), fitted with P 9904 change-over block

G = Internal loop

H = External loop made in part of 18-gauge stainless-steel capillary tubing

J = Sample split heater

K_1 and K_2 = Straight reducing couplings, captive seal type, for ⅜ to ¼ inch o.d. tubing (Drallim, Cat. No. L/50/D/B)

L = Chromatographic column

M = Flame-ionisation detector

N_1 and N_2 = Electrically actuated 3-port pilot valves (Martinair, Type 557C/1Z)

P = Nickel - chromium/nickel - aluminium thermocouple embedded in the split heater

Q = Electronic temperature controller, proportional type

R = Sample-valve time-delay unit, containing three synchronous timers and a semi-conductor, proportional energy controller for the sample heater

FIG. 146. Details of general-purpose instrument for gas chromatographic determination of water and volatiles in polymers.

can be isolated and the sample heated to the required temperature. After an initial heating period the volatile constituents liberated from the sample are 'flushed' on to the chromatographic column by a flow of carrier gas through, or over, the sample, and the required components separated and determined quantitatively. A pneumatic switch valve located in the chromatographic oven to prevent the condénsation of volatile constituents within the valve, and a split heater mounted on a horizontal travel in a plane at right angles, to the sample tube, are essential parts of the apparatus. The instrument is semi-automatic. The carrier gas flows through the copper sulphate in tube D, which imparts a constant amount of water (about 3 ppm w/v) to the helium. The 'wet' carrier gas prevents the gas-flow lines from 'drying out'. Dry pipework tends to adsorb moisture, which can then be desorbed, thus leading to spurious results. The determination of water in a sample is unaffected as the 'wet' carrier gas flows continuously through both the reference and analysis cells of the katharometer.

The pneumatic sample valve, F, operates as a switch valve, directing the carrier-gas flow either around the internal loop, G, or the external loop, H. The pilot valve, N_1, operates sample valve, F. The sample split heater, J, consists of a cylindrical aluminium block 6 in. long with a 9-mm hold through the centre. The block is split axially and the two halves hinged. Each half of the block contains two cartridge-heater elements, each $5\frac{1}{2}$ in. long, $\frac{3}{8}$ in. o.d., and one half contains a thermocouple pocket to accommodate the thermocouple, P. The cartridge heaters are supplied by a semiconductor energy controller contained in R, which is controlled by a galvanometer, two-position, temperature controller, Q. The heater is mounted on a horizontal travel in a plane at right angles to the sample tube. A jig for mounting the sample tubes is also part of the heater assembly. The $\frac{3}{8}$ in. coupling K_1, is brazed to its mounting bracket, which is rigidly attached to the heater base. This coupling is accurately centred, with the central hole through the aluminium block. The coupling, K_2, is brazed to the flexible carrier-gas inlet tube, and rests loosely in the second mounting bracket. The $\frac{3}{8}$-in. couplings are supplied with neoprene, or butyl rubber captive seals.

The sample-heater assembly is placed on top of a Griffin and George oven, so that the L-bend of $\frac{1}{8}$-in. o.d. S/S tubing disappears almost immediately into an opening on the top of the oven. In practice, both the inlet tube (6 in. long, $\frac{1}{4}$-in. o.d. and $\frac{1}{8}$-in. i.d. copper) and the exit tube ($\frac{1}{8}$-in. o.d.) are wrapped with heating tape and lagging to maintain the temperature of the whole assembly at about 100°C. The inlet tube is wrapped to a length of 4–5 in. and the L-bend of the exit tube is wrapped to a point 3 in. inside the oven. Both tubes are wrapped up to, and including, the $\frac{1}{4}$-in. thread of the Drallim coupling, leaving only the centre nut and the

$\frac{3}{8}$-in. coupling nut exposed so that the sample tubes can be readily changed. N_2 acts as a pressure release valve to the external loop H.

The apparatus shown in Fig. 146 is fitted with katharometer and flame-ionization detectors. Although only one detector is necessary for any one specific method, e.g. a katharometer for the determination of water in polymer powder, it is invaluable to have both available (with separate recorders) to establish the conditions, i.e. in the above case to ensure that no organic components are being eluted at the same time as water, and thus contributing to the peak measurement. Figure 147 shows chromato-grams obtained simultaneously from the katharometer and the flame-ionization detectors on a partially dried poly(vinylchloride) powder.

Figures 148 and 149 show some typical results obtained when this method is applied to vinyl polymers and polyolefins. The peaks due to water and monomers are clearly visible.

Jeffs[1085] recommends that before carrying out any quantitative work on the volatile constituents obtained from a polymer powder, a preliminary gas-chromatographic investigation should be carried out, of their com-plexity. For example, Fig. 147 (flame-ionization detector trace) shows nine components other than air, water and the original monomer. These components are chlorinated hydrocarbons, such as 1, 1- and 1, 2-dichloroethane and *cis*- and *trans*-dichloroethylene, that are present as impurities in the original monomer.

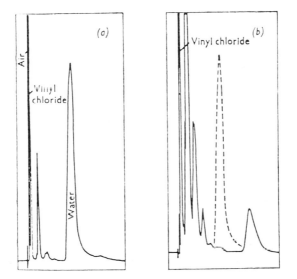

FIG. 147. Gas chromatograms obtained simultaneously with (a) katharometer and (b) flame-ionization detectors on a partially dried PVC powder. In (b) the dotted line represents the portion of the water peak as transposed from (a).

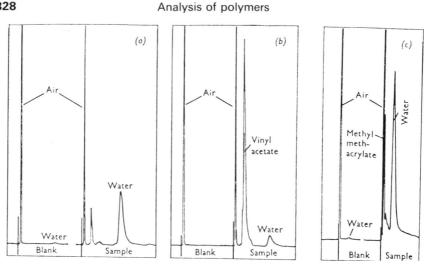

Fig. 148. Typical gas chromatograms obtained with method for (a) PVC powder; (b) PVC–PVA copolymer power; and (c) acrylic moulding powder.

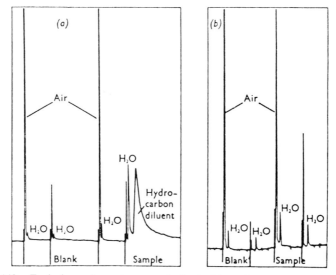

Fig. 149. Typical gas chromatograms obtained with method for (a) polypropylene powder; and (b) high-pressure polyethylene.

Although these impurities are present only in ppm amounts in the original monomer they can be readily detected in the polymer. If water is to be determined, then a suitable column has to be chosen so that the water is eluted free from organic constituents. At this stage the recording of simultaneous chromatograms with the two different detectors is invaluable.

Gas chromatography has also been applied to the determination of water in polyacrylates and PVC.

9.2   Ultraviolet spectroscopy

9.2(a)   *Monomers*

Ultraviolet spectroscopy suffers from several disadvantages in the determination of monomers in polymers, particularly in the case of styrene monomer in polystyrene and its copolymers. In addition to lack of sensitivity, which limits the lower detection limit to about 200 ppm styrene in polymer under the most favourable circumstances, ultraviolet spectroscopic methods are subject to interference by some of the types of antioxidants included in polystyrene formulations. Such interferances can only be overcome by applying a lengthy pre-treatment of the sample to remove antioxidants prior to spectroscopic analysis. In addition to residual styrene monomer, polystyrene may also contain traces of other aromatic hydrocarbons such as benzene, toluene, xylenes, ethyl benzene and cumene, which originate either as impurities in the styrene monomer employed to manufacture the polystyrene or they may have been used in small quantities as dilution solvents at some stage of the manufacturing process. Ultraviolet spectroscopic methods for determining styrene cannot differentiate between the various volatile substances present in polystyrene.

In the case of some types of polymer additives, interference effects can be overcome by the use of a baseline correction techniques. Thus polystyrene contains various non-polymer additives (e.g. lubricants) which result in widely different and unknown background absorptions at the wavelength maximum at which styrene monomer is evaluated (292 nm). The influence of the background absorptions on the evaluation of the optical density due to styrene monomer is overcome by the use of an appropriate baseline technique, claimed to make the method virtually independent of absorptions due to polymer additives. In this technique a straight line is drawn on the recorded spectrum across the absorption peak at 292 nm in such a way that the baseline is tangential to the absorption curve at a point close to the absorption minima occurring at 288 nm and 295–300 nm (Fig. 150). A vertical line is drawn from the tip of the styrene absorption peak at 292 nm to intersect the baseline, and the height of this line is then a measure of the optical density due to the true styrene monomer content of the test solution.

This baseline correction technique can obviously be applied to the determination of styrene monomer in polystyrene only if any other ultraviolet absorbing constituents in the polymer extract (e.g. lubricant, antioxidants) absorb linearly in the wavelength range 288–300 nm). If the

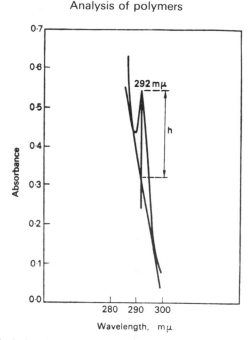

FIG. 150. Typical ultraviolet absorption curve of a polystyrene containing styrene monomer.

polymer extract contains polymer constituents other than styrene with non-linear absorptions in this region, then incorrect styrene monomer contents will be obtained. An obvious technique for removing such non-volatile ultraviolet-absorbing compounds is by distillation of the extract followed by ultraviolet spectroscopic analysis of the distillate for styrene monomer as discussed below.

In the distillation technique[1037] the polystyrene is dissolved in chloroform or ethylene dichloride (20 ml) in a stoppered flask and the solution is poured into an excess of methyl alcohol (110 ml) to reprecipitate dissolved polymer. The polymer is filtered off and washed with methanol (120 ml) and the combined filtrate and washings gently distilled to provide 200 ml of distillate containing styrene monomer and any other distillable component of the original polystyrene sample. Non-volatile polymer components (viz. stabilizers, lubricants and low molecular weight polymer) remain in the distillation residue. The optical density of the distillate is measured at 292 nm or by the baseline method against the distillate obtained in a polymer-free blank distillation. Calibration is performed by applying the distillation procedure to solutions of known weights of pure styrene monomer in the appropriate quantities of methyl alcohol and the chlorinated solvent. Tables 72 and 73 show results obtained for styrene monomer determinations carried out on samples of polystyrene by the

TABLE 72   *Comparison of direct ultraviolet and distillation/ultraviolet methods for the determination of styrene monomer*

		Styrene monomer (% w/w) Polystyrene sample			
Method	Solvent	No. 1		No. 2	
Direct UV method	Chloroform	< 0.05		0.13	
	Carbon tetra-chloride	< 0.05		0.13	0.14
				0.16	0.18
	Ethyl acetate	< 0.05		0.14	0.12
				0.14	0.14
				0.15	
Distillation/UV	Ethylene di-chloride/	0.16	0.18	0.27	0.29
		0.16	0.18	0.26	0.29
	methanol	0.20		0.29	

TABLE 73   *Influence of phenolic antioxidant* on the determination of styrene monomer by direct ultraviolet and by distillation/ultraviolet methods*

	Styrene found (% w/w)				
	Direct UV method[†]			Distillation/UV method	
Styrene added to polystyrene (% w/w)	No phenolic antioxidant* addition		0.5% phenolic antioxidant* added on polymer	No phenolic antioxidant* addition	0.5% phenolic antioxidant* addition on polymer
0.11	0.11	0.11	0.04 0.08	0.12	0.12
0.22	0.22		0.11	—	—
0.27	0.26		0.18	0.29	0.26
0.41	0.41	0.40	0.30 0.27	0.40	0.40

* Wingstay T.
[†] Chloroform used as a sample solvent.

direct ultraviolet method, and by the distillation modification of this method. It is seen that the distillation method gives results that are consistently higher than those obtained by direct spectroscopy, indicating that additives present in the polystyrene are interfering in the latter method of analysis.

9.3   Polarography

9.3(a)   Monomers

Residual amounts of styrene and acrylonitrile monomers usually remain in manufactured batches of styrene–acrylonitrile copolymers. As these

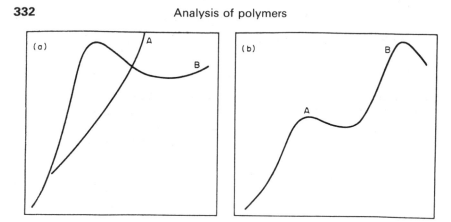

FIG. 151. (a) Cathode-ray polarogram of a synthetic solution of acrylonitrile in tetrabutyl ammonium iodide–dimethylformamide base electrolyte. Curve A, base electrolyte blank solution; curve B, 4 ppm of acrylonitrile in base electrolyte. Start potential − 1.7 V. (b) Cathode-ray polarogram of a synthetic solution of 11.2 ppm of styrene and 9.3 ppm of acrylonitrile monomer in tetrabutyl ammonium iodide–dimethylformamide base electrolyte. Wave A, acrylonitrile wave; wave B, styrene wave. Start potential − 1.7 V.

copolymers have a potential use in the food-packaging field, it is necessary to ensure that the content of both of these monomers in the finished copolymer is below a stipulated level.

In a polographic procedure[1052,1053] for determining acrylonitrile (down to 2 ppm) and styrene (down to 20 ppm) monomers in styrene–acrylonitrile copolymer, the sample is dissolved in 0.2 M tetramethyl-ammonium iodide in dimethyl formamide base electrolyte and polaro-graphed at start potentials of − 1.7 V and − 2.0 V respectively from the two monomers (Fig. 151). Excellent results are obtained by this procedure. Polarography has also been used for the determination of other monomers including acrylonitrile,[1054] methyl methacrylate,[1056] α-methylstyrene,[1055] and styrene.[1055,1052]

9.4   High-performance liquid chromatography

9.4(a)   Monomers

This technique has found a limited number of applications in the determination of monomers including acrylic acid monomer in polyacry-lates[1057] and acrylamide in polyacrylamide.[1058] High-performance liquid chromatography has been used for the determination of oligomers in polyethylene terephthalate[1062,1063] and epoxy resins.[1062]

9.5 Miscellaneous methods

9.5(a) Water

Conventional weight loss methods for determining water in polymers have several disadvantages, not the least of which is that any other volatile constitutents of the polymer such as solvents, dissolved gases, volatile substances produced by decomposition of the polymer or additives therein are included in the determination.

Specific methods for the determination of water in polymers can be based on the principle of heating a weighed amount of polymer in a boat in a tube furnace through which flows a gentle purge of nitrogen.[1086] The wet nitrogen passes into an automated Karl Fisher titration unit which estimates the water as it is released from the polymer. In addition to being absolutely specific for water this method has an additional advantage in that it can provide information on the effects of temperature and time on the release of water from polymers.

9.6 Determination of oligomers

Oligomers are very low molecular weight polymers. This, polystyrene usually contains low concentrations of monomer, dimer, trimer and tetramer of the general formula $(C_6H_6—CH=CH_2)_n$, where $n = 1$ to 4. Various techniques have been used for the determination of oligomers including gel permeation chromatography,[1059,1060] (polyesters), thin-layer chromatography[1061] (polystyrene and poly α-methyl styrene), liquid chromatography p-alkylphenyl formaldehyde oligomers,[1064] polystyrene,[1065,1066] and gas chromatography.[1067]

References

1. Smith, A.J. *Anal. Chem.*, **36**, 944 (1964).
2. Gorsuch, T. *Analyst* (*London*), **87**, 112 (1962).
3. Gorsuch, T. *Analyst* (*London*), **84**, 135 (1959).
4. Nerasaki, H., Umezawa, K. *Kabunshi Kagaku*, **29**, (6), 438 (1972).
5. Mitterberger, W.D., Gross, H. *Kunststofftecknik*, 12, (7) (76), (8), 219, (9), 281, (10), 277, 303 (1973).
6. Bahroni, M.L. Chakravarty, N.K. Chopra, S.C. *Indian J. Technol.*, 13, (12), 576 (1975).
7. Ogure, H. *Bunseki Kagaku*, 24, (12) (1975); *Chem. Abstr.*, **85**, 33845 (1975).
8. Korenaga, T. *Analyst* (*London*), **106**, 40 (1981).
9. Tanaka, K., Morikawa, T. *Kagaku To Kogyo* (*Osaka*), **48**, (10), 387 (1974); *Chem. Abstr.*, **82**, 98702w (1974).
10. Falcon, J.Z. Lone, J.L. Gaeta, L.T. Altenau, A.G. *Anal. Chem.*, **47**, 171 (1975).
11. Narasaki, H. Mijaji, O, Unno, A. *Bunseki Kagaku*, **22**, (5), 541 (1973).
12. Cook, W.S., Jones, C.O., Altenau, A.G. *Canadian Spectroscopy*, **13**, 64 (1968).
13. Kabayashi, O. *J. Polymer Sci.*, A-1, **17**, 293 (1979).
14. Hull, D., Gilmore, J. Division of Fuel Chemistry, 141th Meeting ACS, Washington, DC, March 1962.
15. Bergmann, J.S. Ekhart, C.H. Grantelli, L. Janik, L.J. 153rd National ACS Meeting, Miami Beach, Florida, April 1967.
16. Rowe, W.A., Yates, K.P. *Anal. Chem.*, **35**, 368 (1963).
17. Shott, J.E. Garland, T.J. Clark, R.O. *Anal. Chem.*, **33**, 506 (1961).
18. Korenaga, T. *Analyst* (*London*), **106**, 40 (1981).
19. Crompton, T.R. unpublished work.
20. Crompton, T.R., Reid, V.W. *J. Polymer Sci.*, Part A, **1**, 347 (1963).
21. Albert, N.K. Woodbury Research Laboratory, Shell Chemical Co. Woodbury, U.S.A. Private communication, October 1966.
22. McNeill, I.E. *Polymer*, **4**, 15 (1963).
23. McGuchan, R., McNeill, I.E. *J. Polymer Sci.*, A1, **4**, 2051 (1966).
24. Kuryatnikov, E.C., Vizgert, R.V., Berlin, A.A., Vijn, L'viv. *Politekh Inst.*, *No.* **57**, 36 (1971).
25. Pepper, D.C., Reilly, P.H. *Proc. Chem. Soc.*, **1**, 460 (1961).
26. Gallo, S.G., Weiss, H.K., Nelson, J.F. *Ind. Eng. Chem.*, **40**, 1277 (1948).
27. Lee, T.S., Kalthoff, I.M., Johnson, E. *Anal. Chem.*, **22**, 995 (1950).
28. Hank, R. *Rubber Chem. Technol.*, **40**, 936 (1967).
29. Sloane, H.J., Bramstone-Cooke, R. *Appl. Spectrosc.*, **27**, (3), 217 (1973).
30. Turner, R.R., Carlson, D.W., Altenau, A.G., unpublished work.
31. Sewell, R.P., Skidmore, D.W. *J. Polymer Sci.*, A-1, **6**, 2425 (1968).
32. Shibata, C., Yamazaki, M., Tabeuchi, T. *Bull. Chem. Soc. Japan*, **50**, (1), 311 (1977); *Chem. Abstr.*, **86**. 90540d (1977).
33. Cooper, W., Eaves, D.E., Tunnicliffe, .M.E., Vaughan, G. *European Polymer J.*, **1**, 121 (1965).
34. Tunnicliffe, M.E., Mackillop, D.A., Hank, R. *European Polymer J.*, **1**, 259 (1965).
35. Altenau, A.G., Headley, L.M., Jones, S.O., Ronsaw, H.C. *Anal. Chem.*, **42**, 1280 (1970).
36. Lee, T.C., Koltoff, I.M., Johnson, E. *Anal. Chem.*, **22**, 995 (1950).
37. Kemp, A.R., Peters, H. *Ind. Eng. Chem. Anal. Ed.*, **15**, 52 (1943).
38. Crisp, S., Lewis, B.G., Wilson, A.D.J. *Dental Res.*, **54**, (6), 1238 (1975).
39. Tan, J.S. Gasper, S.P. *Macromolecules*, **6**, (5), 741 (1973).

40. Garrett, S.R., Guile, R.L. *J. Am. Chem. Soc.*, **73**, 4533 (1951).
41. Douglas, J., Timnick, A., Guile, R.L. *J. Polymer Sci.*, A, **1**, 1609 (1963).
42. Mathieson, A.R., McLaren, J.V. *J. Polymer Sci.*, A, **3**, 2555 (1965).
43. Katelhalsky, A., Shavit, N., Eisenberg, H.J. *Polymer Sci.*, **13**, 69 (1954).
44. Mathieson, A.R., McLaren, J.V. *J. Chem. Soc.*, 3581 (1960).
45. Leyte, J.C., Mandel, M. *J. Polymer Sci.*, A, **2**, 1879 (1974).
46. Mandel, M., Leyte, J.C. *J. Polymer Sci.*, **56**, 823 (1962).
47. Johnson, D.E., Lyeria, J.R., Horikawa, T.T., Pederson, L.A. *Anal. Chem.*, **49**, 77 (1977).
48. Majewski, J. *Polimery*, **18**, (3), 142 (1973).
49. Stetzler, R.S., Smullin, C.F. *Anal. Chem.*, **34**, 194 (1962).
50. Spirin, Yu L., Yatsimirskaya, T.A., *Vysokamol. Soedin. Ser.*, A15, (11), 2595 (1973).
51. Fritz, J.S., Schenk, G.H. *Anal. Chem.*, **31**, 1808 (1959).
52. T.R. Crompton, unpublished work.
53. Valenv, V.I., Shiyakhter, R.A., Dimitrieva T.S. *Zh. Khim. Anal. Khim.*, **30**, (6), 1235 (1975), *Chem. Abstr.*, **83**, 132156d (1975).
54. Law, R.D. *J. Polymer Sci.*, A-1, **9**, 589 (1971).
55. Kim, C.S.Y., Dodge, A.L., Lau, S., Kawasaki, A. *Anal. Chem.*, **54**, 232 (1982).
56. Aydin, V., Schulz, R.C. *Makromol Chem.*, **177**, (12), 3537 (1976).
57. Majer, J., Sodomka, J. *Chem. Prumsyl*, **25**, (11), 601 (1975); *Chem. Abstr.*, **84**, 45101 (1976).
58. Munteanu, D., Savu, N. *Rev. Chim. (Bucharest)*, **27**, (10) 902 (1976); *Chem. Abstr.*, **86**, 121973d (1977).
59. Bevington, J.C., Eaves, D.E., Vale, R.L. *J. Polymer Sci.*, **32**, 317 (1958).
60. Bevington, J.C. *Trans. Faraday Soc.*, **56**, 1762 (1960).
61. Esposito, C.G., Swann, M.H. *Anal. Chem.*, **34**, 1048 (1962).
62. Percival, D.F. *Anal. Chem.*, **35**, 236 (1963).
63. Aydin, V., Kaczmar, B.U., Schulz, R.C. *Angew Makromol Chem.*, **24**, 171 (1972).
64. Leukrath, G. *Gummi Asbest. Kunstst*, **29**, (9) 585 (1976); *Chem. Abstr.*, **86**, 44183S (1977).
65. Miller, D.L., Samsel, E.P., Cobler, J.G. *Anal. Chem.*, **33**, 677 (1961).
66. Samsel, E.P., McHard, J.A. *Ind. Eng. Chem. Anal. Ed.*, **14**, 750 (1942).
67. Haslam, J., Hamilton, J.B., Jeffs, A.R. *Analyst (London)*, **83**, 983, 66 (1958).
68. Haslam, J., Jeffs, J. *Anal. Chem.*, **7**, 24 (1957).
69. Barrall, E.M., Porter, R.S., Johnson, J.E. *Anal. Chem.*, **35**, 73 (1963).
70. Porter, R.S., Hoffman, A.S., Johnson, J.F. *Anal. Chem.*, **34**, 1179 (1962).
71. Helmuth, J. *Polym. Vehromarium Plast.*, **3**, (2), 7 (1973).
72. Majer, J., Sodemka, J. *Chem. Prumsyl*, **25**, (11), 601 (1975); *Chem. Abstr.*, **84**, 45101 (1976).
73. Sirynk, A.G., Bulgakova, R.A. *Vysokomol Soedin, Ser B*, **19**, (2), 152 (1977); *Chem. Abstr.*, **86**, 140597a (1977).
74. Czaja, K., Nowadowska, M., Zubek, J. *Polimery (Warsaw)*, **21**, (4), 158 (1976); *Chem. Abstr.*, **85**, 124396h (1976).
75. Porter, R.S., Nicksic, S W , Johnson, J.F. *Anal. Chem.*, **35**, 1948 (1963).
76. Gemanenko, E.N., Perepletchikova, E.M. *Zh. Anal. Khim.*, **29**, (4), 830 (1974).
77. Kato, K. *J. Appl. Polymer Sci.*, **17**, 105 (1973).
78. Scheddel, R.T. *Anal. Chem.*, **30**, 1303 (1958).
79. Tukuchi, T., Tauge, S., Sigimura, Y. *J. Polymer Sci.*, A-1, **6**, 3415 (1968).
80. Neumann, E.W., Nadeau, H.G. *Anal. Chem.*, **10**, 1454 (1963).
81. Swann, W.B., Dux, J.P. *Anal. Chem.*, **33**, 654 (1961).
82. Zeman, I., Novak, L., Mitter, L., Stekla, J., Holendova, O. *J. Chromatogr.*, **119**, 581 (1976).
83. Willbourne, J. *J. Polymer Sci.*, 34, **569** (1959).
84. Brown, J.E., Tyron, M., Mandel, J. *Anal. Chem.*, **35**, 2173 (1963).
85. Tyron, N., Horowicz, E., Mondel, J. *J. Res Nutl Bur. Std.*, **55**, 219 (1955).
86. Cross, L.H., Richards, R.B., Willis, H.A. *Discussions Faraday Soc.*, **9**, 235 (1950).
87. Smith, D.C. *Ind. Eng. Chem. Anal. Ed.*, **48**, 1161 (1956)
88. Newmann, E.W., Nadeau, H.G. *Anal. Chem.*, **10**, 1454 (1963).
89. Gray, A.P. Olin Mathieson Chem. Corp. New Haven Connecticut, unpublished Du Pont of Canada Method (1957).
90. Schlueter, D.D., Siggia, S. *Anal. Chem.*, **49**, 2343 (1977).
91. Schleuter, D.D., Siggia, S. *Anal. Chem.*, **49**, 2349 (1977).
92. Colson, A.F. *Analyst (London)*, **88**, 26 (1963).

93. Colson, A.F. *Analyst* (*London*), **88**, 791 (1963).
94. Colson, A.F. *Analyst* (*London*), **90**, 35 (1965).
95. Salvage, T., Dixon, J.F. *Analyst* (*London*), **90**, 24 (1965).
96. Mann, L.T. *Anal. Chem.*, **35**, 2179 (1963).
97. Johnson, L.A., Leonard, M.A. *Analyst* (*London*), **86**, 101 (1961).
98. Yoshizaki, To. *Anal. Chem.*, **35**, 2177 (1963).
99. Brako, F.D., Wexler, A.S. *Anal. Chem.*, **35**, 1944 (1963).
100. McNeill, I.E. *Polymer*, **4**, 15 (1963).
101. McGuchan, R., McNeill, I.C. *J. Polymer Sci.*, A-1, **4**, 2051 (1966).
102. Fraga, D.W. Emeryville Research Laboratories. Shell Chemical Co. Ltd., Emeryville, U.S.A., private communication.
103. Binder, J.L. *J. Polymer Sci.* A-1, 42 (1963).
104. Natta, G. 15th Annual Technical Conference of the Society of Plastic Engineers, New York, January 1959.
105. Dinsmore, H.C., Smith, D.C. *Rubber Chem. Technol.*, **22**, 572 (1949).
106. Harms, D.C. *Anal. Chem.*, **25**, 1140 (1953).
107. Hummel, D. *Rubber Chem. Technol.*, **32**, 854 (1959).
108. Lermer, M., Gilbert, R.C. *Anal. Chem.*, **36**, 1382 (1964).
109. Tyron, M., Horowicz, E., Mandel, J.J. *J. Res. Nat'l Bur. Stds*, **55**, 219 (1955).
110. McKillop, D.A. *Anal. Chem.*, **40**, 607 (1968).
111. Carlson, D.W., Altenau, A.G. *Anal. Chem.*, **41**, 969 (1969).
112. Mochel, U.D. *Rubber Chem. Technol.*, **40**, 1200 (1967).
113. Carlson, D.W., Ransaw, H.C., Altenau, A.G. *Anal. Chem.*, **42**, 1278 (1970).
114. Binder, J.L. *Anal. Chem.*, **26**, 1877 (1954).
115. Binder, J.L. *J. Polymer Sci.*, A-1, 47 (1963).
116. Braun, D., Canji, E. *Angew Makromol Chem*, **33**, 143 (1973).
117. Brown, D., Canji, E. *Angew Macromol Chem.*, **35**, 27 (1974).
118. Hast, M., Deur Siftar, D. *Chromatographia*, **5**, (9), 502 (1972).
119. Silas, R.D., Yates, J., Thornton, V. *Anal. Chem.*, **31**, 529 (1959).
120. Cornell, S.W., Koenig, J.L. *Makromolecules*, **2**, 540 (1969).
121. Neto, N., diLauro, C. *European Polymer J.*, **3**, 645 (1967).
122. Binder, J.L. *Anal. Chem.*, **26**, 1877 (1954).
123. Clark, J.K., Chen, H.Y. *J. Polymer Sci.*, **12**, 925 (1974).
124. Harwood, H.J., Ritchey, W.M. *J. Polymer Sci.*, **B2**, 601 (1964).
125. Hatada, K., Terewaki, Y., Okuda, H., Tanaka, Y., Sato, H. *J. Polymer Sci. Polymer Letters Ed.*, **12**, (6), 305 (1974).
126. Elgert, K.F., Quack, G., Stutzel, B. *Makromol Chem.*, **176**, (3), 759 (1975).
127. Elgert, K.F., Quack, G., Stutzel, B. *Polymer*, **16**, (3), 154 (1975).
128. Vodchnal, J., Kossler, I. *Coll. of Czech. Communs.*, **29**, 2428 (1964).
129. Vodchnal, J., Kossler, I. *Coll. of Czech. Communs.*, **29**, 2859 (1964).
130. Fraga, D.W., Benson, L.H. Shell Chemical Co. Ltd., Torrance, California, private communication.
131. Binder, J.L. *J. Polymer Sci.* A., 37 (1963).
132. Tanaka, Y., Sato, H., Sumiya, T. *Polymer J.*, **7**, 264 (1975).
133. Tanaka, Y., Sato, H., Ogura, A., Nagoya, I. *J. Polymer Sci. Polymer Chem. Ed.*, **14**, 73 (1976).
134. Morese-Seguela, B., St. Jacques, M., Renaud, J.M., Prud'Lomme, J. *Makromolecules*, **10**, (2), 431 (1977).
135. Gronski, W., Murayamo, N., Carlow, H.J., Miyamato, T. *Polymer*, **17**, 358 (1976).
136. Beebe, D.H. *Polymer*, **19**, (2), 231 (1978).
137. Dalinskaya, E.R., Khachaturov, A.S., Poletaeva, I.A., Kormer, V.A. *Makromol Chem.*, **179**, (2) 409 (1978).
138. Duck, E.W., Grant, D.M. *Marcomolecules*, **3**, 165 (1970).
139. Van, Stratum Dvorak J. *J. Chromatogr*, **71**, (1), 9 (1972).
140. Richardson, W.S., Sacker, A. *J. Polymer Sci.*, **10**, 353 (1953).
141. Plkovskiu, E.I., Volkenstejn, M. *Dokl. Acad. Nauk, S.S.S.R.*, **95**, 301 (1954).
142. Binder, J.L., Ransaw, H.C. *Anal. Chem.*, **29**, 503 (1957).
143. Corish, P.J. *Spectrochim. Acta.*, **15**, 598 (1959).
144. Maynard, J.T., Moobel, W.E. *J. Polymer Sci.*, **13**, 251 (1954).

145. Ferguson, R.C. *Anal. Chem.*, **36**, 2204 (1964).
146. Kossler, I., Vodchnal, J. *Polymer Letters*, **4**, 415 (1963).
147. Binder, J.L., Ransaw, H.C. *Anal. Chem.*, **29**, 503 (1957).
148. Binder, J.L. *Rubber Chem. Technol.*, **35**, 57 (1962).
149. Binder, J.H. *J. Polymer Sci.*, A, **1**, 37 (1963).
150. Golub, M.A. *J. Polymer Sci.*, **36**, 523 (1959).
151. Golub, M.A. *J. Polymer Sci.*, **36**, 10 (1959).
152. Maynard, J.T. Moobel, W.E. *J. Polymer Sci.*, **13**, 251 1954).
153. Luongo, J.P., Solovay, R.J. *Appl. Polymer Sci.*, **7**, 2307 (1963).
154. Barral, M.J. *Polymer Sci.*, **13**, 1515 (1975).
155. Rueda, D.R., Balta-Calleja, F.J. Hidalgo, A. *Spectrochimica. Acta*, **30A**, 1545 (1974).
156. Dankovics, A., *Muanay Gumi*, **11**, (12), 380 (1974); *Chem. Abstr.*, **82**, 86773g (1975).
157. Hackathorne, M.J., Brock, M.J. *J. Polymer Sci. Polymer Chem. Ed.*, **13**, (4), 945 (1975).
158. Furukawa, J., Haga, K., Kobayashi, E., Iseda, Y., Yoshimoto, T., Sakamato, K. *Polymer J.*, **2**, 371 (1971).
159. Hill, R., Lewis, J.R., Simonsen, J.L. *Trans. Faraday Soc.*, **35**, 1067 (1937).
160. Alekseeva, E.N. *J. Gen. Chem. (U.S.S.R.)*, **11**, 353 (1941).
161. Rabjohn, N., Bryan, C.E., Inskip, G.E., Johnston, H.W., Lawson, J.K. *J. Amer. Chem. Soc.*, **69**, 314 (1947).
162. Yakubehik, A.I., Spaaskova, A.J., Zak, A.G., Shotatskaya, I.D. *Zhur. Obshsh Khim.*, **28**, 3080 (1958).
163. Hackathorn, M.J., Brock, M.J. *J. Polymer Sci. Polymer Chem. Ed.*, **13**, (4), 945 (1975).
164. Kawasaki, A. Paper presented at the 27th Autumn meeting, Japan Chemical Society 1672. Pre-prints, **2**, 20 (1972).
165. Hackathorn, M.J., Brock, M.K. *Rubber Chem. Technol.*, **45**, 1295 (1972).
166. Hill, R., Lewis, J.R., Simenson, J.L. *Trans. Faraday Soc.*, **35**, 1073 (1939).
167. Natta, G., Dannusso, F. *J. Polymer Sci.*, **34**, 3 (1959).
168. Bovey, F.A. *Polymer Conformation and Configuration*. Academic Press, New York, N.Y., p.8 (1969).
169. Randall, J.C. *J. Polymer Sci. Polymer Phys. Ed.*, **12**, 703 (1974).
170. Zambelli, A., Locatelli, P., Bajo, G., Bovey, F.A. *Makro-molecules*, **8**, 687 (1975).
171. Stehling, F.C., Knox, J.R. *Macromolecules*, **8**, 595 (1975).
172. Schaefer, J., Natusch, D.F.S. *Macromolecules*, **5**, 416 (1972).
173. Axelson, D.E., Mandelkern, L., Levy, G.C. *Macromolecules*, **10**, 557 (1977).
174. Provasoli, A., Ferre, D.R. *Macromolecules*, **10**, 874 (1977).
175. Randall, J.C. ACS Symposium Series No. 103. *Carbon 13 NMR Polymer Science*. Wallace M. Pasika (ed.), American Chemical Society (1979).
176. Dovey, F.A. *Polymer Conformation and Configuration*. Acedemic Press, New York, N.Y., p.16 (1969).
177. Randall, J.C. *J. Polymer Sci. Polymer Phys. Ed.*, **14**, 2083 (1976).
178. Stehling, F.C. *J. Polymer Sci.* A-2, 1815 (1964).
179. Randall, J.C. *J. Polymer Sci.*, **13**, 889 (1975).
180. Inone, Y., Nishioka, A., Chujo, R. *Macromolecular Chem.*, **156**, 207 (1972).
181. Matsuzaki, K., Urya, T., Osada, K., Kawamura, T. *Macromolecules*, **5**, 816 (1972).
182. Lafeyre, W., Cheradame, H., Spaasky, N., Sigwait, P.J. *Chem. Phys. Physio Chim. Biol.*, **70**, (5), 838 (1973).
183. Uryu, T., Shimazu, H., Matsuzak, K. *J. Polymer Sci. Polymer Lett. Ed.*, **11**, 275 (1973).
184. Schaefer, J. *Macromolecules*, **5**, (5), 590 (1972).
185. Kawamura, I.H., Uryu, T. *J. Polymer Sci.*, A-1, **11**, 971 (1973).
186. Matsuzaki, K., Okazono, S., Kanai, T. *J. Polymer Sci.* A-1, **17**, 3447 (1979).
187. Baker, C., Maddams, W.F., Park, G.S., Robertson, H. *Makromol. Chem.*, **165**, 231 (1973).
188. Millan, J.L., De la Pena, J.I. *Rev. Plast. Mod.*, **26**, (206), 232 (1973).
189. Kelen, T., Galambos, G., Tudos, F., Balint, G. *European Polymer J.*, **5**, 617 (1969).
190. Bezadea, E., Braun, D., Buriana, E., Caraculacu, A., Istrate-Robila, G. *Angew Makromol Chem.*, **37**, 35 (1974).
191. Schroeder, E., Byrdy, M. *Plaste Kautsch*, **24**, (11), 757 (1977).
192. Keller, F. *Plaste Kautsch*, **22**, 8 (1975).
193. Keller, F., Findeisen, G., Raetzsch, M., Roth, H. *Plaste Kautsch*, **22**, (9), 722 (1975).

194. Elgert, K.F., Ritter, W. *Makromol. Chem.*, **177**, (7), 2021 (1976).
195. Laiber, M., Opera, H., Toader, M., Laiber, V. *Rev. Chim.* (*Bucharest*), **28**, (9) 881 (1977); *Chem. Abstr.*, **88**, 23539p (1978).
196. Solovay, R., Pascale, J.W. *J. Polymer Sci.*, **2**, 2041 (1964).
197. D-ASTM-2283-64T. Tentative Methods of Test for the Absorbance of Polyethylene due to Methyl Groups at 1378 cm-1. Method A.
198. Neilson, O., Holland, O. *J. Mol. Spect.*, **4**, 448 (1960).
199. Boyle, D.A., Simpson, W., Waldron, J.D. *Polymer*, **2**, 323 (1961).
200. Willbourne, A.H. *J. Polymer Sci.*, **34**, 509 (1959).
201. Nerheim, A.G. *Anal. Chem.*, **47**, 1128 (1975).
202. Barral, M. *J. Polymer Sci.*, **13**, 1515 (1975).
203. Nishoika, A., Ando, I., Matsumoto, J. *Bunseki Kayaku*, **26**, (5), 308 (1977); *Chem. Abstr.*, **87**, 102729h (1977).
204. Kraus, G., Stacy, C.J. *J. Polymer Sci.* Symposium No. 43, 329 (1973).
205. Randall, J.C. *J. Polymer Sci., Polymer Phys. Ed.*, **11**, 275 (1973).
206. Bovey, F.A., Schilling, F.G., McCrackin, F.L., Wagner, H.L. *Macromolecules*, **9**, 76 (1976).
207. Foster, G.N. *Polymer Preprints*, **20**, 463 (1970).
208. Barlow, A., Wild, A., Ranganath, R. *J. Appl. Polymer Sci.*, **21**, 3319 (1977).
209. Wild, L., Ranganath, R., Barlow, A. *J. Appl. Polymer Sci.*, **21**, 3331 (1977).
210. Randall, J.C. *J. Polymer Sci., Polymer Phys. Ed.*, **11**, 275 (1973).
211. Randall, J.C. *J. Appl. Polymer Sci.*, **22**, 585 (1978).
212. Grant, D.M., Paul, E.G. *J. Amer. Chem. Soc.*, **86**, 2984 (1964).
213. Axelson, D.E., Levy, G.C., Mandelkern, L. *Macromolecules*, **12**, 41 (1979).
214. Kamath, P.M., Barlow, A. *J. Polymer Sci.*, A-1, **5**, 2023 (1967).
215. Bellamy, L.J. *Infrared Spectra of Complex Molecules*. Wiley, New York, **27**, 13 (1958).
216. Willbourn, A.J. *J. Polymer Sci.*, **34**, 569 (1959).
217. Schacfer, J. *Natusch DFS Macromolecules*, **5**, 416 (1972).
218. Inoue, Y., Nishoika, A., Chujo, R. *J. Polymer Sci. Polymer Phys. Ed.*, **11**, 2234 (1973).
219. Porter, N., Nicksie, O., Johnson, P. *Anal. Chem.*, **35**, 1948 (1963).
220. Wilkes, C.E., Carmen, C.J., Harrington, R.A. *J. Polymer Sci.*, **43**, 237 (1973).
221. Carmen, C.J., Harrington, R.A., Wilkes, C.E. *Macromolecules*, **10**, 536 (1973).
222. Ray, G.J., Johnson, D.E., Knox, J.R. *Macromolecules*, **10**, 773 (1977).
223. Sanders, J.M., Kamoroski, R.A. *Macromolecules*, **10**, 1214 (1977).
224. Paxton, J.R., Randall, J.C. *Anal. Chem.*, **50**, 1777 (1978).
225. Randall, J.C. *Macromolecules*, **11**, 33 (1978).
226. Voigt, J. *Kunststoffe*, **54**, 2 (1964).
227. Voigt, J. *Kunststoffe*, **51**, 18 (1961).
228. Voigt, J. *Kunststoffe*, **51**, 314 (1961).
229. Van Schooten, J., D.E.W. Berkenbosch, R. *Polymer* (*London*), **2**, 357 (1961).
230. Van Schooten, J., Mostert, S. *Polymer* (*London*), **4**, 135 (1963).
231. Barlow, A., Lehrie, R.S., Robb, J.C. *Polymer* (*London*), **2**, 27 (1961); S.C.I. Monogr. No. 17, 'Techniques of Polymer Science', p. 267, London (1963).
232. Van Schooten, J., Mostert, S. *Polymer* (*London*), **4**, 135 (1963).
233. Van Schooten, J., Evenhuis, J.K. *Polymer* (*London*), **6**, 561 (Nov. 1965).
234. Van Schooten, J., Evenhuis, J.K. *Polymer* (*London*), **561**, Nov. 1965 and *Polymer* (*London*), **6**, 343 (1965).
235. Lehmann, F.A., Brauer, G.M. *Anal. Chem.*, **33**, 676 (1961).
236. Strassburger, J., Brauer, C.M., Tyron, M., Forziati, A.F. *Anal. Chem.*, **32**, 454 (1960).
237. Brauer, G.M. *J. Polym. Sci.*, C, **8**, 3 (1965).
238. Neumann, E.W., Nadeau, H.G. *Anal. Chem.*, **10**, 1454 (1963).
239. Madorsky, S.L., Straus, S. *J. Res. Nat'l. Bur. Std.*, **53**, 361 (1954).
240. Wall, L.A., Madorsky, S.L., Brown, D.W., Straus, S., Simka, R. *J. Amer. Chem. Soc.*, **76**, 3430 (1954).
241. Kolb, B., Kaiser, H. *J. Gas Chromatog.*, **2**, 233 (1964).
242. Seeger, M., Exner, J., Cantow, H.J. Paper presented at 23rd International Congress of Pure and Applied Chemistry, Boston, 1971, *Macromolecular Preprints*, **11**, 739 (1971).
243. Exner, J., Seeger, M., Cantow, J.J. *Angew Chem.* (*International Ed.*), **10**, 346 (1971).
244. Seegner, M., Barrall, E. *J. Polymer Sci.*, A-1, **13**, 1515 (1975).

245. Tsuchiya, J., Sumi, K. *J. Polymer Sci.*, B, **6**, 356 (1968).
246. Tsuchiya, Y., Sumi, K. *J. Polymer Sci.*, A-1, **6**, 415 (1968).
247. Ahlstrom, D.H., Leibman, S.A., Abbas, K.P. *J. Polymer Sci. Polymer Chem. Ed.*, **14**, 2479 (1976).
248. Barrall, E.M., Porter, R.S., Johnson, J.F. *J. Chromatog.*, **11**, 177 (1963).
249. Van Schooten J., Evenhuis, J.K. Private communication.
250. Wall, L.A., Strauss, S. *J. Polymer Sci.*, **44**, 313 (1960).
251. Cobler, J.G., Samsel, E.P. *SPE Transactions*, 145, April (1962).
252. Ohtani, S., Ishikawa, T. *Kogyo Kagaku Zasshi*, **65**, 617 (1962).
253. O'Mara, M.M.J. *Appl. Polymer Sci.*, **13**, 1887 (1970).
254. Watson, J.T., Biemann, K. *Anal. Chem.*, **36**, 1135 (1964).
255. Stromberg, R.R., Straus, S., Achhammer, B.G. *J. Polymer Sci.*, **35**, 355 (1959).
256. Tsuchiya, F., Sumi, K. *J. Appl. Chem.*, **17**, 364 (1967).
257. Ohta, M. *Kogyo Kagaku Zasshi*, **55**, 31 (1952).
258. Boettner, E.A., Ball, G., Weiss, B. *J. Appl. Polymer Sci.*, **13**, 377 (1969).
259. Noffz, D., Benz, W., Pfab, W. *Z. Anal. Chem.*, **235**, 121 (1968).
260. Liebman, S.A., Ahlstrom, D.H., Quinn, E.J., Geighley, A.G., Meluskey, J.T. *J. Polymer Sci.*, A-1, **9**, 1921 (1971).
261. Boettner, E.A., Weiss, B. *Amer. Ind. Hyg. Assoc. J.*, **28**, 535 (1967).
262. Van Schooten, J., Evenhuis, J.K. *Polymer (London)*, **6**, 353 (1965).
263. Galin, M. *J. Macromol. Sci.-Chem.*, A, **7**, 783 (1973).
264. Hackathorn, M.J., Brock, M.J. *J. Polymer Sci. Polymer Lett. Ed.*, **8**, 617 (1970).
265. Tanaka, Y., Sato, H., Ogura, A., Nagoya, I. *J. Polymer Sci., Polymer Chem. Ed.*, **14**, 73 (1976).
266. Seeger, M., Barrall, E. II, *J. Polymer Sci.*, A-1, **13**, 1515 (1975).
267. Willbourn, A.H. *J. Polymer Sci.*, **34**, 569 (1959).
268. Hank, R. *Rubber Chem. Technol.*, **40**, 936 (1967).
269. Hoer, H., Kooyman, *Anal. Chim. Acta.*, **5**, 550 (1951).
270. Cotman, J.D., Jr, *J. Amer. Chem. Soc.*, **77**, 2790 (1955).
271. Carrega, M., Bonnebat, C., Zednik, G. *Anal. Chem.*, **42**, 1807 (1970).
272. de Vries, A.J., Bonnebat, C., Carrega, M. *Pure Appl. Chem.*, **26**, 209 (1971).
273. Bovey, F.A., Abbas, K.B., Schilling, F.C., Starnes, W.H. *Macromolecules*, **8**, 437 (1975).
274. Chang, E.P., Salovey, R. *J. Polymer Sci., Polymer Chem. Ed.*, **12**, 2927 (1974).
275. Suzuki, M., Tsuge, S., Takeuchi, T. *J. Polym. Sci.*, A-1, **10**, 1051 (1972).
276. McMurray, H.L., Thornton, V. *Anal. Chem.*, **24**, 318 (1952).
277. Van Schooten, J., Duck, E.W., Berkenbosch, R. *Polymer*, **2**, 357 (1961).
278. Bucci, G., Simonazzi, T. *J. Polymer Sci.*, C-7, 203 (1964).
279. Bucci, G., Simonazzi, T. *Chimica Industria*, **44**, 262 (1962).
280. Natta, G., Mazzanti, G., Valvassori, A., Sartori, G., Merero, D. *Chim. Ind. (Milan)*, **42**, 125–132 (1960).
281. Veerkamp, Th. A., Veermans, A. *Makromol. Chem.*, **50**, 147 (1961).
282. Van Schooten, J., Mostert, S. *Polymer*, **4**, 135 (1963).
283. Bucci, G., Simonazzi, T. *J. Polymer Sci.*, C-7, 203 (1964).
284. McMurray, H.L., Thornton, V. *Anal. Chem.*, **24**, 138 (1952).
285. Natta, G., Dall'Asta, G., Mazzanti, G., Ciampelli, F. *Kolloid Z.*, **182**, 50 (1962).
286. Sheppard, N., Sutherland, G.B. *Nature (London)*, **159**, 739 (1947).
287. Rugg, F.M., Smith, J.J., Martman, L.H. *Ann. N.Y. Acad. Sci.*, **57**, 398 (1953); *J. Polymer Sci.*, **11**, 1 (1953).
288. Sutherland, G.B. *Disc. Faraday Soc.*, **9**, 279 (1950).
289. Liang, C.Y., Lytton, M.R., Boone, C.J. *J. Polymer Sci.*, **54**, 523 (1961).
290. Liang, C.Y., Watt, W.R. *J. Polymer Sci.*, **51**, 514 (1961).
291. Natta, G., Mazzanti, G., Valvassori, A., Pajaro, A. *Chim. Ind. (Milan)*, **39**, 733 (1957).
292. Drushell, H.V., Iddings, F.A. *Anal. Chem.*, **35**, 28 (1963).
293. Drusbell, H.V., Iddings, F.A. Division of Polymer Chemistry 142nd Meeting American Chemical Society Atlantic City, September 1962.
294. Wei, P.E. *Anal. Chem.*, **33**, 215 (1961).
295. Gossl, T. *Makromol. Chem.*, **42**, 1 (1961).
296. Corish, P.J. *Anal. Chem.*, **33**, 1798 (1961).
297. Tosi, G., Simonazzi T. *Die Angewandte Makromolecular Chemie*, **32**, 153 (1973).

298. Ciampelli, F., Bucci, G., Simonazzi, A., Santambrogio, A. *La Chimica l'Industria*, **44**, 489 (1962).
299. Bucci, G., Simonazzi, T. *La Chemica l'Industria*, **44**, 262 (1962).
300. Lomonte, J.N., Tirpak, G.A. *J. Polymer Sci.*, A-2, 705 (1964).
301. Schneider, B., Stokr, J., Doskocilova, D., Kolinsky, M., Sykora, S., Lim, D. Paper presented at International Symposium on Macromolecular Chemistry, Prague (1965).
302. Schneider, J.S.B., Kalinsky, M., Ryska, M., Lim, D. *J. Polymer Sci.*, A-1, **5**, 2013 (1967).
303. Peraldo, M. *Gazz. Chem. Ital.*, **89**, 798 (1959)
304. McDonald, M.P., Ward, I.M. *Polymer*, **2**, 341 (1961).
305. Brada, J.J. *J. Polymer Sci.*, **3**, 370 (1960).
306. Luongo, J.P. *J. Appl. Polymer Sci.*, **3**, 302 (1960).
307. Liang, C.V., Pearson, F.G. *J. Mol. Spect.*, **5**, 290 (1960).
308. Sibilia, J.P., Wincklhofer, R.C. *J. Appl. Polymer Sci.*, **6**, 557 (1962).
309. Kissin, Yu B., Gol'dfarb, Yu Ya, Novoderzhkin, Yu V, Krentsel, B.A. *Vysokmol Soedin. Ser. B*, **18**, (3), 167 (1976).
310. Tosi, C. *Makromol. Chem.*, **170**, 231 (1973).
311. Popov, V.P., Duvanova, A.P. *Zh Prikl Spektrosk.*, **18**, (6), 1077 (1973).
312. Seeger, M., Exner, J., Cantow, H.J. *Quad. Ric. Sci.*, **84**, 102 (1973); *Chem. Abstr.*, **81**, 50225 (1973).
313. Seno, J., Tsuge, S., Takeuchi, T. *Makromol. Chem.*, **161**, 195 (1972).
314. Bakuyutov, N.G., Kissin, Yu V., Vavilova, I., Arkhipova, Z.V. *Vysokmol. Soedin. Ser., A*, **17**, 2163 (1975).
315. Modric, I., Holland-Moritz, K., Hummel, D.O. *Colloid Polymer Sci.*, **254**, (3), 342 (1976).
316. Kamida, K., Yamaguchi, K. *Makromol. Chem.*, **162**, 205 (1972).
317. Simak, P., Ropte, E. *Makromol. Chem. Suppl.*, **1**, 507 (1975).
318. Oi, N., Moriguchi, K. *Bunseki Kagaki*, **23**, (7) 798 (1974); *Chem. Abstr.*, **82**, 17308x (1974).
319. Oi, N., Miyazaki, K., Moriguchi, K., Shimada, H. *Kabunski Kagaki*, **29**, (6), 388 (1972).
320. Ebdon, J.J., Kandil, S.H., Morgan, K.J. *J. Polymer Sci.*, A-1, **17**, 2783 (1979).
321. Shirakawa, H., Yamagaki, N., Kambara, S. Personal communication for Dr. N. Yamazaki, Tokyo Institute of Technology.
322. Kalal, J., Honska, M., Seycek, O., Adamek, P. *Makromol. Chem.*, **164**, 249 (1973).
323. Arg, T.L., Harwood, H.J. *J. Amer. Chem. Soc. Polymer Preprints*, **5**, (1), 306 (1964).
324. Yaragisawa, K. *Chem. Hisg. Polymers (Tokyo)*, **21**, 312 (1964).
325. Germar, M. *Makromolecular Chem.*, **84**, 36 (1965).
326. Enomoto, S. *J. Polymer Sci.*, **55**, 95 (1961).
327. Oswald, H.J. Kubu, E.T. *Soc. Plastics Eng. Trans.*, **3**, 168 (1963).
328. Bruck, D., Hummel, D. *Makromolecular Chem.*, **163**, 245 (1973).
329. Kumpanenko, I.I., Kazananskii, K.S. *J. Polymer Sci., Polymer Symp. No. 42.* Pt. 2, 973–80 (1973).
330. Reilly, G.A. Shell Chemical Co., Emeryville, California, private communication (1964).
331. Mitani, K. *J. Makromol. Sci. Chem.*, A, **8**, (6), 1033 (1974).
332. Inoue, Y., Mishoika, O. Chuka, R. *Makromolecular Chem.*, **152**, 15 (1972).
333. Zambelli, A., Dorman, D.E., Brewster, A.I.R. Bovey, F.A. *Makromolecules*, **6**, 925 (1973).
334. Randall, J.C. *J. Polymer Sci. Polymer Phys. Ed.*, **12**, 703 (1974).
335. Randall, J.C. *J. Polymer Sci. Polymer Phys. Ed.*, **14**, (11), 283 (1976).
336. Randall, J.C. *J. Polymer Sci. Polymer Phys. Ed.*, **14**, 208 (1976).
337. Randall, J.C. *J. Polymer Sci.*, **14**, 1693 (1976).
338. Cavelli, L. *Relaz Corso Teor-Prat Risonenza Magh. Nuci.*, 351 (1973).
339. Brosio, E., Delfini, M., Conti, F. *Nuova Chim.*, **48**, (11), 35 (1972).
340. Asakurin, T., Ando, I., Nishoika, A., Doi, Y., Keii, T. *Makromol. Chem.*, **178**, (3), 791 (1977).
341. Porter, R.S. *J. Polymer Sci.*, A-1, **4**, 189 (1966).
342. Porter, R.S., Nicksie, J.W., Johnson, J.F. *Anal. Chem.*, **35**, 1948 (1963).
343. Stehling, F.C. *J. Polymer Sci.*, A-1, **4**, 189 (1966).
344. Barrall, E.M., Porter, R.S., Johnson, J.F. Paper presented to Division of Polymer Chemistry 148th National Meeting, American Chemical Society, Chicago, September 1964. *Preprints* **5**, No. 2, 816 (1964); *J. Appl. Polymer Sci.*, **9**, 3061 (1965).
345. Satoh, S.R., Chujo, T., Nagai, E. *J. Polymer Sci.*, **62**, 510 (1962).

346. Tanaka, T., Hatada, K. *J. Polymer Sci.*, **11**, 2057 (1973).
347. Grant, D.M., Paul, E.C. *J. Amer. Chem. Soc.*, **86**, 2984 (1964).
348. Crain, W.O., Zambelli, A., Roberts, J.P. *Makromolecules* **4**, 330 (1971).
349. Cannon, C.J., Wilkes, C.E. *Rubber Chem. Technol.*, **44**, 781 (1971).
350. Ray, G.J., Johnson, R.E., Knox, J.R. *Makromolecules*, **10**, (4), 773 (1977).
351. Randall, J.C. *Makromolecules*, **11**, (3), 592 (1978).
352. Carmen, C. *Macromolecules*, **7**, 789 (1974).
353. Mauzac, M., Vairon, J.P., Sigmalt, P. *Polymer*, **18**, (11), 1193 (1977).
354. Ritter, W., Elgert, K.F., Cantow, H.J. *Makromol. Chem.*, **178**, (2), 557 (1977).
355. Chen, H.Y., *J. Polymer Sci. Polymer Letters Ed.*, **12**, 85 (1974).
356. Elgert, K.F., Cantow, H.J., Stutzel, B. *Frenzel Angew Chem. Int. Ed.*, **12**, 427 (1973).
357. Brame, E.G., Khan, A.A. *Rubber Chem. Technol.*, **50**, (2), 272 (1977).
358. Lindsay, G.A., Santee, E.R., Harwood, H.J. *Polymer Preprints*, American Chemical Society. Divn. Polym. Chem., **14**, (2), 646 (1973).
359. Araki, K. *Fukui-ken Kogyo Shinenjo Nempo*, **50**, 63 (1975); *Chem. Abstr.*, **87**, 118514 (1977).
360. Wilkes, C.E. *J. Polymer Sci. Polymer Symp.*, **60**, 161 (1977); *Chem. Abstr.*, **89**, 24952a (1978).
361. Okada, T., Ikushige, T. *J. Polymer Sci. Polymer Chem. Ed.*, **14**, (8), 2059 (1976).
362. Keller, F., Opitz, H., Hoesselbarth, B., Beckert, D., Reicherdt, W. *Faserforsch Textiltech.*, **26**, (7), 329 (1975).
363. Keller, F., Hoesselbarth, B. *Faserforsch Textiltech*, **29**, (2), 152 (1978).
364. Keller, F., Zepnik, S., Hoesselbarth, B. *Faserforsch Textiltech.*, **28**, (6), 287 (1977).
365. Abe, A., Nishoika, A. *Kobunski Kagaku*, **29**, (6), 402, 448 (1972).
366. Bovey, F.A., Schilling, F.C., Kwei, T.K., Frisch, H.L. *Macro-molecules*, **10**, (3), 559 (1977).
367. Okuda, K.J. *Polymer Sci.*, A-2, 1749 (1964).
368. Chujo, R., Satoh, S., Nagai, E. *J. Polymer Sci.*, A-2, 895 (1964).
369. McClanalan, J.L., Privitera, S.A *J. Polymer Sci.* A-3, 3919 (1965).
370. Carman, C.J., Tarpley, A.R., Goldstein, J.H. *Macromolecules*, **4**, 445 (1971).
371. Carman, C.J. *Macromolecules*, **6**, 725 (1973).
372. Bovey, F.A., Anderson, E.W., Douglas, D.C., Mason, J.A. *J. Chem. Phys.*, **39**, 1199 (1963).
373. Ramey, K.C. *J. Phys. Chem.*, **70**, 2525 (1966).
374. Bovey, F.A., Hood, F.P., Anderson, E.W., Kornegay, R.L. *J. Polymer Chem.*, **71**, 312 (1967).
375. Heatley, F., Bovey, F.A. *Macromolecules*, **2**, 241 (1969).
376. Cavelli, L., Borsini, G.C., Carraro, G., Confalonieri, G. *J. Polymer Sci.*, A-1, **8**, 801 (1970).
377. Wu, T.K. *Macromolecules*, **6**, (5), 737 (1973).
378. Shipman, J.J., Golub, M.A. *J. Polymer Sci.*, A-1, 832 (1963).
379. Isa, I.A. *J. Polymer Sci.*, A-1, **10**, 881 (1972).
380. Fsa, I.A., Meyers, M.E. *J. Polymer Sci.*, A-1, **11**, 2125 (1973).
381. Mitani, K., Ogata, T., Iwasaki, M. *J. Polymer Sci. Polymer Chem. Ed.*, **12**, 1653 (1974).
382. Brame, E.G. *J. Polymer Sci.*, A-1, **9**, 2051 (1971).
383. Elgart, K F. Cantow, H.J., Stutzel, B. *Frenzel Angew Chem. Int. Ed.*, **12**, 427 (1973).
384. Inoeu, Y., Nishoika A., Chujo, R. *Makromol Chem.*, **156**, 207 (1972).
385. Jasse, B., Laupretre, F., Monnerie, L. *Makromol. Chem.*, **178**, (1), 1987 (1977)
386. Borsa, F., Lanzi, G. *J. Polymer Sci.*, A-2, 2623 (1964).
387. Randall, J.C. *J. Polymer Sci. Polymer Phys. Ed.*, **14**, 283 (1976).
388. Collins, G.C.S., Lowe, A.C., Nicholas, D. *European Polymer J.*, **9**, (11), 1173 (1973).
389. Buchak, B.E., Ramey, K.C. *J. Polymer Sci. Polymer Lett. Ed.*, **14**, (7), 401 (1976).
390. Yakata, K., Hirabayashi, T. *J. Polymer Sci.*, A-1, **14**, 57 (1976).
391. Elgert, K.F., Stuetzel, B. *Polymer*, **16**, (10), 758 (1975).
392. Sandner, B., Keller, F., Roth, H. *Faserforsch Textiltech.*, **26**, (6), 278 (1975).
393. Yamashita, Y., Yoshida, M., Kawasc, J., Ito, K. *Aschi Garasu Kogyu Gijutsu, Shoreeikal Kenkyo Hukoku*, **25**, 87 (1974); *Chem. Abstr.*, **84**, 60070s (1974).
394. Regel, W., Westfeld, L., Cantow, H.J. *Angew Chem. Int. Ed. Engl.*, **12**, (5), 434 (1973).
395. Abe, A., Nishoika, N. *Kobunshi Kagaku*, **29**, (6), 402, 448 (1972).
396. Ibrahim, R., Katritzky, A.R., Smith, A., Weiss, D.E. *J. Chem Soc. Perkin Trans.*, **2**, (13), 1537 (1974).
397. Keller, F. *Plaste Kautsch*, **22**, (1), 8 (1975).
398. Delfini, M., Segre, A.C., Conti, F. *Macromolecules*, **6**, 645 (1973).

399. Wu, T.K., Ovenall, D.W., Reddy, G.S. *J. Polymer Sci. Polymer Phys. Ed.*, **12**, 901 (1974).
400. Kusakov, M.M., Koshevnick, A., Yu, Mekenitskaya, L.I., Shulipine, L.M., Amerik, Yu. B., Golova, L.K. *Vysokomol. Soedin. Ser. B.*, **15**, (3), 150 (1973).
401. Suzuki, T., Horwood, H.J. *Polymer. Prec. Amer. Chem. Soc. Div. Polymer Chem.*, **16**, (1), 638 (1975).
402. Cavelli, L., Relaz, O., *Corso Teor. Prat. Rizonenza Magn. Nucl.*, 351 (1973).
403. Spevacek, J., Schneider, B. *Makromol. Chem.*, **176**, (3), 729 (1975).
404. Suzuki, T., Santee, E.R., Harwood, H.J., Vogl, O., Tanaka, T. *J. Polymer Sci. Polymer Lett. Ed.*, **12**, 635 (1974).
405. Matsuzaki, K., Kanai, T., Kawamura, T., Matsumoto, S., Uryu, T. *J. Polymer Sci. Polymer Chem. Ed.*, **11**, 961 (1973).
406. Hateda, K., Ohta, K., Okamoto, Y., Kitayama, T., Umemuru, Y., Yuki, H. *J. Polymer Sci. Polymer Lett. Ed.*, **14**, (9), 531 (1976).
407. Heublein, G., Boerner, R., Schnetz, H. *Faserforsch Textiltech.*, **29**, (5), 317 (1978); *Chem. Abstr.*, **88**, 191695c (1978).
408. Johnson, A., Klesper, E., Wirthlin, T. *Makromol Chem.*, **177**, (8), 2398 (1976).
409. Klesper, E., Johnson, A., Gronski, W., Wehrilr, F.W. *Makromol. Chem.*, **176**, (4), 1071 (1975).
410. Mori, Y., Ueda, A., Tamzawa, H., Matsuzaki, K., Kobayoshi, H. *Makromol. Chem.*, **176**, (3), 699 (1975).
411. Suzuki, T., Mitani, K., Takegami, Y., Furukawa, J., Kobayashi, E., Arai, Y. *Polymer J.*, **6**, (6), 496 (1974).
412. Ebdon, J.R. *J. Macromol. Sci. Chem.*, (2), 417 (1974).
413. Katritzky, A.R., Smith, A., Weiss, D.E. *J. Chem. Soc. Perkin Trans.*, **2**, (13), 1547 (1974).
414. Wang, A., Suzuki, T., Harwood, H.J. *J. Polymer Prep. Amer. Chem. Soc. Div. Polymer Chem.*, **16**, (1), 644 (1975).
415. Pogorel'skii, K.Y., Asanov, A., Akhmedov, K.S. *Dokl. Akad. Nauk. Uzh. S.S.R.*, **27**, (3), 28 (1970).
416. Keller, F., Fukrmann, C., Roth, H., Findetsen, G., Raezsch, M. *Plaste Kautsch*, **24**, (9); 626 (1977); *Chem. Abstr.*, **88**, 38232g (1978).
417. Roussel, R., Galin, J.C. *J. Macromol. Sci. Chem.*, A, **11**, (2), 347 (1977).
418. Heublein, G., Freitag, W., Schuetz, H. *Faserforsch Textil.*, **26**, (10), 498 (1975).
419. Okada, T., Otsurn, M. *J. Polymer Sci. Polymer Lett. Ed.*, **14**, (10), 595 (1976).
420. Matsuzaki, K., Ito, H., Kawamura, T., Uryu, T. *J. Polymer Sci. Polymer Chem. Ed.*, **11**, 971 (1973).
421. Whipple, E.B., Green, P.J. *Macromolecules*, **6**, 38 (1973).
422. Wu, T.K., Ovenall, D.W. *J. Polymer Prep. Amer. Chem. Soc. Div. Polymer Chem.*, **17**, (2), 693 (1976).
423. Cross, L.H., Richards, R.B., Willis, H.A. *Disc. Faraday Soc. (London)*, **9**, 235 (1950).
424. Rugg, R.M., Smith, J.J., Bacon, R.C. *J. Polymer Sci.*, **13**, 535 (1954).
425. Pross, A.W., Black, R.M. *J. Soc. Chem. Ind. (London)*, **69**, 115 (1950).
426. Beachell, A.C., Nemphos, S.P. *J. Polymer Sci.*, **21**, 113 (1956).
427. Beachell, A.C., Tarbet, G.W. *J. Polymer Sci.*, **45**, 451 (1960).
428. Luongo, J.P. *J. Polymer Sci.*, **42**, 139 (1960).
429. Cooper, G.D., Prober, X. *J. Polymer Sci.*, **44**, 397 (1960).
430. Thompson, H.W., Tarkington, P. *Proc. Royal Soc. (London)*, 184, A, **3**, (1945).
431. Pross, A.W., Black, R.M. *J. Soc. Chem. Ind. (London)*, **69**, 113 (1950).
432. Kveder, H., Ungar, G. *Nafta (Zagreb)*, **24**, (2), 85 (1973).
433. Pentin, Yu A., Tarasevich, B.N., El'tsefon, B.S. *Vestn. Mosk Univ. Khim.*, **14**, (1), 13 (1973).
434. Miller, P.J., Jackson, J.F., Portor, R.S. *J. Polymer Sci. Polymer Phys. Ed.*, **11**, 2001 (1973).
435. Tabb, D.L., Sevcik, J.J., Koenig, S.L. *J. Polymer Sci. Polymer Phys. Ed.*, **13**, (4), 815 (1975).
436. Chan, M.G., Allara, D.I. *Polymer Eng. Sci.*, **14**, (1), 12 (1974).
437. Cooper, G.D., Prober, M. *J. Polymer Sci.*, **44**, 397 (1960).
438. Heacock, J.F. *J. Appl. Polymer Sci.*, **7**, 2319 (1963).
439. Adams, J.H. *J. Polymer Sci.*, A-1, **8**, 1279 (1970).
440. Wood, R.P., Statton, W.O. *J. Polymer Sci. Polymer Phys. Ed.*, **12**, 1575 (1974).
441. Goldstein, M., Seeley, M.E., Willis, H.A., Zichy, V.J.I. *Polymer*, **14** (11), 530 (1973).
442. Grassie, N., Weir, N.A. *J. Appl. Polymer Sci.*, **9**, 963 (1965).
443. Grassie, N., Weir, N.A. *J. Appl. Polymer Sci.*, **9**, 975 (1965).

444. Schole, R.G., Bednorszyk, J., Tamanei, T. *Anal. Chem.*, **38**, 331 (1966).
445. Shaw, J.N., Marshall, M.C. *J. Polymer Sci.*, A-1, **6**, 449 (1968).
446. Fritz, D.F., Sahil A., Keller, H.P., Kovat, E. *Anal. Chem.*, **51**, 7 (1979).
447. Bezdadea, E.C. private communication.
448. Bezdadea, E.C. *J. Polymer Sci. Polymer Phys. Ed.*, **15**, (3), 611 (1977).
449. Motorina, M.A., Kalinina, L.S., Metalkina, E.I. *Plast. Massy.*, **6**, 74 (1973).
450. Smirnova, O.V., Kirshak, V.V., Solonim, I. Ya, Urman, Ya G., Alikseeva, S.G., Bayamov, V.A. *Vysokomol. Soedin. Ser. A.*, **17**, (11), 2415 (1975).
451. Law, R.D. *J. Polymer Sci.*, A-1, **9**, 589 (1971).
452. Tsefanov, K.L.B., Panalotov, I.M., Erusalimskii, B.L. *European Polymer J.*, **10**, (7), 557 (1974).
453. Manaff, S.L., Ingham, J.D., Miller, J A. *Org. Magn. Reson.*, **10**, 198 (1977).
454. Valuev, V.I., Shiykhter, R.A., Dmitrieva, T.S., Tsvetovskii, I.B. *Zhur Anal. Khim.*, **30**, (6), 1235 (1975); *Chem. Abstr.*, **83**, 131156d (1975).
455. Nissen, D., Rossback, V., Zahn. H. *J. Appl. Polymer Sci.*, **18**, 1953 (1974).
456. Ghosh, P., Chadha, S.C., Mukoyee, A.R., Palit, R. *J. Polymer Sci.*, A-2, 4443 (1964).
457. Palit, S.R., Ghosh, P. *Microchem. J. Symp. Ser.*, **2**, 663 (1961).
458. Bartlett, P.D., Nozaki, K. *J. Polymer Sci.*, **3**, 216 (1948).
459. Ghosh, P., Mukherjee, A.R., Palit, S.R. *J. Polymer Sci.*, A-2, 2807 (1964).
460. Maiti, S., Saha, M.K. *J. Polymer Sci.*, A-1, **5**, 151 (1967).
461. Palit, S.R. *Makromol. Chem.*, **36**, 89 (1959).
462. Palit, S.R. *Makromol. Chem.*, **38**, 96 (1960).
463. Maiti, S., Ghosh, A., Saha, M.K. *Nature (London)*, **210**, 513 (1966).
464. Ghosh, P., Sengupta, P.K., Pramanick, A. *J. Polymer Sci.*, A, **3**, 1725 (1965).
465. Ghosh, P., Chadha, S.C., Mukherjee, A.R , Palit, S.R. *J. Polymer Sci.*, A, **2**, 4433 (1964).
466. Saha, M.K., Ghosh, P., Palit, S.R. *J. Polymer Sci.*, A, **2**, 1365 (1964).
467. Palit, S.R. *Makromol. Chem.*, **36**, 89 (1959).
468. Palit, S.R. *Makromol. Chem.*, **38**, 96 (1960).
469. Stetnagel, W.J., Palit, S.R. *J. Polymer Sci.*, A-1, **15**, 945 (1977).
470. Ghosh, P., Chadha, S.C., Mukherjee, A.R., Palit, S.R. *J. Polymer Sci.*, A-2, 4443 (1964).
471. Banthia, A.K., Mandal, B.M., Palit, S.R. *J. Polymer Sci. Polymer Chem. Ed.*, **15**, (4), 945 (1977).
472. Kanjilal, C., Mitra, B.C., Palit, S.R. *Makromol. Chem.*, **178**, (6), 1707 (1977).
473. Mukhopadhyay, S., Mitra, B.C., Palit, S.R. *J. Polymer Sci.*, A-1, **7**, 2442 (1969).
474. Palit, S.R., Ghosh, P., Palit, S.R. *J. Polymer Sci.*, A-2, 1365 (1964).
475. Saha, M.K., Ghosh, P., Palit, S.R. *J. Polymer Sci.*, A-2, 1365 (1964).
476. Slovakhotova, N.A., II'iccheva, Z.F., Vasiliev, L.A., Kangin, V.A. Karpov Physicochemical Scientific Research Institute, unidentified publication 1963.
477. McMurry, H.L., Thornton, V. *Anal. Chem.*, **24**, 318 (1952).
478. Binder, Z.Z., Ransow, H C *Anal. Chem.*, **29**, 303 (1957).
479. Fisher, N., Hellwege, R.H. *J. Polymer Sci.*, **56**, 33 (1962).
480. Browning, H.L., Ackermann, H.D., Patton, H.W. *J. Polymer Sci.*, A4, 1433 (1966).
481. Tsuji, K. *J. Polymer Sci. Polymer Chem. Ed.*, **11**, 1407 (1973).
482. Tsuji, K. *J. Polymer Sci. Polymer Chem. Ed.*, **11**, (2), 467 (1973).
483. Cambell, D., Araki, K., Turner, D.T. *J. Polymer Sci.*, A-1, **4**, 2597 (1966).
484. Cambell, D., Turner, D.T. *J. Polymer Sci.*, A-1, **5**, 2199 (1967).
485. Hama, Y., Hosano, Y., Shinohara, K. *J. Polymer Sci.* A9, 1411 (1971).
486. Seiki, T., Takeshita, T. *J. Polymer Sci.*, A-1, **10**, 3119 (1972).
487. Tsuji, K. *J. Polymer Sci.*, A-1, **11**, 467 (1973).
488. Tsuji, K. *J. Polymer Sci.*, A-1, **11**, 1407 (1973).
489. Hori, Y., Shimaka, S., Kashiwabara, H. *Polymer*, **18**, (6), 567 (1977).
490. Hori, Y., Shimada, S., Kashiwabara, H. *Polymer*, **18**, (6), 1143 (1977).
491. Olnishi, S.I., Sugimato, S.I., Nitta, I. *J. Polymer Sci.*, Part A., **1**, 605 (1963).
492. Loy, B.R. *J. Polymer Sci.*, A-1, 225 (1963).
493. Wall, L.A. *J. Polymer Sci.*, **17**, 141 (1955).
494. Buck, T. PhD thesis, North Western University, 1960. Microfilms MIC 60–4761, Ann Arbor. Michigan.
495. Forrestal, L.J., Hodgson, W.G. *J. Polymer Sci.*, A-2, 1275 (1964).

496. Ohnishi, S., Sugimoto, S., Nitta, I. *J. Polymer Sci.*, A-1, 625 (1963).
497. Kusumoto, N., Matsumoto, K., Tabayagni, M. *Polymer Sci.*, A-1, **7**, 1773 (1969).
498. Ooi, T., Schiotsubo, M., Hama Y., Shinokara, K. *Polymer*, **16**, (7), 510 (1975).
499. Florin, R.E., Wall, L.A., Brown, D.W. *J. Polymer Sci.*, A-1, 1521 (1963).
500. Bullock, A.T., Cameron, G.G., Smith, P.M. *Polymer*, **14**, 525 (1973).
501. Florin, R.E., Wall, L.A. *J. Chem. Phys.*, **57**, (4), 1791 (1972).
502. Ouchi, I. *J. Polymer Sci.*, A-3, 2685 (1965).
503. Liebman, S.A., Renwer, J.F., Gollatz, K.A., Nauman, C.D. *J. Polymer Sci.*, A-1, **9**, 1823 (1971).
504. Bowden, M.J., O'Donnell, J.H. *J. Polymer Sci.* A-1, 7, 1665 (1969).
505. Sakai, Y., Iwasaki, M. *J. Polymer Sci.*, A-1, **7**, 1749 (1969).
506. Harris, J.A., Horojosa, O., Author, J.C. *J. Polymer Sci.*, A-1, **11**, 3215 (1973).
507. Gueskens, G., David, C. *Makromol. Chem.*, **165**, 273 (1973).
508. Yashoika, H., Matsumotoh, H., Uno, S., Higashide, F. *J. Polymer Sci.*, A-1, **14**, 1331 (1976).
509. Sakaguchi, M., Kodama, S., Ediund, O., Sohma, J. *J. Polymer Sci. Polymer Lett. Ed.*, **12**, 609 (1974).
510. Bullock, A.T., Cameron, G.G., Elson, J.M. *Polymer*, **15**, 74 (1974).
511. Michel, R.E., Chapman, F.W., Mao, T.J. *J. Polymer Sci.*, A-1, **5**, 1077 (1967).
512. Sakaguchi, M., Yamakawa, H., Sohmo, J. *J. Polymer Sci. Polymer Lett. Ed.*, **12**, 193 (1974).
513. Hama, Y., Shinohara, K. *J. Polymer Sci.*, A-1, **8**, 651 (1970).
514. Clay, M.R., Charlesby, A. *European Polymer J.*, **11**, (2), 187 (1975).
515. Placek, J., Szocs, F., Borsig, E.J. *J. Polymer Sci.*, A-1, **14**, 1549 (1976).
516. Shioji, J., Ohnishi, S., Nitta, I. *J. Polymer Sci.*, A-1, 3373 (1963).
517. Fisher, H., Hellwege, R.H. *J. Polymer Sci.*, **56**, 33 (1962).
518. Fisch, M.H., Dannenberg, J. *J. Anal. Chem.*, **49**, (9), 1405 (1977).
519. Yur'eva, F.A., Guseinova, F.O., Portyanskii, A.E., Seidov, N.M., Mamedova, V.M., Abasov, A.T., Malova, H.G. *Vysokomol Soedin. Ser A.*, **19**, (10) 2401 (1977); *Chem. Abstr.*, **87**, 202315p (1977).
520. Popov, V.P., Duvanova, A.P. *Zh. Prikl Spektrosk.*, **22**, (6), 1115 (1975); *Chem. Abstr.*, **83**, 115221d (1976).
521. Gol'benberg, A.L. *Zh. Prikl. Spektrosk.*, **19**, (3), 510 (1973).
522. Manlus, G.G. *J. Polymer Sci.*, C8, 137 (1965).
523. Barrall, E.M., Porter, R.S., Johnson, J.F. *J. Chromatog.*, **11**, 177 (1963).
524. Barrall, E.M., Porter, R.S., Johnson, J.F. *Anal. Chem.*, **35**, 73 (1963).
525. Porter, R.S., Hoffman, A.S., Johnson, J.F. *Anal. Chem.*, **34**, 1179 (1962).
526. Grasley, M.H., Barnum, E.R. Hartinez Research Laboratory, Shell Chemical Co. Ltd, private communication.
527. Wasilewska, W., Grzywa, E., Rajkiewicz, M. *Chem. Anal. (Warsaw)*, **19**, (1), 89 (1974).
528. Kubic, I., Singbar, M., Navratil, M. *Petrochemia*, **17**, (1-2) 10 (1977); *Chem. Abstr.*, **88**, 51308f (1978).
529. Kubic, I., Singbar Navratil, M. *Petrochemia*, **17**, (1-2), 15 (1977).
530. Lucas, R., Kolinsky, M., Doskocilova, D. *J. Polymer Sci. Polymer Chem. Ed.*, **16**, (5), 889 (1978).
531. Chujo, S., Satoh, S., Ozeka, T., Nagai, E. *J. Polymer Sci.*, **61**, 512 (1962).
532. Chujo, R., Satoh, S., Nagai, E. *J. Polymer Sci.*, A-2, 895 (1964).
533. Paul, S., Ranby, B. *Anal. Chem.*, **47**, (8), 1428 (1975).
534. Johnson, D.E., Lyerla, J.R., Horikawa, T.T., Pederson, L.A. *Anal. Chem.*, **49**, (1), 77 (1977).
535. Cross, C.K., Mackay, A.C. *J. Amer. Oil Chem. Soc.*, **50**, (7), 249 (1973).
536. Biyer, P., Padhye, M.R. Silk Rayon Ltd, India, **16**, (5), 174 (1973).
537. Munteanu, D., Toader, M. *Mater. Plast. (Bucharest)*, **13**, (2) 97 (1976); *Chem. Abstr.*, **85**, 19345v (1976).
538. Manett, S., Horowicz, S. *Anal. Chem.*, **52**, 1529 (1980).
539. Manlus, G.G. *J. Polymer Sci.*, **62**, 263 (1962).
540. Nishoika, A., Kato, Y., Ashikari, N. *J. Polymer Sci.*, **62**, 510 (1962).
541. Nishoika, A., Kato, Y., Mitsuoka, H. *J. Polymer Sci.*, **62**, 59 (1962).
542. Bovey, F.A. *J. Polymer Sci.*, **62**, 197 (1962).
543. Ito, K., Yashimata, Y. *J. Polymer Sci.*, B-3, 625 (1965).
544. Harwood, H.J., Ritchey, W.M. *J. Polymer Sci.*, B-3, **419** (1965).

545. Kranz, D., Dinges, K., Wendling, P. *Angew Makromol. Chem.*, **51**, (1), 25 (1976).
546. Ozawa, T. *Bull. Chem. Soc. Japan*, **38**, 1881 (1965).
547. Flynn, J.H., Wall, L.A. *Polyme Lett.*, **4**, 323 (1966).
548. Doyle, C.D. *J. Appl. Polymer Sci.*, **5**, 285 (1961).
549. Zsako, J., Zsako, J. Jr. *J. Therm. Anal.*, **19**, 333 (1980).
550. Basch, A., Lewin, M. *J. Polymer Sci. Polymer Chem. Ed.*, **11**, 3071 (1973).
551. Basch, A., Lewin, M. *J. Polymer Sci. Polymer Chem. Ed.*, **11**, 3095 (1973).
552. Basch, A., Lewin, M. *J. Polymer Sci. Polym.*, **11**, 1707 (1973).
553. Dollimore, D., Holt, B. *J. Polymer Sci. Polymer Phys. Ed.*, **11**, 1703 (1973).
554. Varma, D.S., Narasimhan, V. *J. Appl. Polymer Sci.*, **16**, 3325 (1972).
555. Funt, J.M., Magill, J.H. *J. Polymer Sci. Polymer Phys. Ed.*, **12**, 217 (1974).
556. Kokta, B.V., Valade, J.L., Martin, W.N. *J. Appl. Polymer Sci.*, **17**, 1 (1973).
557. Judd, M.D., Norris, A.C. *J. Therm. Anal.*, **5**, (2), 179 (1973).
558. Tulupov, P.E., Karpov, O.N. *Zh. Fiz. Khim.*, **47**, (6), 1420 (1973).
559. Joesten, B.L., Johnston, N.W. *J. Macromol. Sci. Chem.*, **8**, (1), 83 (1974).
560. Halip, V., Stan, V., Biro, A., Radovici, R. *Mater. Plast. (Bucharest)*, **10**, (11), 601 (1973).
561. Cooper, D.R., Sutton, G.J., Tighe, B.J. *J. Polymer Sci. Polymer Chem. Ed.*, **11**, 2045 (1973).
562. Patterson, A., Sutton, G.J., Tighe, B.J. *J. Polymer Sci. Polymer Chem. Ed.*, **11**, 2343 (1973).
563. Sutton, G.J., Tighe, B.J. *J. Polymer Sci. Polymer Chem. Ed.*, **11**, 1069 (1973).
564. Varma, D.S., Ravisankar, S. *Angew, Makromol. Chem.*, **28**, 191 (1973).
565. Kromalte, R., Malegina, N.D., Kotov, B.V., Oksent'evich, L.A., Pravednikov, A.N. *Vysokomol. Soedin, Ser. A.*, **14**, (10), 2148 (1972).
566. Barrales-Rienda, J.M., Ramos, J.G. *J. Polymer Sci. Symp., No. 42*, 1249 (1973).
567. Hodd, K.A., Holmes-Walker, W.A. *J. Polymer Sci. Symp. No. 42*, 1435 (1973).
568. Gedemer, T.J. *J. Macromol Sci. Chem.*, **8**, (1), 95 (1974).
569. Kiran, E., Gillham, J.K., Gipstein, E.J. *Makromol. Sci-Phys. B*, 9, (2), 341 (1974).
570. Ahlstrom, D.II., Foltz, C.R. *J. Polymer Sci.*, A-1, **16**, 2703 (1978).
571. Gedemer, T.J. *J. Macromol. Sci. Chem.*, **8**, (1), 95 (1974).
572. Dassanayake, N.L., Phillips, R.W. *Anal. Chem.*, **56**, 1753 (1984).
573. Clampett, B.H. *Anal. Chem.*, **35**, 1834 (1963).
574. Clampett, B.H. *J. Polymer Sci.*, A-3, 671 (1965).
575. Bambaugh, K.J., Clampett, B.H. *J. Polymer Sci.*, A-3, 805 (1965).
576. Geacinfov, C., Schotland, R.S., Miles, R.B. *J. Polymer Sci.* C-6, 197 (1964).
577. Schwenken, F.R.F., Zuccarello, R.K. *J. Polymer Sci.*, C-61 (1964).
578. Donald, H.J., Humes, E.S., White, L.W. *J. Polymer Sci.*, C-6, 93 (1964).
579. Marx, C.L., Cooper, S.L. *Makromol. Chem.*, **168**, 339 (1973).
580. Wunderlich, B., Bodily, D.M. *J. Appl. Polymer Sci.*, C-6, 137 (1964).
581. Kamida, K., Yamaguchi, K. *Makromol. Chem.*, **162**, 205 (1972).
582. Duswalt, A.A., Cox, W.W. *Polymer Characterization: Inter disciplinary Approaches*, (Craven C.D., ed.). Plenum Press, New York, p. 147 (1971).
583. Marchetti, A., Martuscelli, E. *J. Polymer Sci. Polymer Phys. Ed.*, **12**, 1649 (1974).
584. Miller, G.W. *Thermochimica Acta*, **8**, (1–2), 129 (1974).
585. Smith, O.F. *Anal. Chem.*, **35**, 1835 (1963).
586. Holden, H.W. *J. Polymer Sci.*, C-6, 209 (1964).
587. Clampitt, B.H., Hughes, R.H. *J. Polymer Sci.*, C-6, 43 (1964).
588. Holden, H.W. *J. Polymer Sci.*, C-6, 53 (1964).
589. Dick, N., Westerberg, C.J. *Macromol Sci. Chem. A*, **12**, (3), 455 (1978).
590. Poshet, G. *Kunstst-Plast. (Solothurn, Switz.)*, **25**, (1), 24 (1978).
591. Reich, L. *Thermochimica Acta*, **5**, 433 (1973).
592. Kotoyori, T. *Thermochimica Acta*, **5**, 51 (1972).
593. Guseinov, T.I., Seldov, N.M., Abasov, A., Kasumov, K. *Mater. Vses. Soveshch Relaksatsionnym Yaulenlyum Polim.* 2nd, 1971, **2**, 218–24 (1974).
594. Illers, K.H. *Europ. Polymer J.*, **10**, (10), 911 (1974).
595. Peppas, N.A. *J. Appl. Polymer Sci.*, **20**, (6), 1715 (1976).
596. Bosch, K. *Beitr Gerichtl Med.*, **33**, 280 (1975).
597. Caillot, C., Fournie, R., Alidoux, C. *Briver-Rev. Gen. Therm.* **11**, (125), 461 (1972).
598. Lemstra, P.J., Schouten, A.J., Challa, G. *J. Polymer Sci. Polymer Phys Ed.*, **10**, 2301 (1972).
599. Lemstra, P.J., Schouten, A.J., Challa, G. *J. Polymer Sci. Polymer Phys. Ed.*, **12**, 1565 (1974).

600. Lety, A., Noel, C. *J. Chim. Phys. Physicochim. Biol.*, **69**, (5), 875 (1972).
601. Gedemer, T.J. *Plast. Eng.*, **31**, (2), 28 (1975).
602. Duswalt, A.A. *Thermochimica Acta*, **8**, 57 (1974).
603. Kimuru, M., Hatakeyama, T., Nakano, J. *J. Appl. Polymer Sci.*, **18**, 3069 (1974).
604. Era, V., Venalainen, H. *J. Polymer Sci. Symp., No.* **42**, 879 (1973).
605. Heyns, H., Heyer, S., in *Thermal Analysis*, vol. 3 (Wiedermann H.G., ed.). Birkhaeuser, Basel, Switzerland p. 341 (1972).
606. Johnsen, U., Spilgies, G. *Kolloid-Z Z. Polym.*, **250**, (11–12), 1174 (1972).
607. Gilbert, M., Hybart, F. *J. Polymer*, **15**, 407 (1974).
608. Ceccorulli, G., Manescaichi, F. *Makromol Chem.*, **168**, 303 (1973).
609. Savolainen, A. *J. Polymer Sci. Symp., No.* **42**, 885 (1973).
610. Ebdon, J.R., Hunt, B.J. *Anal. Chem.*, **45**, 1401 (1973).
611. Godard, P., Mercier, J.P. *J. Appl. Polymer Sci.*, **18**, 1493 (1974).
612. Malavasic, T., Vizovisek, I., Lananje, S., Moze, A. *Makromol. Chem.*, **175**, 873 (1974).
613. Moze, A., Vizovisek, I., Malavasic, T., Cernec, F., Lapanje, S. *Makromol. Chem.*, **175**, 1507 (1974).
614. Heinen, K.U., Hummel, D.O. *Kolloid-Z. Z. Polym.*, **251**, 901 (1973).
615. Sebenik, A., Vizovisek, I., Lapanje, S. *Europ. Polymer J.*, **10**, 273 (1974).
616. Barton, J.M. *J. Macromol Sci. Chem.*, **8**, (1), 25, (1974).
617. Barton, J.M. *Makromol Chem.*, **171**, 247 (1973).
618. Crane, L.W., Dynes, P.J., Kaelble, D.H. *J. Polymer Sci. Polymer Lett. Ed.*, **11**, 533 (1973).
619. Sacher, E. *Polymer*, **14**, 91 (1973).
620. Goldfarb, L., Foltz, C.R., Messersmith, D.C. *J. Polymer Sci. Polymer Chem. Ed.*, **10**, 3289 (1972).
621. Van, K.V., Malhotra, S.L., Blanchard, L.P. *J. Macromol. Sci. Chem.*, **8**, 843 (1974).
622. Slysh, R., Hettinger, A.C., Guyler, K.E. *Polymer Eng. Sci.*, **14**, 264 (1974).
623. Willard, P.E. *J. Macromol. Sci. Chem.*, **8**, (1), 33 (1974).
624. Willard, P.E. *SPEJ*, **29**, 38 (1973).
625. Dunlap, L.H., Foltz, C.R., Mitchell, A.G. *J. Polymer Sci. Polymer Phys. Ed.*, **10**, 2223 (1972).
626. Gervais, M., Gallot, B. *Makromol. Chem.*, **171**, 157 (1973).
627. Johnston, N.W. *J. Macromol Sci. Chem.*, **7**, (2), 531 (1973).
628. Johnston, N.W. *Macromolecules*, **6**, 453 (1973).
629. Frosini, V., Magagnini, P.L., Newman, B.A. *J. Polymer Sci. Polymer Phys. Ed.*, **12**, 23 (1974).
630. Suwa, T., Takehisa, M., Machi, S. *J. Appl. Polym. Sci.*, **17**, 3253 (1973).
631. Ahad, E. *J. Appl. Polymer Sci.*, **18**, 1587 (1974).
632. Radhakrishnan, N.G., Padhye, M.R. *Angew Makromol. Chem.*, **43**, (1), 177 (1975).
633. Coppola, G., Filippini, R., Pallesi, B. *Polymer*, **16**, (7), 546 (1975).
634. Kamide, K., Imanaka, A. *Kobunshi Ronbunshu*, **32**, (9), 537 (1975).
635. Peppas, N.A., Merrill, E.W. *J. Appl. Polymer Sci.*, **20**, (6), 1457 (1976).
636. Era, V.A., Jauhiainen, T. *Angew Makromol. Chem.*, **43**, (1), 157 (1975).
637. Sickfield, J., Heinze, B. *J. Therm. Anal.*, **6**, (6), 689 (1974).
638. Sourour, S., Kamal, M.R. *Thermochim. Acta*, **14**, (1–2), 41 (1976).
639. Kay, R., Westwood, A.R. *Europ. Polymer J.*, **11**, (1), 25 (1975).
640. Frederick, W.J. Jr, Mentzer, C.C. *J. Appl. Polymer Sci.*, **19** (7), 1799 (1975).
641. Haberfeld, J.L., Reffner, J.A. *Soc. Plast. Eng. Tech. Pap.*, **21**, 585 (1975).
642. Pope, D.P. *J. Polymer Sci. Polymer Phys. Ed.*, **14**, (5), 811 (1976).
643. Macallum, J.R. *Makromol. Chem.*, **83**, 137 (1965).
644. McNeill, I.C. *J. Polymer Sci.*, A-1, **4**, 2479 (1966).
645. Mehmet, Y., Roche, R.S. *J. Appl. Polymer Sci.*, **20**, (7), 1955 (1976).
646. Grasselli, J.G. *Anal. Chem.*, **55**, 1220 (1983).
647. Zelenev, Yu. V., Kardanov, Kh. K. *Zavod Lab.*, **38**, (10), 1277 (1972).
648. Maruta, M., Ito, K., Kunimatsu, Y., Yamada, K. *Therm. Anal. (Proc. Int. Conf.)*, 483 (1977).
649. Gillen, K.T.J. *Appl. Polymer Sci.*, **22**, (5), 1291 (1978).
650. Yano, S. *J. Appl. Polymer Sci.*, **21**, (10), 2645 (1977).
651. O'Mara, M.M. *J. Polymer Sci.*, A-1, **8**, 1887 (1970).
652. O'Mara, M.M. *J. Polymer Sci.*, A-1, **9**, 1387 (1971).
653. Watson, S.T., Biemann, K. *Anal. Chem.*, **36**, 1135 (1964).
654. Iida, T., Nakánishi, M., Goto, K. *J. Polymer Sci.*, **12**, 737 (1974).

655. Coloff, S.G., Vanderborgh, N.E. *Anal. Chem.*, **45**, 1507 (1973).
656. Schmitt, C.R. *J. Fire Flammability*, **3**, 303 (1972).
657. Risby, T.H., Yergey, J.A. *Anal. Chem.*, **54**, 2228 (1982).
658. Foti, S., Liguori, A., Moravigra, P., Montando, G. *Anal. Chem.*, **54**, 647 (1982).
659. Moe, G.J. *Thermochimica Acta*, **10**, (3), 259 (1974).
660. Hughes, J.C., Wheals, B.B., Whitehouse, M.J. *Analyst (London)*, **102**, 143 (1977).
661. Joseph, K.T., Browner, R.F. *Anal. Chem.*, **52**, 1083 (1980).
662. Abney, G., Head, B.C., Poller, R.C. *J. Polymer Sci. Macro-mol. Rev.*, **8**, 1 (1974).
663. Danforth, J.D., Takeuchi, T.J. *Polymer Sci.*, A-1, **11**, 2091 (1973).
664. Guyot, A., Bert, A., Spitz, R. *J. Polymer Sci.*, A-1, **8**, 1596 (1970).
665. Shibazaki, Y., Kamebe, H. *Kabish Kayaku*, **21**, 65 (1964).
666. Beckewitz, F., Housinger, H. *Angew Makromol. Chem.*, **46**, (1), 143 (1975).
667. Wall, L.A. *J. Elastoplast*, **5**, 36 (1973).
668. Mitera, J., Kubelka, V., Novak, J., Mosteckey, J. *Plasty Kavc*, **14**, (1), 18 (1977); *Chem. Abstr.*, **86**, 172232u (1977).
669. Chaigneau, M. *Analysis* **5** (5), 223 (1977).
670. Grassie, N., Bain, D.R. *J. Polymer Sci.*, A-1, **8**, 2683 (1970).
671. Grassie, N., Bain, D.R. *J. Polymer Sci.*, A-1, **8**, 2669 (1970).
672. Schmitt, C.R. *J. Fire Flammability*, **3**, 303 (1972).
673. Beachel, C., Beck, D.L. *J. Polymer Sci.*, A-3, 457 (1965).
674. Seeger, M., Gritter, R.J. *J. Polymer Sci.*, A-1, **15**, 1393 (1977).
675. Tsuchiya, A., Sumi, K. *J. Polymer Sci.*, A-1, **7**, 1599 (1969).
676. Moiseeva, U.D., Nieman, M.H. *Vu Soed*, **3**, 1383 (1961).
677. Wall, L.A. *Soc. Petrol Engrs. J.*, **16**, 1 (1960).
678. McGaugh, M.C., Kottle, S. *J. Polymer Sci.*, A-1, **6**, 1243 (1968).
679. Grassie, N., Torrance, B.D.J. *J. Polymer Sci.*, A-1, **6**, 3303 (1968).
680. Grassie, N., Torrance, B.D.J. *J. Polymer Sci.*, A-1, **6**, 3315 (1968).
681. Nagasawa, M., Holtzer, A. *J. Amer. Chem. Soc.*, **86**, 538 (1964).
682. Troitskii, B.B., Varyukhin, V.A., Khokhlova, L.V. *Trudy Khim, Khim Tekhnol*, **2**, 115 (1974).
683. Gilland, J.C., Lewis, J.S. *Angew Makromol. Chem.*, **54**, (1), 49 (1976).
684. Ehlers, G.F.C., Fisch, K.R., Powell, W.R. *J. Polymer Sci.*, A-1, **7**, 2969 (1969).
685. Ehlers, G.F.C., Fisch, K.R., Powell, W.R. *J. Polymer Sci.*, A-1, **7**, 2955 (1969).
686. Ehlers, G.F.C., Fisch, K.R., Powell, W.R. *J. Polymer Sci.*, A-1, **7**, 2931 (1969).
687. Clark, J.E., Jellinek, H.H.G. *J. Polymer Sci.*, A, **3**, 1171 (1965).
688. Ny, T.H., Williams, H.L. *Makromol. Chem.*, **182**, 3323 (1981).
689. Bambough, K.J., Clampitt, B.H. *J. Polymer Sci.*, A, **3**, 805 (1965).
690. Schole, R.G., Bednarczyk, J., Yamanchi, T. *Anal. Chem.*, **38**, 331 (1966).
691. Neiman, M.P. *Ageing and Stabilization of Polymers* (translated from Russian). Consultants Bureau, New York, chapter 4 (1965)
692. Paulik, J., Paulik, F. *O.I.I. (Glas-Ontrum-Tech) Fachz Lab.* **16**, (9), 1043 (1972).
693. Meyer, C.S. *Ind. Eng. Chem.*, **44**, 1095 (1952).
694. Carlsson, P.J., Kato, Y., Wiles, D.N. *Macromolecules*, **1**, 459 (1968).
695. Scheehan, W.C., Cole, T.B. *J. Appl. Polymer Sci.*, **8**, 2359 (1964).
696. Ross, S.E. *J. Appl. Polymer Sci.*, **9**, 2729 (1965).
697. Adams, J.H., Goodrich, J.E. *J. Polymer Sci.*, A-1, **8**, 1269 (1970).
698. Purdon, J.R., Mate, R.D. *J. Polymer Sci.*, B, **1**, 451 (1963).
699. Purdon, J.R., Mate, R.D. *J. Polymer Sci.*, A-1, **8**, 1306 (1970).
700. Rogozinski, M., Kramer, M. *J. Polymer Sci.*, A-1, **10**, 3110 (1972).
701. Turner, R.R., Carlson, P.W., Altenau, A.G. *J. Elastomers Plast.*, **9**, 94 (1974).
702. Nakajima, A. *Chem. High Polymers Japan, Kobunski Kagaku*, **7**, 64 (1950).
703. Nakajima, A., Fujiwara, H. *High Polymers Japan, Kobunski Kaguku*, **37**, 909 (1964).
704. Wijga, P.W.O., Van Schooten, J., Beerma, J. *Makromol. Chemie*, **36**, 115 (1960).
705. Kenyon, A.C., Salyer, C.O., Kirz, J.E., Brown, O.R. *J. Polymer Sci.*, C-8, 205 (1965).
706. Luvric, L.J., Grubisic-Gallot, Z. Kunst, B. *Europ. Polymer J.*, **12**, (3), 189 (1976).
707. Kolke, V., Billmeyer, F.W. *J. Polymer Sci.*, C-8, 217 (1965).
708. Akutin, M.S., Goldberg, V.M., Lavruchin, F.G. *Vysokmore. Soedin. Ser., A.*, **19**, (5), 1113 (1977).

709. Ogawa, T., Hoshino, S. *J. Appl. Polymer Sci.*, **17**, 2235 (1973).
710. Ogawa, T., Tanaka, S., Inaba, T. *J. Appl. Sci.*, **18**, 1351 (1974).
711. Loconti, J.D., Cahill, J.W. *J. Polymer Sci.*, **49**, 152 (1961).
712. Loconti, J.D., Cahill, J.W. *J. Polymer Sci.*, A-1, 3163 (1963).
713. Ruskin, A.M., Parravano, G. *J. Appl. Polymer Sci.*, **8**, 565 (1964).
714. Bryson, A.P., Hawke, J.G., Parts, A.G. *J. Polymer Sci.*, **12**, 1323 (1974).
715. Kalal, J., Marousek, V., Svec, F. *Sb Vys, Sk Chem. Technol. Praze Org. Chem. Technol.*, **C22**, 57 (1975); *Chem. Abstr.*, **83**, 132443v (1975).
716. Gal'perin, V.M., Zak, A.G., Kuznetsov, N.A., Roganova, Z.A., Smolyanskii, A.L. *Vysokomol. Soedin. Ser. A.*, **17**, (3), 575 (1975).
717. Kossler, L.D., Vodchnal, J. *J. Polymer Sci.*, A-3, 2511 (1965).
718. Baker, C.A., Williams, R.J.P. *J. Chem. Soc. (London)*, **2**, 852 (1965).
719. Sutherland, J.E. *Research Laboratories Polymer Prep.*, R **17**, (2) (1976).
720. Teramachi, S., Fukao, T. *Polymer J.*, **6**, (6), 532 (1974).
721. Kuzaev, A.I., Susiova, E.N., Entelis, S.G. *Dokl. Akad. Nauk. S.S.R.*, **208**, (1), 142 (1973).
722. Chang, M.S., French, D.M., Rogers, P.L. *J. Macromol. Sci. Chem.*, **7**, (8), 1727 (1973).
723. Law, R.D., *J. Polymer Sci. Polymer Chem. Ed.*, **11**, 175 (1973).
724. Hori, F., Ikada, Y., Sakurada, J.J. *J. Polymer Sci. Polymer Chem. Ed.*, **13**, 755 (1975).
725. Peebles, L.H., Huffman, M.W., Ablett, C.T. *J. Polymer Sci.*, A-1, 479 (1969).
726. Repina, L.P., Khalatur, P.G. *Zavod Lab.*, **41**, (3), 287 (1975); *Chem. Abstr.*, **83**, 60334n (1975).
727. Katada, T., White, J.L. *Macromolecules*, **7**, (1), 106 (1974).
728. Gloeckner, G., Kahle, D. *Plast. Kautsch*, **23**, (8), 577 (1976).
729. Gloeckner, G., Kahle, D. *Plast Kautsch*, **23**, (5), **338** (1976).
730. Vakhtina, I.A., Khrenova, R.F., Tarankov, O.G. *Zh-Anal. Khim.*, **28**, (8), 1625 (1973).
731. Hori, F., Ikada, Y., Sakurada, I.J. *J. Polymer Sci. Polymer Chem. Ed.*, **13**, 755 (1955).
732. Buter, R., Tan, Y.Y., Challa, G. *Polymer*, **14**, 171 (1973).
733. Belen'skii, B.G., Gankina, E.S., Nefedov, P.P., Lazareva, M.A., Savitskaya, T.S., Voichikhina, M.D. *J. Chromatogr.*, **108**, (1), 61 (1975).
734. Inagaki, H., Kamiama, F. *Macromolecules*, **6**, (1), 107 (1973).
735. Kotaka, T., Uda, T., Tanaka, T., Inagaku, H. *Makromol. Chem.*, **176**, (5), 1273 (1975).
736. Geymer, D.O. Shell Chemical Co. Ltd., Emeryville, California, private communication.
737. Inagaki, H., Miyamoto, T., Kamiyama, F.J. *Polymer Sci.*, B-7, 329 (1969).
738. Benningfield, L.V. 'A dielectric constant detector for liquid chromatography and its applications.' Paper presented at the 30th Pittsburgh Conference on Analytical Chemistry and Applied Spectroscopy, Cleveland, Ohio, 5–9 March, 1979, p. 123 (1979).
739. Benningfield, L.V., Mowry, R.A. *J. Chromatogr. Sci.*, **19**, 115 (1981).
740. Fuller, E.N., Porter, G.T., Roof, L.B. *J. Chromatogr. Sci.*, **17**, 661 (1979).
741. Roof, L.B., Porter, G.T., Fuller, E.N., Mowrey, R.A. 'On-line Polymer Analysis by Liquid Chromatography'. Instrumentation and Automation in the Paper, Rubber, Plastics and Polymerization Industries. 4th TFAC Conference, Ghent, Belgium, 3–5 June 1980. (A. Van Canwenberghe, ed.) Pergamon Press, New York, pp. 47–53 (1980).
742. Ross, J.H., Shank, R.C. *Adv. Chem. Ser.*, **125**, 108 (1971).
743. Bode, R.K., Benningfield, L.V., Mowrey, R.A., Fuller, E.N. *International Laboratory*, **40**, Nov./Dec. (1981).
744. Grubisic, Z., Rempp, P., Benoit, H. *J. Polymer Sci. B*, **5**, **753** (1967).
745. Moore, J.C., Hendrickson, J.G. *J. Polymer Sci. C*, **8**, 233 (1985).
746. Mori, S.J. *Chromatography*, **157**, 75 (1978).
747. Mori, S. *J. Appl. Polymer Sci.*, **18**, 2391 (1974).
748. Balke, S.T., Hamielec, A.E., Leclair, B.P., Pearce, S.L. *Ind. Eng. Chem. Prod. Res. Dev.*, **8**, 54 (1969).
749. Loy, B.R. *J. Polymer Sci. Polymer Chem. Ed.*, **14**, 2321 (1976).
750. Vrijbergen, R.R., Soeteman, A.A., Smit, J.A.M. *J. Appl. Polymer Sci.*, **22**, 1267 (1978).
751. McCrackin, F.L. *J. Appl. Polymer Sci.*, **21**, 191 (1977).
752. Mahabadi, H.K., O'Driscoll, K.F. *J. Appl. Polymer Sci.*, **21**, 1283 (1977).
753. Mori, S., Suzuki, T. *J. Liq. Chromatogr.*, **3**, 343 (1980).
754. Weiss, A.R., Cohn-Ginsberg, E. *J. Polymer Sci. B*, **7**, 379 (1969).
755. Belinskii, B.G., Nefedov, P.F. *Vysokomol. Soedin*, A-14, 1568 (1972).
756. Kalinsky, M., Janca, J. *J. Polymer Sci.*, A-1, **12**, 1181 (1974).

757. Morris, M.C. *J. Chromatogr.*, **55**, 203 (1971).
758. Hamielec, A.E., Omoridion, S.N.E. *ACS Symposium Series No. 138*, 183 (1980).
759. Mori, S. *Anal. Chem.*, **53**, 1817 (1981).
760. Mori, S., Yamakawa, A. *Liquid Chromatogr.*, **3**, 329 (1980).
761. Ogawa, T., Suzuki, Y., Inaba, T. *J. Polymer Sci.*, A-1,**10**, 737 (1972).
762. Crouzet, P., Fine, O., Mangin, P. Paper presented at the 5th International Seminar, London (1968). Preprint 10, *J. Appl. Polymer Sci.*, **13**, 1205 (1969).
763. Williamson, G.R, Cervanka, A. *Europ. Polymer J.* **8**, (8), 1009 (1972).
764. Nakajima, N. *Adv. Chem. Ser.*, 125, **98**, (1971) (unpublished 1973).
765. Eldarov, E.G., Goldberg, V.M., Parkratova, G.V., Akutin, M.S., Toptygin, D. Ya. *Zavod Lab.*, **40**, (3), 269 (1974).
766. Lovric, L. *Nafta (Zagreb)*, **23**, (12), 606 (1972).
767. Lovric, L. *Nafta (Zagreb)*, **24**, (1), 47 (1973).
768. Ross, J.H., Shank, R.L. *Adv. Chem. Ser.*, **125**, 108 (1971).
769. Maley, J. *Polymer Sci.*, C-8, 253 (1965).
770. Starck, P., Kantola, P. *Kem-Kemi*, **3**, (2), 100 (1976).
771. Gianotti, G., Gaita, A., Romanini, D. *Polymer*, **21**, 1087 (1980).
772. Maley, I.E. *J. Polymer Sci.*, C-8, 253 (1965).
773. Cooper, W., Vaughan, G., Eaves, D.E., Madden, R.W. *J. Polymer Sci.*, **50**, 159 (1961).
774. Ambler, M.R., Mate, R.D., Durden, J.R. *J. Polymer Sci.*, **12**, 1771 (1974).
775. Ogawa, T., Inaba, T. *J. Polymer Sci.*, **21**, (11), 2979 (1977).
776. Vaughan, M.F. *Ind. Polym. Charact. Mol. Weight Proc. Meeting*, **111**, (1973).
777. Winns, A.M., Swarin, S.J. *J. Appl. Polymer Sci.*, **19**, (5), 1243 (1975).
778. Mori, S. *Anal. Chem.*, **53**, 1813 (1981).
779. Otaka, E.P. *J. Chromatogr.*, **76**, (1), 149 (1973).
780. Uglea, C.V. *Makromol. Chem.*, **166**, 275 (1973).
781. Beden'kii, B.G., Gankina, E.S., Nefedov, P.D., Kuznetsova, M.A., Valchikhina, M.D. *J. Chromatogr.*, **77**, 209 (1973).
782. Kranz, D., Rahl, Hu, Baumann, H. *Angew Makromol. Chem.*, **26**, 67 (1972).
783. Tung, L.H., Runyon, J.R. *J. Appl. Polymer Sci.*, **17**, 1589 (1973).
784. Moore, J.C., Hendrickson, J.G. *J. Polymer Sci.*, C-8, 233 (1965).
785. Zhdanov, S.P., Belenkii, B.G., Nekdov, P.P., Koromal'di, E.V. *J. Chromatogr.*, **77**, (1), 149 (1973).
786. Moore, J.G. *J. Polymer Sci.*, A-2, 835 (1964).
787. Belen'kii, B.G., Gankina, E.S., Nefedov, P.P., Kuznetsova, M.A., Valchikhina, M.D. *J. Chromatogr.*, **77**, (1), 209 (1973).
788. Guenet, I.M. Gallot, Z., Picot, C., Bennett, D. *J. Appl. Polymer Sci.*, **21**, (8), 2181 (1977).
789. Chang, F.S.C. *Adv. Chem. Ser.*, **125**, 154 (1971) (published 1973).
790. Hoffmann, M., Urban, H. *Makromol. Chem.*, **178**, (9), 268 (1977).
791. Amber, M.R., Mate, R.D., Durdon, J.R. *J. Polymer Sci.*, **12**, 1771 (1974).
792. Stojanov, K., Shirazi, Z.H., Audu, T.O.K. *Chromatographia*, **11**, (5), 274 (1978).
793. Stojanov, K., Sharaji, Z.H., Audu, T.O.K. *Ber. Bunsenges Phys. Chem.*, **81**, (8), 767 (1976).
794. Regnier, F.E., Noel, R. *J. Chromatogr. Sci.*, **14**, 316 (1976).
795. Janca, J., Kolinsky, M. Ustav Makromol. Chem. Lesk. Akad. Ved. (Prague, Czeckoslovakia), *Plasty Kauc.*, **13**, (5), 138 (1976).
796. Daley, L.E. *J. Polymer Sci.*, C-8, 253 (1965).
797. Lin, F.T., Tung, L.Z., Liu, F.Y., Hsu, W.W. *J. Chim., Inst. Chem. Eng.*, **4**, 43 (1973).
798. Nakao, K., Kuramoto, K. *Nippon Secchaku Kyokal Shi.*, **8**, (4), 186 (1972).
799. Mori, S. *Anal. Chem.*, **53**, 1817 (1981).
800. Janca, J., Kolinsky, M. *J. Appl. Polymer Sci.*, **21**, (1), 83 (1977).
801. Revillon, A., Dumont, R., Guyot, A. *J. Polymer Sci. Polymer Chem. Ed.*, **14**, 2263 (1976).
802. Fritzsche, P., Klug, P., Grobe, V. *Nuova Chem.*, **49**, (6), 39 (1973).
803. Krans, D., Pohl, Hu., Baumann, H. *Angew Makromol. Chem.*, **26**, 67 (1972).
804. Mori, S. *Anal. Chem.*, **55**, 2414 (1983).
805. Cooper, D.R., Semylen, J.A. *Polymer*, **14**, 185 (1973).
806. Paschke, E.E., Bidingmeyer, B.A., Bergmann, J.C. *J. Polymer Sci. Polymer Chem. Ed.*, **15**, (4), 983 (1977).
807. Minarzk, M., Sir, Z., Coupek, J. *Angew Makromol. Chem.*, **64**, (1), 147 (1977).

808. Shiono, S. *Anal. Chem.*, **51**, 2398 (1979).
809. Meyerhoff, G. *J. Chromatogr. Sci.*, **9**, (10), 596 (1971).
810. Miyamoto, Y., Tomoshige, S., Inagaka, H. *Polymer J.*, **6**, (6), 564 (1974).
811. Grubisic, Z., Rempp, P., Benoit, H.J. *Polymer Sci.*, B, **5**, 753 (1967).
812. Weiss, A.R., Cohn-Ginsberg, E. *J. Polymer Sci.*, A-2, **8**, 148 (1970).
813. Belinski, B.G., Nefedov, P.P. *Vysokomol. Soedin.*, A-14, 1568 (1972).
814. Kolinsky, M., Janca, J. *J. Polymer Sci.*, A-1, **12**, 1181 (1974).
815. Janca, J., Vlcek, P., Trekoval, J., Kolinsky, M. *J. Polymer Sci. Polymer Chem. Ed.*, **13**, 1471 (1975).
816. Robertson, A.P., Cook, J.A., Gregory, J.T. *Kinetics Symposium Boston, Massachusetts*, pp. 258–273 (1972).
817. Moore, J., Hillman, D.E. *Brit. Polymer J.*, **3**, 259 (1971).
818. Kato, Y., Sasaki, H., Aiuru, M., Hashimato, T. *J. Chromatogr.*, **153**, (2), 546 (1978).
819. Berck, D., Nova, L. *Chem. Prumsyl*, **23**, (2), 91 (1973).
820. Vakhtina, I.T., Tarakanov, O.G. *Plast. Kaut*, **21**, (1), 28 (1974).
821. Taylor, W.C., Tung, L.H. Paper presented at 140th Meeting of American Chemical Society, Chicago, Illinois, September 1961 (also SPE transactions, p. 119. April 1962).
822. Morey, D.R., Tamblyn, J.W. *J. Appl. Phys.*, **16**, 419 (1945).
823. Claesson, S. *J. Polymer Sci.*, **16**, 193 (1955).
824. Gamble, L.W., Nipke, W.T., Lane, T.L. *J. Appl. Polymer Sci.*, **9**, 1503 (1965).
825. Wesslau, H. *Makromol. Chem.*, **20**, 111 (1956).
826. Beattie, W.H., private communication.
827. Gooberman, G. *J. Polymer Sci.*, **40**, 469 (1959).
828. Tanaka, S., Nakamura, A., Morikawa, H. *Die Makromol. Chem.*, **85**, 164 (1965).
829. Gruber, U., Elias, H.G. *Die Makromol. Chem.*, **78**, 58 (1964).
830. Taylor, W.C., Graham, J.P. *Polymer Letters*, **2**, 169 (1964).
831. Taylor, W.C., Tung, L.H. *SPE Society Plastics Engineers Trans.*, **2**, 119 (1963).
832. Gamble, L.W., Wipke, W.T., Lane, T. *J. Amer. Chem. Soc. Polymer Chem. Preprints*, **4**, No. 2, 162 (1963).
833. Wesslau, H. *Makromol. Chem.*, **20**, 111 (1956).
834. Klenin, V.I., Schegolev, S. Yu. *J. Polymer Sci. Polymer Symp.*, **42**, part 2, 965 (1973).
835. Hall, O. *Techniques of Polymer Characterization* (P.M. Allen, ed.). Academic Press, New York, chapter II (1959).
836. Beattie, W.H. *J. Polymer Sci.* A, **3**, 527 (1965).
837. Klenin, V.I., Padol'skii, A.F., Shchegolev, S. Yu., Shvortsburd, B.I., Petrova, N.E. *Vysokomol. Soedin. Ser.*, A, **16**, (5), 974 (1974).
838. Cantow, H.H., Kowalski, M., Krozer, C. *Angew Chem. Int. Ed. Engl.*, **11**, (4), 336 (1972).
839. Tashmuklamedov, S.A., Khasankhanova, M.N., Tillaev, R.S. *Uzb. Khim Zh.*, **17**, (2), 35 (1973).
840. Grechanovskii, V.A. *Vysokomol. Soedin.*, **17**, 2721 (1975).
841. Lanikova, J., Hlensek, H. *Chem. Prumsyl*, **23**, (6), 10 (1973).
842. Benicka, E., Ciganekova, V Sh Prednasek. *Makrotest 1973*, **2**, 64 (1973).
843. Vasile, C., Onu, A., Popa, O., Matel, T. *Mater. Plast. (Bucharest)*, **10**, (12), 631 (1973).
844. Case, L.C. *Makromol. Chem.*, **51**, 61 (1960).
845. Nakagawa, T., Nakata, I. *Kogyo Kagaku Zasshi*, **59**, 710 (1956); *Chem. Abstr.*, **52**, 4418d (1958).
846. Burger, K. *Z. Anal. Chem.*, **196**, 259 (1963).
847. Mikkelsen, L. *Characterization of High Molecular Weight Substances.* Pittsburgh Conference, Anal. Chem., 5–9 March (1962).
848. Puschmann, O. *Fette Seife Anstrichmittel*, **65** (1963).
849. Celedes Pacquot, C. *Rev. France Corps. Gras.*, **9**, 145 (1962).
850. Sweeley, C.C., Bentley, R., Makita, M., Wells, W.W. *J. Amer. Chem. Soc.*, **85**, 2497 (1963).
851. Fletcher, J.P., Persinger, H.E. *J. Polymer Sci.*, A-1, **6**, 1025 (1968).
852. Myers, L.W. Shell Research Ltd., Carrington, Cheshire, U.K., private communication.
853. Platonov, J.P., Belyaev, V.M., Grigor'eva, F.P. *Plast. Massy*, **4**, 77 (1975).
854. Stark, P., Kantola, P. *Kem-Kemi*, **3**, (2), 100 (1976).
855. Lanikova, J., Hlousek, M. *Chem. Prumsyl*, **27**, (12), 628 (1977).
856. Ross, J.H., Shank, R.L. *Ad. Chem. Ser.*, **125**, 108 (1971).

857. Chiang, R. *J. Polymer Sci.*, A-3, 3679 (1965).
858. Cuesta de la, Billmeyer, F.W. *J. Polymer Sci.*, A-1, 1721 (1963).
859. Das, N., Palit, S.R. *J. Polymer Sci.*, A-1, 11, 1025 (1973).
860. Nakajuima, A. *Chem. High Polymers, Japan, Kobunski Kagaku,* 7, 64 (1950).
861. Nakajuma, A., Fujiwara, H. *High Polymers Japan Kobunski Kagaku,* 37, 909 (1964).
862. Welsh, T., Engewald, W., Kowash, E. *Plaste u Kaut,* 23, 584 (1976).
863. Wozniak, T. *Pezegl Wlok,* 30, (1), 18 (1976).
864. Feldman, P., Ughea, C.W., Nuta, V., Popa, M. *Rev. Roum Chim (Bucharest),* 17, 1033 (1972).
865. Simonova, M.I., Alzenshtein, E.M. *Zavod Lab.,* 40, (4), 435 (1974).
866. Schmalz, E.O. *Fas Forsch Tex. Tech.,* 21, 209 (1970).
867. Ambler, M.R., Mate, R.D. *J. Polymer Sci.,* A-1, 10, (9), 2677 (1972).
868. Sakurada, I., Ikada, Y., Kawahara, T.J. *Polymer Sci.,* 11, 2329 (1973).
869. Berens, A.R. Paper delivered to the 168th American Chemical Society Meeting, Atlantic City, New Jersey, entitled 'The solubility of vinyl chloride in polyvinylchloride'. September 1974.
870. Gilbert, M., Hylart, F.J. *J. Polymer Sci., A-1,* 9, 227 (1971).
871. Takashima, K., Nakae, K., Shibata, M. *Macromolecules,* 7, (5), 641 (1974).
872. Kallistov, O.V. *Zavod Lab.,* 38, (6), 711 (1972).
873. Kamata, T., Nakahara, T. *J. Colloid Interface Sci.,* 43, (1), 89 (1973).
874. Dautzenberg, H. *J. Polymer Sci., C.,* 39, 123 (1972).
875. Pogorel'skii, K.V., Asanov, A., Akhmedov, K.S. *Doklady Akad. Nauk. Uzh. S.S.R.,* 27, (3), 28 (1970).
876. Bradley, J.H. *J. Polymer Sci., C-8,* 305 (1965).
877. Blair, W.E. *J. Polymer Sci.,* C-8, 287 (1965).
878. Suzuki, H., Leonis, C.G. *Brit. Polymer J.,* 5, (6), 485 (1973).
879. Kalinina, L.S., Motorina, H.A. U.S.S.R. Patent 340, 962 (6/5/72).
880. Mattern, D.E., Hercules, D.M. *Anal. Chem.,* 57, 2041 (1985).
881. Lattimer, R.P., Harmon, D.J. *Anal. Chem.,* 52, 10808 (1980).
882. Crompton, T.R. *Europ. Polymer J.,* 4, 473 (1968).
883. Lorenz, O., Scheele, W., Dummer, W. *Kautschuk, Gummi,* 7, 273 (1954).
884. Campbell, R.H., Wise, R.W. *J. Chromatog.,* 12, 178 (1963).
885. Slonaker, D.F., Sievers, D.C. *Anal. Chem.,* 36, 1130 (1964).
886. Van der Heide, R.F., Wouters, O. *Lebensm. Untersuch. Forsch.,* 117, 129 (1962).
887. Schroder, E., Rudolph, G. *Plaste Kautschuk,* 10, 22 (1963).
888. Metcalf, K., Tomlinson, R. *Plastics (London),* 25, 319 (1960).
889. Yushkevichyute, S.S., Shlyapnikov, Yu. A. *Plasticheskie Massy,* 1, 54 (1967).
890. Stafford, C. *Plasticheskie Massy,* 34, 794 (1962).
891. Korn, O., Woggon, H. *Plaste Kautschuk,* 11, 278 (1964).
892. Zilio-Grandi, F., Libralesso, G., Sassu, G., Sveglidao, G. *Mater. Plaste. Elast.,* 30, 643 (1964).
893. Brock, M.J., Louth, G.D. *Anal. Chem.,* 27, 1575 (1955).
894. Wandel, M., Tengler, H. *Fette Seifen Anstrichmittel,* 66, 815 (1964).
895. Varma, J.P., Suryanaraya, N.P., Sircar, A.K. *J. Sci. Ind. Res. India,* 20, 79 (1961).
896. Yuasa, T., Kamiya, K. *Japan Analyst,* 13, 966 (1964).
897. Varma, J.P., Suryanaraya, N.P., Sircar, A.L. *J. Sci. Ind. Res. India,* 21, 49 (1962).
898. Hilton, C.L. *Anal. Chem.,* 32, 1554 (1960).
899. Luongo, J.P. *Appl. Spectrosc.,* 19, (4), 117 (1965).
900. Miller, R.G.J., Willis, H.A. *Spectrochim Acta,* 14, 119 (1959).
901. Guichon, G., Henniker, J. *Brit. Plast.,* 37, 74 (1964).
902. Drushel, H.V., Sommers, A I.. *Anal. Chem.,* 36, 836 (1964).
903. Albarino, R.V. *Appl. Spectrosc.,* 27, 47 (1973).
904. Murakami, S., Tukumori, T., Tsurugi, J., Murata, N. *J. Chem. Soc. Japan, Ind. Chem. Sect.,* 67, 1161 (1964).
905. Hilton, C.L. *Anal. Chem.,* 32, 383 (1960).
906. British Standard 2782 Part 4, Method 405 D (1965).
907. Mayer, H. *Deut. Lebensm Rundshau,* 57, 170 (1961).
908. Glavind, J. *Acta Chem. Scand.,* 17, 1635 (1963).
909. Blois, M.S. *Nature (London),* 181, 1199 (1958).

910. Morgenthaler, L.P., unpublished work.
911. Cornish, P.J. *J. Appl. Polymer Sci.*, **7**, 727 (1963).
912. Fiorenza, A., Bonomi, G., Saredi, A. *Mater. Plast. Elast.*, **31**, 1045 (1965).
913. Auler, H. *Asbest. Kunststoffe*, **14**, 1024 (1961).
914. Kawaguchi, T., Ueda, K., Koga, A. *J. Soc. Rubber Ord. (Japan)*, **28**, 525 (1955).
915. Ruddle, L.H., Wilson, J.R. *Analyst (London)*, **94**, 105 (1969).
916. Hilton, C.L. *Rubber Age*, **84**, 263 (1958).
917. Burchfield, H.P., July, J.N. *Anal. Chem.*, **19**, 383 (1960).
918. Cieleszky, C., Nagy, F. *Z. Lebensm. Unter-Forsch*, **114**, 13 (1961).
919. Wexler, A.S. *Anal. Chem.*, **35**, 1926 (1963).
920. Parker, C.A., Barnes, W.J. *Analyst (London)*, **82**, 606 (1957).
921. Parker, C.A. *Anal. Chem.*, **37**, 140 (1961).
922. Kirkbright, G.F., Narayanswamy, R., West, T.S. *Anal. Chim. Acta*, **52**, 237 (1970).
923. Mocker, F. *Kautschuk Gummi*, **11**, 1161 (1964).
924. Mocker, F. *Kautschuk Gummi*, **12**, 155 (1959).
925. Mocker, F. *Kautschuk Gummi*, **13**, 91 (1960).
926. Mocker, F., Old, J. *Kautschuk Gummi*, **12**, 190 (1959).
927. Adams, R.N. *Rev. Polarog.*, **11**, 71 (1963).
928. Zweig, A., Lancaster, E., Neglia, M.T., Jura, W.H. *J. Amer. Chem. Soc.*, **86**, 413 (1964).
929. Vodzinskii, Yu, Semchikova, G.S. *T. Po Khim i Khim Teknol.*, **272** (1963).
930. Vermillion, F.J., Pearl, T.A. *J. Electrochem. Soc.*, **111**, 1392 (1964).
931. Barendrecht, E. *Anal. Chim. Acta*, **24**, 498 (1961).
932. Gaylor, V.F., Elving, P.J., Conrad, A.L. *Anal. Chem.*, **25**, 1078 (1953).
933. Gaylor, V.F., Conrad, A.L., Landerl, J.H. *Anal. Chem.*, **29**, 228 (1957).
934. Gaylor, V.F., Conrad, A.F., Landerl, J.H. *Anal. Chem.*, **29**, 224 (1957).
935. Hedenberg, J.F., Freiser, H. *Anal. Chem.*, **25**, 1355 (1953).
936. Hamilton, J.W., Tappel, A.L. *J. Amer. Oil Chem. Soc.*, **40**, 52 (1963).
937. Analytical Applications Report No. 637D, Southern Analytical, Camberley, Surrey, U.K.
938. Analytical Applications Report No. 651/2, Southern Analytical Ltd., Camberley, Surrey, U.K.
939. Budyina, V.V., Marinin, V.G., Vodzinskii, Yu.V., Kalinina, A.I., Korschunov, I.A. *Zavod. Lab.*, **36**, 1051 (1970).
940. Vasil'eva, A.A., Vodzinskii, Yu.V., Korshunov, I.A. *Zavod. Lab.*, **34**, 1304 (1968).
941. Budke, C., Bannerjee, D.K., Miller, F.D. *Anal. Chem.*, **36**, 523 (1964).
942. Ward, G.A., *Talanta*, **10**, 261 (1963).
943. Elving, P.J., Krivis, A.F. *Anal. Chem.*, **30**, 1645 (1958).
944. Elving, P.J., Krivis, A.F. *Anal. Chem.*, **30**, 1648 (1958).
945. Lintner, C.J., Schleif, R.H., Higuchi, T. *Anal. Chem.*, **22**, 534 (1950).
946. Yamaji, O., Yamashina, T. *J. Soc. Rubber Ind. (Japan)*, **35**, 284 (1962).
947. Sawada, M., Yamagi, I., Yamashina, T. *J. Soc. Rubber Ind. (Japan)*, **35**, 284 (1962).
948. Spell, H.L., Eddy, R.D. *Anal. Chem.*, **32**, 1811 (1960).
949. Schroder, E., Hagen, E. *Plaste Kautsch.*, **15**, 625 (1968).
950. Roberts, C.B., Swank, J.D. *Anal. Chem.*, **36**, 271 (1964).
951. Nosikov, Yu.D., Vetchinkina, U.N. *Neftekhimya*, **5**, 284 (1965).
952. Helf, C., Bockwan, D. *Plaste Kautschuk*, **11**, 624 (1964).
953. Pocaro, P.J. *Anal. Chem.*, **36**, 1664 (1964).
954. Freedman, R.W., Charlier, G.O. *Anal. Chem.*, **36**, 1880 (1964).
955. Duvall, A.H., Tully, W.F. *J. Chromatog.*, **11**, 38 (1963).
956. Brooks, V.T. *Chem. Ind.*, 1317 (1959).
957. Knight, H.S., Siegel, H. *Anal. Chem.*, **38**, 1221 (1966).
958. Grant, D.W., Vaughan, G.H. *Gas Chromatography*, p. 305. Butterworths, London (1962).
959. Long, R.E., Guvernator, G.C. *Anal. Chem.*, **39**, 1493 (1967).
960. Roberts, C.B., Swank, J.D. *Anal. Chem.*, **36**, 271 (1964).
961. Schroder, E., Rudolph, G. *Plaste Kautschuk*, **1**, 22 (1963).
962. Buttery, R.G., Stuckey B.N. *J. Agr. Food Chem.*, **9**, 283 (1961).
963. Styskin, E.L., Gurvich, Ya.A., Kumak, S.T. *Khim Prom.*, **5**, 359 (1973); *Zavod Lab.*, **39**, 27 (1973).
964. Hilton, C.L. *Rubber Chem. Technol.*, **32**, 844 (1959).

965. British Standard 2782, Part 4, Method 405B (1965).
966. Kabota, T., Kuribayashi, S., Furuhama, T.S. *Soc. Rubber Ind. (Japan)*, **35**, 662 (1962).
967. Gaeta, L.J., Schleuter, E.W., Altenau, H.G. *Rubber Age*, **101**, 47 (1969).
968. Wize, R.W., Sullivan, A.B. *Rubber Chem. Technol.*, **35**, No. 3, July–September (1962).
969. Wize, R.W., Sullivan, A.B. *Rubber Age*, **91**, 773 (1962).
970. *Bundesgesundheitschlatt*, **16**, (10), 155 (1973).
971. Lappin, G.R., Zannucci, J. *Anal. Chem.*, **41**, 2076 (1969).
972. Denning, J.A., Marshall, J.A. *Analyst (London)*, **97**, 710 (1972).
973. Nawakowski, A.C. *Anal. Chem.*, **30**, 1868 (1958).
974. Brandt, H.J. *Anal. Chem.*, **33**, 1390 (1961).
975. Kellum, G.E. *Anal. Chem.*, **43**, 1843 (1971).
976. Sedlar, J., Feniokova, E. Pac., J. *Analyst (London)*, **99**, 50 (1974).
977. Udris, J. *Analyst (London)*, **96**, 130 (1971).
978. Fossy, H., Lalet, P. *Chim. Analyst*, **52**, 1281 (1970).
979. Groagova, A., Pribl, M. *Z. Analyst Chem.*, **234**, 423 (1968).
980. Belpaire, F. *Revue Belg. Mat. Plast.*, **6**, 201 (1965).
981. Simpson, D., Currell, B.R. *Analyst (London)*, **96**, 515 (1971).
982. Mal'kova, L.N., Kalanin, A.I., Perepletchckova, E.M. *Zhur Analit. Khim*, **27**, 1924 (1972).
983. *Verfungungen und Mittalungen des Ministeriums for Gesund-heitwesan*, No. 2 (1965).
984. Schroeder, E., Thinius, K. *Deutsch Farben Z.*, **14**, 146 (1960).
985. Schroeder, E., Malz, S. *Deutsch Farben Z.*, **5**, 417 (1958).
986. Robertson, M.M., Rowley, R.M. *Brit. Plast.*, 26 January (1960).
987. Haslam, J., Soppet, W.W. *J. Chem. Soc.*, **67**, 33 (1948).
988. Doebring, H. *Kunststoffe*, **28**, 230 (1938).
989. Thinius, K. *Analytische Chemie der Plaste*. Springer, Berlin (1952).
990. Haslam, J., Squirrel, D.C.M. *Analyst (London)*, **80**, 871 (1955).
991. Wake, W.C. *The Analysis of Rubber and Rubber like Polymers*. McLaren, London, 1958.
992. Clarke, A.D., Bazill, G. *Brit. Plast.*, **31**, 16 (1958).
993. Haslam, J., Hamilton, J.B., Jeffs, A.R. *Analyst (London)*, **83**, 66 (1958).
994. Miller, D.F., Samsel, E.P., Cobler, J.G. *Anal. Chem.*, **33**, 677 (1961).
995. Masfield, P.B. *Chemy. Ind.*, **28**, 792 (1971).
996. Kuta, E.J., Quackenbush, F.W. *Anal. Chem.*, **32**, 1069 (1960).
997. Bukata, S.W., Zabrocki, L.L., McLaughlin, M.F. *Anal. Chem.*, **35**, 886 (1963).
998. Hyden, S. *Anal. Chem.*, **35**, 133 (1963).
999. Brammer, J.A., Frost, S., Reid, V.W. *Analyst (London)*, **92**, 91 (1967).
1000. Cook, W.S., Jones, C.O., Altenau, G. *Can. Spectrosc.*, **13**, 64, 71 (1968).
1001. Hank, W.W., Silverman, L. *Anal. Chem.*, **31**, 1069 (1959).
1002. Mitchell, B.J., O'Hear, H.J. *Anal. Chem.*, **34**, 1621 (1962).
1003. Bergmann, J.S., Ehrhart, C.H., Granatulli, L., Janik, J.L. 153rd National American Chemical Society Meeting, Miami Beach, Florida, April (1967).
1004. Krishen, A. *Anal. Chem.*, **43**, 1130 (1971).
1005. Smuts, T.W., Van Niekerk, F.A., Pretorius, V.J. *Gas Chromatog.*, **5**, 190 (1967).
1006. Giddings, J.C. *Anal. Chem.*, **35**, 2215 (1963).
1007. Giddings, J.C. *Anal. Chem.*, **37**, 61 (1965).
1008. Locke, D.C. *Anal. Chem.*, **39**, 921 (1967).
1009. Snyder, L.R. *Anal. Chem.*, **39**, 698, 705 (1967).
1010. Schroder, E. *Pure Anal. Chem.*, **36**, 233 (1973).
1011. Kirkland, J.J. *Anal. Chem.*, **41**, 218 (1969).
1012. Kirkland, J.J. *J. Chromatogr. Sci.*, **7**, 7, (1969).
1013. Halasz, I., Walking, P. *J. Chromatogr. Sci.*, **7**, 129 (1969).
1014. Mcdonell, H.L. *Anal. Chem.*, **40**, 221 (1968).
1015. Haulein, A. Corning Glass Works, personal communication (1970).
1016. Waters Associates, *New Development in Chromatography*, No. 1 (1970).
1017. Kirkland, J.J. *J. Chromatogr. Sci.*, **7**, 361 (1969).
1018. Majors, R.E. *J. Chromatogr. Sci.*, **8**, 338 (1970).
1019. Griddle, W.J. *Brit. Plast.*, **242**, May (1963).
1020. Campbell, R.H., Wize, R.W. *J. Chromatog.*, **12**, 178 (1963).
1021. Schabrom, J.F., Fenska, L.E. *Anal. Chem.*, **52**, 1411 (1980).

1022. Dohmann, K. *Lab. Prac.* July (1965).
1023. Slonaker, D.F., Sievers, D.C. *Anal. Chem.*, **36**, 1130 (1964).
1024. Van der Neut, J.H., Maagdenberg, A.C. *Plastics*, **31**, 66 January (1966).
1025. Van der Heide, R.F., Wouters, O. *Z. Fur Lebenmittelforschung*, **17**, 129 (1962).
1026. Waggon, H., Jehle, D. *Die Nahrung*, **4**, 495 (1965).
1027. Braun, D. *Chimica Ind.*, **19**, 77 (1965).
1028. Wandel, M., Tengler, H. *Kunststoffe*, **55**, 11 (1965).
1029. Haalpaap, H. *Chem. Org. Tech.*, **35**, 488 (1963).
1030. Braun, D., Geenen, I.T. *J. Chromatog.*, **7**, 56 (1962).
1031. Knappe, E. Peteri, D. *Z. Anal. Chem.*, **188**, 184 (1961).
1032. Lynes, A.J. *Chromatography*, **15**, 108 (1964).
1033. Lane, E.S. *J. Chromatog.*, **18**, 426 (1965).
1034. Crump, G.B. *Anal. Chem.*, **36**, 2447 (1964).
1035. Dallas, H.S.J. *J. Chromatog*, **17**, 267 (1965).
1036. McCoy, R.N., Fiebig, E.C. *Anal. Chem.*, **37**, 593 (1965).
1037. Crompton, T.R., Myers, L.W., Blair, D. *Brit. Plast.*, December (1965).
1038. Crompton, T.R., Myers, L.W. *Europ. Polymer J.*, **4**, 355 (1968).
1039. Pfab, W., Noffz, D. *Z. Anal. Chem.*, **37**, 195 (1963).
1040. Shapras, P., Claver, G.C. *Anal. Chem.*, **36**, 2282 (1964).
1041. Schwoetzer, G. *Z. Anal. Chem.*, **10**, 260 (1972).
1042. Pozdeeva, R.M., Lukhovitskii, U.I., Korpov, V.L. *Zavod Lab.*, **37**, 160 (1971).
1043. Ragelis, E.P., Gajan, R.J.F. *Assoc. Office. Agric. Chem.*, **45**, 918 (1962).
1044. Tweet, O., Miller, W.K. *Anal. Chem.*, **35**, (7), 852 (1963).
1045. Cobler, J.G., Samsel, E.P. *SPE (Society for Petroleum Engineers) Transactions*, April (1962).
1046. Streichen, R.J. *Anal. Chem.*, **48**, 1398 (1976).
1047. Shiryaev, B.V. Kozhbukhova, E.B. *Zavod Lab.*, **38**, 1303 (1972).
1048. Mel'nikova, S.L., Tishchenko, U.T., Sazonenko, U.V. *Lakokras Mater. Ikh. Primen*, **4**, 56 (1977); *Chem. Abstr.*, **87**, 118118u (1977).
1049. Rosenthal, R.W., Schwartzman, L.H., Greco, N.P., Proper, P. *J. Org. Chem.*, **28**, 2835 (1963).
1050. Rohrschneider, L. *Z. Anal. Chem.*, **255**, 345 (1971).
1051. Shanks, R.A. *Pye Unicorn Newsletter*, 1975.
1052. Clover, G.C., Murphey, M.E. *Anal. Chem.*, **31**, 1682 (1959).
1053. Crompton, T.R., Buckley, D. *Analyst (London)*, **90**, 76 (1965).
1054. Mayo, F.R., Lewis, F.M., Walling, C. *J. Amer. Chem. Cos.*, **70**, 1529 (1948).
1055. Novak, V. *Chem. Prumsyl*, **22**, (6), 298 (1972).
1056. Katelin, A.I., Komleva, V.N., Mol'kova, L.N. *Metody Anal. Kontrolya Proizvod Khim. Prom-Sti*, **11**, 62 (1977); *Chem. Abstr.*, **88**, 153329f (1978).
1057. Brown, L. *Analyst (London)*, **104**, 1165 (1979).
1058. Husser, E.R., Stehl, R.H., Price, D.R., Dehap, R.A. *Anal. Chem.* **49**, 154 (1977).
1059. Shiono, S.J. *Polymer Sci.*, A-1, **17**, 4120 (1979).
1060. Zaborsky, L.M. *Anal. Chem.*, **49**, 1166 (1977).
1061. Gankina, E.S., Val'chikhina, M.D., Belen'kii, G. *Vysokomol. Soedin. Ser.*, A, **18**, (5), 1170 (1976).
1062. Van der Maeden, F.P.B., Biemond, M.E.F., Janssen, P.C.G.M. *J. Chromatog.*, **149**, 539 (1978).
1063. Zaborsky, L.M. *Anal. Chem.*, **49**, (8) 1166 (1977).
1064. Ludwig, J.F., Bailie, A.G. *Anal. Chem.* **56**, 2081 (1984).
1065. Mouney, T.H., Smith, G.A. *Anal. Chem.*, **56**, 1773 (1984).
1066. Mourey T.H. *Anal. Chem.*, **56**, 1777 (1984).
1067. Utterback, D.F., Millington, D.S., Gold, A. *Anal. Chem.* **56**, 470 (1984).
1068. Davies, J.T., Bishop, J.R. *Analyst (London)*, **96**, 55 (1971).
1069. Schmidt, W. *Beckman Report, Beckman Associates*, **4**, 6 (1961).
1070. Schmidt, W. *Beckmann Report, Beckmann Associates*, **3**, 13 (1962).
1071. Shapras, P., Claver, G.C. *Anal. Chem.*, **36**, 2282 (1964).
1072. Shanks, R.A. *Scan*, **6**, 20 (1975).
1073. Berens, A.R., Crider, L.B., Tomamek, C.M., Whitney, J.M., B.P. Goodrich Tyne Centre. Brecksville Ohio. Manuscript circulated to members of the Vinyl Chloride Safety Association. 14 November 1974, entitled 'Analysis of Vinyl Chloride in PVC Powders by Head Space Chromatography'.

1074. Berens, A.R. Paper delivered to the 168th American Chemical Society Meeting. Atlantic City, New Jersey, September 1974, entitled 'The Solubility of Vinyl Chloride in Polyvinyl Chloride'.
1075. Steichen, R.J. *Anal. Chem.*, **48**, (9), 1398 (1976).
1076. Hartley, A.J., Lueng, Y.K., McMahon, J., Booth, C., Shepherd, I.W. *Polymer*, **18**, (4), 336 (1977).
1077. Crompton, T.R., Myers, L.W. *Plastics and Polymers*, 205 June (1968).
1078. Jeffs, A.R. *Analyst (London)*, **94**, 249 (1969).
1079. Maltese, P., Clementini, L., Panizza, S. *Mater. Plaste. Elastomerie*, **35**, 1669 (1969).
1080. Takashima, S., Okada, F. *Himeji Kogyo Daigaku Kenkyu Hokoku*, **12**, 34 (1960).
1081. Haslam, J., Jeffs, A.R., Willis, H.A. *J. Oil Col. Chem. Assoc.*, **45**, 325 (1962).
1082. Haslam, J. *Chem. Age*, **82**, 169 (1959).
1083. Haslam, J., Jeffs, A.R. *J. Appl. Chem.*, **7**, 24 (1957).
1084. Haslam, J., Jeffs, A.R. *Analyst (London)*, **83**, 455 (1958).
1085. Jeffs, A.R. *Analyst (London)*, **94**, 249, (1969).
1086. Maltese, P., Clementini, L., Panizza, S. *Mater. Plaste. Elastomerie*, **35**, 1669 (1969).

Index